CHARL...

Arctic Ocean

North Atlantic Ocean

NORTH AMERICA

1930 First flight from east to west across the Atlantic Ocean

Harbour Grace

New York City

North Pacific Ocean

Oakland

1930 Completion of the world circumnavigation

Honolulu

HAWAII

1928 First trans-Pacific flight

1934 First flight across the Pacific Ocean in a single-engine aircraft

FIJI

...ane

South Pacific Ocean

SOUTH AMERICA

28 ...rans- ...flight

Christchurch

NEW ZEALAND

South Atlantic Ocean

N

0   5000 Km

Scale at the equator

180°E   150°W   12...

60°N

30°N

30°S

60°S

180°E   150°W   120°W   90°W   60°W

King of the Air

King of the Air

THE TURBULENT LIFE OF
CHARLES KINGSFORD SMITH

ANN BLAINEY

Published by Black Inc.,
an imprint of Schwartz Publishing Pty Ltd
Level 1, 221 Drummond Street
Carlton VIC 3053, Australia
enquiries@blackincbooks.com
www.blackincbooks.com

9781760641078 (hardback)
9781743820711 (ebook)

 A catalogue record for this
book is available from the
National Library of Australia

Cover design by Tristan Main
Text design and typesetting by Marilyn de Castro
Index by Kerry Anderson
Cover photograph by Aviation History Collection/Alamy Stock Photo
Back cover photograph courtesy of the Museum of Applied Arts and Sciences
(photographer unknown; gift of Austin Byrne, 1965)

Printed in China by 1010 Printing International Limited.

CONTENTS

To my father,
Commander Frances William Heriot

PREFACE

I have known about Charles Kingsford Smith since I was eleven years old. On my schoolroom wall there was a map of the Pacific with the pioneering route of his plane, the *Southern Cross*, painted in red. That year in my history class I also learned about Magellan and Columbus, but my enthusiasm centred on Smithy. He was an Australian, and about the same age as my male cousins, which gave me a sense of kinship. Moreover, my mother had made a short trip in the *Southern Cross*. And my father, a navigator with the Royal Australian Navy, had charted parts of the coastline over which Smithy had flown, and could explain what extraordinary flights his had been. I began to wonder what Smithy was really like.

Perhaps that is why my book tries to focus on the inner as well as the outer man. Having been world-famous for so long, his career highlights are widely known, but his hopes and doubts, his impetuosity and patience, his courage and fears, are less well understood. And yet his states of mind play crucial roles in his dazzling success and his final disaster.

My book also examines his relationship with fame, for Smithy was revered as the greatest aviator of his generation, and is today accounted one of the most famous Australians of all time. Fame is a tricky companion. How did Smithy cope with such a conspicuous friend and enemy? How did it shape his short life? In an earlier book of mine – a biography of the celebrated soprano Dame Nellie Melba – I tried as best I could to fathom her surrounding fame and its effects. While Smithy rejoiced in his fame, he could never manage to master it.

Half a dozen people have written accounts of his life, starting with Beau Sheil and Norman Ellison in 1937, less than two years after Smithy and his plane vanished near Myanmar. Ellison did not publish his biography for another twenty years – and even then it was uncompleted – but it is still a valuable source for biographers. His collection of letters and documents, many given to him by Smithy's family, reside today in the Ellison Collection of the National Library of Australia, and also remain a precious resource for a biographer.

Since then several large biographies have appeared, such as those by Ian Mackersey and Peter FitzSimons, to both of whom I am indebted. For his widely read book, Mackersey secured interviews, in the 1990s, with Kingsford Smith's two wives and a number of his associates, none of whom is still living. Thanks to his foresight, much valuable information has been bequeathed to future biographers and historians, and I have gratefully drawn on it. To Mr FitzSimons I also owe a debt: his sparkling account of Smithy's exploits has renewed public interest in the great aviator's achievements.

Another debt is to Pedr Davis, the author of books on both Smithy and his fellow aviator Keith Anderson. Pedr has provided encouragement and practical help, reading my typescript and picking up errors. I sincerely thank him.

In my quest to find the man beneath the image, I have been much helped by the excellent reportage in the newspapers of the time. Smithy gave many interviews, and often spoke to reporters with a spontaneity and candour that is missing in his official writings. The Sydney *Sun* chronicled his career in useful detail, but it is in such outback West Australian newspapers as Carnarvon's *The Northern Times* and

The Pilbara Goldfield News that one comes across episodes and attitudes that have so far been little examined.

While out-of-the-way newspapers have been of considerable help, I could not have written this book without the assistance of librarians and archivists around Australia. I thank particularly those of the National Library of Australia, the National Archives of Australia, the Mitchell Library in the State Library of New South Wales, the State Archives of New South Wales and the John Oxley Library in the State Library of Queensland.

I am also deeply grateful for the help of friends. John Day checked my page proofs; John Drury loaned me books; Nicole Ritzdorf translated a difficult German text for me; Gaynor Sinopolis arranged for me to climb up to the cramped cockpit of the immortal *Southern Cross*, now housed proudly at Brisbane Airport; Lady (Marie) Fisher took me to Smithy's birthplace in Brisbane.

Martin Price unearthed obscure facts for me in British archives, and Kathryn Lindsay unearthed others in Vancouver. In Sydney, Charles and Amelia Slack-Smith gave me details of Smithy's parents' house in Arabella Street; Dr Allan Beavis obtained a copy of Smithy's classroom report at the Cathedral school; and Kitty Greenwood described her mother's romance with the young Kingsford Smith. Nor should I forget Smithy's grandniece, Kristin King, and Smithy's son, Charles Arthur Kingsford Smith, who lives on the west coast of the United States. They answered questions and gave encouragement. I thank them all.

Before writing about Smithy's panic disorder, I gained from discussions in Melbourne with Ms Roslyn Glickfeld, a psychoanalyst, and Professor Nicolino Paoletti, a psychiatrist, both well known in their fields. Moreover, Nicolino took time from his busy schedule to read and correct crucial chapters in my typescript.

Lastly, I must thank my publisher, Chris Feik, and my husband, Geoffrey Blainey. My husband travelled with me to many of Smithy's airfields and flight paths, including to outback Australia, New Zealand, Fiji, North America and Asia, including the steamy coastline of

Myanmar, where Smithy met his end. I also thank my daughter, Anna, and my son-in-law, Tim Warner, whose knowledge of navigation has been most welcome. They have been obliged to live with Smithy for a long time, and their support has never flagged.

ANN BLAINEY
Melbourne, 2018

SUNNY FACED,
SUNNY HEARTED,
SUNNY TEMPERED

It was dawn on the foggy morning of 9 June 1928, and people were swarming over the frosty paddocks of Eagle Farm, which served as Brisbane's aerodrome. Some had been there since well before midnight; others, with coats and scarves over their evening clothes, had arrived later, coming from dances and parties. As the sun rose, the long road skirting the river was dotted with walkers, while trams brought thousands more to the nearest stop. They walked the rest of the way in the cold. And certainly it was cold – 'but not cold enough', one woman said, 'to stop Australian hearts beating warmer than ever'.

At half past six, the Governor of Queensland arrived, along with the Premier, who was clutching a movie camera, being determined to record the historic events that they hoped soon to see. Numerous photographers and cameramen had also set up their equipment around the flying field. With each hour excitement grew, because it promised to be, as one news reporter put it, 'the most spectacular flying feat in history'. Everyone was praying that it would 'end in a win for the men who were game enough to play the odds at victory or certain death'.

By eight o'clock the winter sun was banishing the frost, and the roads into Eagle Farm were lined with parked automobiles. Fences had been erected around the field, and beside them stood expectant people of all ages and conditions, waiting for news. Nothing had been heard for hours of the plane they were hoping to see, and the spectators were growing restless. 'Give us some news!' voices cried impatiently. In a vacant hangar, technicians had set up an impromptu radio station, where two men sat with headphones clamped to their ears. 'No messages are coming through,' they informed their listeners. 'However, we can hear that the plane is in the air.'

At about half past eight, the men in the headphones suddenly grew tense and began fingering the dials on their receivers. 'They're safe,' they called to the crowd. 'They've been blown off course, but they've reached the coast near Ballina.' A sigh of relief went up and the waiting continued.

At a little after ten, a leather-coated man with field glasses picked out a distant object in the sky. 'Here she comes!' he shouted. A babble of voices cried in reply, 'Where, where?' Gleaming in the sun came a large blue and silver monoplane, escorted by two small biplanes that looked, in contrast, almost like toy machines. Twice the monoplane circled the field 'like a giant blue hawk', gradually dropping lower, until finally she passed close to the treetops and landed with the grace of a bird.

As she touched down, a cheer went up from 15,000 throats, while a thousand car horns honked their welcome. 'Smithy, Smithy!' shouted the welcomers. They were hailing the pilot, Charles Kingsford Smith, who, with three companions, had just flown 7000 miles from California to Brisbane. By completing the longest transoceanic flight so far made – a feat widely considered impossible – he may well have been, at that moment, the most newsworthy man in the Western world. But what mattered equally to the watching crowd was his birthplace. He was Brisbane-born, and the city was giving him a hero's welcome.

Charles Kingsford Smith had been born in a house within walking distance of the Eagle Farm aerodrome, and on both sides of his family he had strong Queensland roots. His maternal grandfather, Richard Kingsford, had settled in Brisbane in 1854, subsequently becoming a wealthy draper, the city's mayor and eventually a member of the Queensland Legislative Assembly. The aviator's father, William Charles

Smith, had come to Brisbane to join his brother, an officer in the Queensland customs and the son-in-law of Brisbane's first mayor. Thus it was that William, a humble bank clerk, came to mix in Brisbane's high society, and fell in love with the daughter of the rich and powerful Richard Kingsford.

On 17 April 1878, at the age of twenty-six, William Smith married twenty-year-old Catherine Kingsford at the Baptist Church in South Brisbane. Four years later, William was appointed manager of the Queensland National Bank in Cairns, in North Queensland, and he and Catherine and their three small children sailed to their new home. Their steamship, the *Ranelagh*, struck a reef near the mouth of the Johnstone River, and its 125 passengers, with 'nothing but their night clothes', waited many hours to be rescued. The shipwreck was a sign of hardships to come, but Catherine had inherited her father's energy and determination. Luckily also, although she was slightly built and barely more than five feet tall, she had the physical stamina to withstand the enervating humidity and tropical illnesses that were part of North Queensland life. She would need these resources, because in the first ten years of her marriage she gave birth to six children: Harold, Winifred, Wilfrid, Elsie, Leofric and Eric.

In a port like Cairns, the bank manager had considerable power. The money he loaned shaped businesses, and since head office was a long way away, he had wide powers of discretion. William took his responsibilities seriously, and his courtesy and fairness earned the town's respect, while his gregarious nature won him many friends. He was honorary treasurer, and sometime president, of most of the clubs and committees in Cairns. He belonged to cricket, boating, jockey and dramatic clubs, the Anglican Church committee and the School of Arts Library committee, and he was an alderman on the Municipal Council. In the words of *The Cairns Post*, William Smith was 'a valuable and public spirited citizen', and one of the 'half dozen gentlemen in the town who take an active interest in matters pertaining to the public'.

Yet this sober banker also had a frivolous side that made him the town's 'leading amateur comedian'. His performance as the Widow

Twankey in the pantomime *Aladdin* 'fairly convulsed the house', *The Cairns Post* reported; William was 'far away above the average for an amateur'. Catherine and the children were also capable performers, and a family production of *Mother Goose* was long remembered.

Successful as William was, he felt unsettled by Cairns' accelerating economy. The discovery in the early 1880s that the region's soil could grow some of the finest sugar in Queensland was transforming this sleepy little port into a 'sugaropolis' of 1500 people. Beyond the town stretched mile upon mile of cane fields, which attracted investors from cities far to the south. The Melbourne firm of Swallow & Derham acquired a large plantation and crushing mill at Hambleton. In a few years' time, Catherine's sister, Caroline, would marry William Swallow, and the Hambleton Mill would become a second home to the Smith family.

When Catherine's father, Richard Kingsford, lost his seat in the Legislative Assembly, he turned his entrepreneurial gaze towards Cairns. In June 1884 he and his family moved to a house on the Esplanade, and soon he dominated the town's institutions, being unanimously elected its first mayor. Meanwhile, William was transferred to the Queensland National Bank at the port of Mackay. It was described as a promotion, but to the Smith family it seemed more like an exile. After a few months William quit the bank, returned to Cairns and went into business, putting to use the commercial expertise he had acquired during his years in banking. In January 1886 he joined his older brother, Edward Eldridge Smith, in running what they rather grandly called 'a big importing establishment known as Smith Brothers'. In the buoyant economic climate of that decade, the store flourished and diversified, opening a branch in Brisbane, but in the economic slump of the 1890s, the business did not survive.

In February 1889 Richard Kingsford was elected mayor of Cairns for a third time, but his term of office did not last long. His wife, Sarah, became gravely ill. Her best hope, according to her doctors, was to leave the tropics and move to the cooler climate of Tasmania. There Richard paid £5000 for a two-storey mansion by the Tamar River, in Launceston.

In a corner of the garden stood a brown weatherboard cottage of thirteen rooms called *Rose Lyn*. To faraway *Rose Lyn* came the Smith family for long holidays, Catherine to comfort her dying mother and William to regain his health. In May 1889 he had suffered 'a severe indisposition', and ten months later 'a sharp attack of fever'. After both illnesses he was advised by his doctor to take a long holiday in one of the cooler southern cities. His illness was almost certainly malaria, for which swampy, mosquito-infected Cairns was becoming notorious, the long wet seasons of 1889 and 1890 provoking severe outbreaks of the disease.

In September 1890 Sarah Kingsford died, and at the age of seventy-one the indomitable Richard embarked on a new life, remarrying and fathering a daughter. He kept in touch with his business interests in Launceston and Cairns by making the long coastal voyage between them. He finally resettled in Cairns, building a grand house on the outskirts of town. Meantime, William Smith, only partially restored to health, resolved to return to banking. In 1892 he became manager of the Cairns branch of the small Bank of North Queensland. It was the profession he did best, and his courtesy, attention to detail and integrity were again widely respected.

Four years later William was made the assistant manager of the bank's Brisbane office. The move to the cooler climate of Brisbane was probably a response to his continuing illness, for it was not unusual for the Bank of North Queensland to transfer a senior employee to a southern branch in such circumstances. The previous year William had been forced to spend two more months in Tasmania, recovering from another attack of fever. A severe case of malaria could bequeath lifelong recurrences of fever and anaemia, and this seems to have been William's fate.

The move to Brisbane saddened William's friends: he was 'a man who will be greatly missed', wrote *The Cairns Post*. It saddened William, too. Even so, Brisbane had its compensations, being the home of relatives and many acquaintances. What troubled him most was Catherine's health. In her fortieth year, and after an interval of ten years, she was once again pregnant. In the medical opinion of the time, a pregnancy at this age and in these circumstances was risky for both mother and child.

When this late-in-life baby went on to become Australia's most famous aviator, to the family it seemed a miracle.

The Smiths needed to find a suitable home before the baby's birth, and they chose Hamilton, a semi-rural suburb of large houses and old money on the north bank of the Brisbane River. Little is known of the house they rented, other than that it stood at the corner of Riverview Terrace and what was then Hamilton Road but is now called Kingsford Smith Drive. The house has long since disappeared and no photograph of it seems to exist, but popular belief insists that it was a modest cottage of humble appearance.

If this were so, it must have been an unusually large humble cottage, because it housed the Smith parents, seven children, a maid, a nurse and a large dog. And it was unlikely to have been conspicuously modest, because William, as assistant manager at the city office, would have been expected to live in a certain style. It is relevant to note that when the Smiths moved the following year to O'Connell Terrace, in Brisbane's Bowen Hills, they chose a substantial villa called Kingsdown, which had ten rooms and servants' quarters, and was very much the residence of a bank manager. However, Australians – like Americans – seem to prefer their national heroes to be born in the proverbial cramped hut or log cabin, thus making a humble cottage an appropriate place for a celebrated aviator's birth.

Whatever its size, the house stood in a handsome position, virtually on the riverbank. This pleased the Smith children, who loved to boat and swim. They enjoyed the carefree summer days, but Catherine did not, for her thoughts were on her unborn baby. On 8 February she felt so near her time that she kept twelve-year-old Leofric home from school in case she needed him to summon the doctor. It was a wise decision, because that same day her labour started. Between five and six o'clock the following morning – 9 February 1897 – Catherine's fifth son and seventh child came safely into the world.

The Smiths do not seem to have chosen any names for the baby in advance. Perhaps they believed, in view of the danger, that it was unlucky to select a name before the birth. However, six months later, when the baby was baptised at St Andrew's Anglican Church at Lutwyche, the names bestowed on him showed careful thought. All the previous Smith children had received their mother's maiden name of Kingsford as their final Christian name, but none so far had received a Christian name taken from their father's side. This new baby was given his father's second name, his paternal uncle's first name and his mother's maiden name. Thus he became Charles Edward Kingsford Smith.

At the time of the baby's christening, his brothers, Harold, Wilfrid, Leofric and Eric, were eighteen, fifteen, twelve and ten, and his sisters, Winifred and Elsie, were seventeen and fourteen. With such a throng of older brothers and sisters, it was inevitable that the baby would be indulged. The boys were said to have regarded him as 'a delightful little plaything', and treated him rather like one of their puppies: at the age of two, he was often to be seen 'squatting on his knees with chubby hands hanging down, paw-wise, while the boys tossed him sweets'. The girls mothered him and marvelled at his beauty. Although at birth he had been 'wrinkled, flat-nosed and bald', by his first birthday he was 'really pretty, with golden hair, blue eyes, rose-leaf skin, and the reddest lips imaginable curling over pearly teeth'. 'Sunny-faced, sunny-hearted, sunny-tempered', Charles was the type of child who wins baby competitions. Catherine could scarcely credit that this son of her middle age would bring her so much joy.

It was just as well that the family had something to cheer them, because William's health was causing anxiety. According to Catherine's diary, he was often too unwell to work. Nevertheless, he was considered sufficiently well to be sent to the Rockhampton office of the bank in December 1897, to relieve the manager over Christmas. Back in Brisbane the following January, he received the unexpected news that he had been placed in charge of the Sydney office. That he should have been given such a position of trust despite his poor record of attendance suggests how much his superiors valued him.

Early in 1898, the family moved to Sydney. William had grown up on Sydney's North Shore and been educated at the Fort Street School, so the city was familiar to him. It was less so to Catherine, but with her usual zeal she set about renting a suitable house, no easy task with so large a family. In February they moved to the suburb of Randwick, where a nurse named Ruby Irwin was engaged to care for the baby, at 'ten shillings a month and clothes'. Before the end of the year they had moved again, to the suburb of Ashfield.

In the new year of 1899, on Tuesday, 3 January, Catherine took Charles, now a lively toddler, on his first visit to the zoo. She arrived home to find William back from work and ill in bed. Since this was a common occurrence, she was not immediately alarmed, but as they talked, certain facts emerged that filled her with dismay. Something had happened that day at the office that made his position as the manager untenable. Two days later – 'Black Thursday', according to Catherine's diary – William resigned.

Possibly his resignation arose from his ill health, which was an almost constant cause for concern. But there might also have been a more dramatic reason. In the 1990s a daughter of the Smiths' fourth son, Eric, told the biographer Ian Mackersey that William had approved an unsecured loan to a trusted friend, and that the friend, finding himself unable to repay it, had committed suicide.

In the eyes of the bank, the granting of an unsecured loan was a serious offence, and once discovered, would have incurred immediate dismissal. Was this a risk that William was likely to have taken on behalf of a friend? Would a manager of his experience and probity have behaved so carelessly? As the relevant records of the Bank of North Queensland have long since disappeared, it is impossible to know. If William were guilty and had wished to keep the vestige of his good name, he would, according to the custom of the time, have had no option but to resign, and to accompany his resignation with a promise to restore the missing money as soon as possible. Perhaps on this one occasion William's kind heart overcame his good sense, and sadly cut short his career in a profession that he loved.

William was now unemployed, and being forty-six years old – and presumably lacking a reference – he had little hope of gaining work. He needed help, and his first thoughts seem to have been of Richard

Kingsford, who, of all his relatives, had the influence and money to solve his problems. A week after William's resignation, Catherine boarded the steamship *Bulimba* for Cairns. Her companion was Charles – aged one year and eleven months, and 'mischievous as a kitten' – whom she judged too young and too unruly to leave behind. Sick at heart, and seasick as well, she dreaded the demands of the voyage, but the sailors took the little boy off her hands, returning him only to sleep and to eat. Hour upon hour, down in the crew's quarters, the child sang his nursery rhymes and babbled his baby chatter. His blue eyes and golden curls, along with his endearing unselfconsciousness, captured the sailors' hearts. It was an early sign of Charles's extraordinary capacity to charm.

Catherine must have been profoundly relieved to see her father. Money was her most pressing need and Richard Kingsford supplied it, although not in such quantities that she did not need to economise. At the end of the month she and baby Charles returned to Sydney. Almost immediately she moved the family to a cheaper house, 'a quaint stone cottage with red tiles, called Bromage', on Kangaroo Hill in the ocean-facing suburb of Manly. To save money, the servants were dismissed, the exception being little Charles's nurse.

Catherine was the powerhouse of the family, resourceful, efficient and energetic. With the servants gone, she undertook most of the household tasks herself, and refused to bow to weakness. In the evenings, when, rightly, she should have been almost exhausted, she would change into a formal dark dress with a starched white collar and cuffs and preside over the dinner table. She was very much, as her daughter Winifred recalled, the lady of the house.

Fortunately, the Smiths did not need to spend much on entertainment. The parents' love of music and acting had been passed down to the children, especially to the eldest boy, Harold, who had taken acting and elocution lessons from the distinguished Shakespearean actor Walter Bentley. Few evenings went by without the whole family gathering in the living room to act scenes from plays, compose poems and riddles, and sing around the piano. Catherine, a competent pianist, was in charge of the musical evenings, and expected every member of the family to contribute an item. Winifred remembered her baby brother, 'standing on the big polished table, surrounded by an admiring family group, lustily singing in quite good time and tune'. The family soon realised that the little boy had a quick and retentive musical ear, which easily picked up popular songs. They also recognised that he was a born performer who loved to be the centre of attention.

On Sunday evenings the family sang hymns together. The boys loved 'Onward, Christian Soldiers', while the girls favoured 'There Is a Green Hill Far Away'; little Charles liked 'There Is a Happy Land Far, Far Away'. When the hymns were over, William would read a chapter aloud from the Bible, for without being overly zealous, the Smiths were a religious family. They were regular attenders at the local Anglican church, and the children were regular pupils at the Sunday school.

Infant Charles, with his quaint ways and sayings, played an important role at the Smith family gatherings. His mother refused to allow discussion of his quaintness in his hearing: she had seen too many children, encouraged by over-fond elders, grow into spoiled little exhibitionists. For the same reason she was careful to whip him when he was naughty, however amusing or endearing his naughtiness might be. In fact, she kept a small riding whip especially for the purpose. Charles accepted his punishments philosophically. Once, when he put his nurse's hat, sewn with artificial poppies and cornflowers, through the family mangle, and it emerged as a squashed mixture of red and blue, he fetched the whip himself and offered it to his mother.

With no employment in sight, William settled into a desultory life, attempting to enliven it with trips to relatives. The following year he sailed to Brisbane to visit his mother at New Farm, and to see his son Harold, who, notwithstanding his father's abrupt departure, was doing well as a promising accountant at the Bank of North Queensland. In March 1900, still without a job, William made a protracted visit to Cairns, sailing from Sydney with his brother-in-law, William Swallow. Their joint father-in-law, Richard Kingston, now ailing and almost eighty, greeted them warmly, but they could see that the old man's days were numbered. The question of their wives' inheritance must have been foremost in their minds.

On 2 January 1902 Richard Kingsford died. Catherine mourned her father, but on hearing the terms of his will she must also have rejoiced. His fortune of just over £7000 pounds was to be divided equally among his four children. Catherine's share was substantial: the equivalent of the wages of a skilled worker for fifteen or sixteen years. After three years of anxiety, she could breathe easily. The family's financial future at last seemed assured.

VANCOUVER
BECKONS

In 1901, a year before coming into her inheritance, Catherine moved the family from their Manly cottage to a house in Longueville. Although she may have moved for reasons of economy, this picturesque suburb, set on steep slopes above the blue waters of Sydney's Lane Cove, caught Catherine's fancy, and retained it. In years to come it would be the favourite suburb of the Smith family, and the setting for Charles's earliest and fondest memories.

There, four-year-old Charles discovered that he adored dogs, especially Sir Toby, the champion English setter that the Smiths had brought with them from Cairns. Exhibited at the Sydney and Brisbane shows, it twice won 'best sporting dog' trophies. 'Toby is a nice good Toby dog,' Charles told his father in one of his earliest letters.

Machines also began to fascinate him. He spent hours on the verandah of his home, watching the ferries dart about the harbour. When not watching ships, he was drawing pictures of them, and of trains and engines, and telling himself stories about them. His sister Winifred remembered a tale about a 'dood twain' that 'didn't twy when its face was washed and didn't call its nurse a bwute'. When Charles wrote to Santa Claus that Christmas, it was to beg for a special train, with piston rods to make its 'wheels go around'.

Of aeroplanes he knew nothing, for they had not yet been invented, but he was familiar with hot air balloons, and these too fired his imagination. At about the age of five, he substituted his father's open umbrella for a balloon: gripping the handle, he leaped from the roof of the family shed, shouting to his brother Leofric, 'Leffy, I'm going to fly!' Fortunately, he crashed into Leffy's shoulders, and so saved himself from harm. The family was fond of recounting this anecdote during Charles's years of fame.

By the age of five he was enrolled at Miss Clifford's nearby infants' school. He loved Miss Clifford; indeed, realising she was unmarried, he promised to marry her, becoming henceforth her 'chivalrous body guard'. Winifred remembered him setting out for school, a determined figure 'in a cream sweater, little pants and a schoolbag on his shoulders'. At Miss Clifford's, he learned to write grown-up letters, several of which survive, penned in large and wobbly capitals. 'Ullo Daddy,' he wrote to his father. 'You are a good Daddy. Ill write something nice. Im good Charlie.' He finished his letter with: 'Mummy and Elsie starves me. Ullo Daddy.'

Meanwhile, the Smiths were coming to an important decision. By the end of 1902, Catherine's inheritance was in their hands, and while they had enough for their present needs, they had no idea what the future held. Their three elder sons, on the threshold of adulthood, would need assistance to start their life journeys. William believed that he must somehow increase his capital.

William now had the means to set up a business of his own, but the Australian economy did not encourage him. Australia's prosperity rode, as it was said, 'on the sheep's back', and the worst drought in the history of white settlement had destroyed nearly half the country's flocks. Unemployment was high, and tens of thousands of Australians were migrating. Canada, in particular, was hungry for settlers, and was conducting an aggressive publicity campaign to attract migrants to its shores.

Vancouver, on Canada's west coast, beckoned Australians most persuasively. Dense forest only twenty years before, Vancouver now had 26,000 inhabitants, was ranked the fourth-largest port in Canada, and boasted a booming commercial hub that had connections to the Atlantic seaboard along the new Canadian Pacific Railway. Fishing and lumber provided much of its wealth, sparking the creation of banks and investment companies, and wholesale and retail businesses. Somewhere in this expanding market, William believed, he could surely find a niche. Moreover, thanks to the warm sea current close to its shores, the temperatures in Vancouver were neither too high nor too low in summer and winter, making it an ideal climate for fever-prone William.

Not the least of its advantages was the presence of a number of North Queensland friends who had already migrated and were urging William to join them. He had merely to board a steamship, subsidised by the New South Wales and Canadian governments, and in three weeks he would be there. And this is what he decided to do. Late in 1902, at Sydney's Circular Quay, he boarded the steamship *Aorangi*, having promised that, if all seemed well, he would send for his wife and children at the first opportunity.

The spirit that had brought their ancestors to Australia still ran in the family veins, so Catherine and most of the children – Wilfrid and Harold excepted – were prepared to migrate. The main stay-at-home was Harold, who, now aged twenty-two, was a newly appointed accountant at the Cairns branch of the Bank of North Queensland.

Overjoyed to be back in his home town, Harold had thrown himself into its public life, making much of the fact that he was Richard Kingsford's grandson. He also made much of the fact that he had been taught acting by the distinguished Walter Bentley, offering to give lessons to fellow townsfolk. But a promising actor and accountant who was a grandson of the great Richard Kingsford should not, he decided, carry the humdrum name of Smith. Kingsford was a far better choice – and Kingsford Smith sounded even better. Thus ran the reasoning – or so it seems – that ultimately led to the family's change of name. Harold had been christened Richard Harold Kingsford Smith, and by January 1901 was calling himself R.H. Kingsford Smith. It was under this name that he became engaged to Sydney-born Elsie St Claire Johnson, and it was under this name that he hoped to take over management of the Bank of North Queensland in the town of Atherton. He had no intention of migrating anywhere.

Meanwhile, Catherine, awaiting William's summons, sold most of their possessions and moved herself and children into Mrs Latchford's boarding house in the Sydney suburb of Greenwich. Charles – or Chilla, as he called himself, for his infant tongue had difficulty saying Charles – was now aged six, and was excited at the thought of living in Canada. He had never seen snow and marvelled when his mother explained that he would soon be making snowballs. He missed his father and wrote him loving letters, describing life at Mrs Latchford's: 'We have a table to our selfes and a room. We have sold a lot of furnechur. We have sold the penow.' In a postscript, he admitted: 'I don't like liveing this country and I want to see you … I sepose you have that toy ready for me, I do like toys. I do like train toys.'

To Catherine's relief, the summons from Vancouver came early in 1903. On the basis of William's accounting experience, the powerful Canadian Pacific Railway had appointed him a voucher clerk in its claims department. To be trusted once more, and to be able to perform work that required honesty, efficiency and financial skills – the very work for which he had been trained – must have lifted a burden from his shoulders. Thereafter, the family joined him at intervals, Catherine, Charles and his sisters sailing in July and August, having waited behind for Harold's wedding.

For Charles, the three weeks on the *Aorangi* must have seemed like heaven. As well as providing exotic ports, the ship was a child's adventure playground, and he found himself with readymade playmates, something he often lacked at home. Winifred once remarked that because Charles grew up in a largely adult household, he was often 'the odd man out', and since he was naturally gregarious, he must have often felt lonely. On the ship, he quickly attached himself to a couple of older lads and followed them about like an eager puppy. Some of their escapades, such as climbing into the hawsehole that housed the anchor cable, were wildly unsafe for a six-year-old, but the adoring little boy did whatever the older boys told him. Even at that tender age Charles seems to have had a liking for daredevil pranks. Once Catherine heard about their mischief, she informed the captain, and Charles and the older boys were summoned to the bridge. They were sternly warned that further misbehaviour would see them 'clapped in irons'.

The first sight of Vancouver raised the family's hopes. The placid waters of the harbour and the snow-capped mountains rising above it gave promise of that outdoor life that the Smith family relished. Over the following summers there would be numerous camping expeditions to those mountains, and numerous explorations of the inland waterways. Catherine's initial energies, however, centred on finding a house that could accommodate them all, and by the end of the year they had moved into 1334 7th Avenue West. According to *Henderson's City of Vancouver* directories, it would remain their home for the next three years. This contradicts the later stories that William was a 'rolling stone', and that he and his family 'moved about like gypsies'.

Catherine was something of a feminist and believed that her daughter as well as her sons should have trained occupations. Winifred and Elsie gained jobs as stenographers; Eric, the resourceful son, who had worked his passage over on a ship, became a junior accountant at the British Columbia Telephone Company; and nineteen-year-old Leofric joined the Real Canadian Property Management Company, where he trained as a draughtsman. Charles seems to have attended two schools: the Sir William Dawson School and the Lord Roberts School.

Living together in 7th Avenue West, the Smith family re-created the happy, convivial life they had so enjoyed in Sydney. To Charles this was an undoubted blessing, for so major a move at so young an age could have disturbed him. That he remained secure is shown by an incident that occurred not long after the family's arrival. Sailing back from a camping trip, the Smiths were caught in a fierce wind, and Eric anchored the boat close to the shore to ride out the storm. There, through the wild night, the family watched anxiously – all but young Charles, who showed no fear and slept peacefully at his mother's feet.

By September 1904 the Smiths were describing their Canadian life so rosily in letters home that Harold summarily abandoned the banking profession and arrived in Vancouver with his wife and baby daughter. Interviewed as he left the ship, he told reporters that he planned to establish a hardwood trade between western Canada and his native North Queensland. As time passed, no more was heard of the hardwood trade, but Harold threw himself into the life of the city and took his family with him. Having found work as an accountant with the leading booksellers Clarke and Stuart, he began to seek out fellow Australians, with an eye to forming a social and business network.

One of his first projects was the setting up of an Australian Club, designed to 'welcome all Australians who come to Vancouver', and give 'all prospective settlers every possible assistance'. It opened its club rooms in October 1905 with 100 members; William was its first vice-president and Harold its first honorary secretary. The following year, William and Harold, who were keen Masons, took leading roles in founding an Australian Masonic Lodge, which later came to be called the Southern Cross Lodge. Father and son were early office bearers.

Harold's ultimate aim was to join his father in running an import business, along the lines of the firm of Smith Brothers, set up in Cairns twenty years before. However, 1905 was an unfortunate year to start a business in western Canada: a poor harvest the previous year had tilted the economy into recession, and most businesses were struggling. Harold and William nevertheless went ahead, renting premises in North Vancouver, and arranging to import Australian luxury goods.

They called the business 'Smith Brothers' in memory of its Australian counterpart, but the family was now known by the double surname of Kingsford Smith. Why did they adopt this new name? The reason usually given is one of utility: there were so many Smiths in Vancouver that to ensure that mail was delivered correctly they needed to particularise the name. This may have been a reason, but the name had been adopted in Australia by Harold, who had been calling himself Kingsford Smith since 1901; and it had possibly been used also by Catherine, who had been named on the passenger list of the *Aorangi* as Mrs Kingsford Smith. A double-barrelled name had a distinctive, even aristocratic ring to it, and that, most likely, was why they adopted it.

What of young Charles in these years? At the age of nine he was an irrepressible extrovert, and as fond of songs and jokes and banter as his brothers and sisters. His daredevil pranks still led him into trouble, but by now he knew his winning smile could lead him out again. He continued to adore animals; he had recently brought home a stray puppy and insisted on having it sleep on his bed. Best of all he loved machines, especially trains, to which his father's work in the railway gave easy access. Once, when Charles vanished during a family excursion into the Rocky Mountains, he was found at a railway turntable, in charge of a train engine. The driver had allowed him to take the controls, and was decidedly impressed. He declared that 'the lad shaped up mighty well'.

Engrossed in the Australian Club, in the Masonic Lodge and, above all, in the firm of Smith Brothers, William and Harold were more than content in their Vancouver life. At first the firm had been seen as only a sideline, and father and son had wisely retained their regular jobs, but now, encouraged by success and the rebounding Canadian economy, their plans grew bolder. In 1906 they decided to leave off clerking and accounting and devote themselves entirely to their new business.

The Kingsford Smith women, on the other hand, were homesick and longed to return to Sydney, if only for a few months. Winifred was the first to leave, sailing on the *Miowera* in July 1906. Catherine, Charles, Elsie and Harold's wife and two small children followed on the *Moana* four months later. Much as she was looking forward to it, the ensuing visit proved unexpectedly unsettling for Catherine. Throughout the hot Sydney summer, her thoughts kept returning to her husband and sons on the wintry side of the Pacific.

Charles's affections and loyalties also remained in Vancouver. He scarcely remembered Australia, and thought and spoke like a Canadian schoolboy. Fortunately, he was gregarious by nature, and, having been encouraged always to speak up, he had little hesitation in approaching new people. Of course, this did not please everyone. Some thought he had been indulged and was far too sure of himself but he had no difficulty making friends. Raymond, Phillip and Godfrey Kingsford, the sons of Catherine's brother Arthur, and Robin and Rupert Swallow, the sons of Catherine's sister, Caroline Swallow, quickly absorbed him into their circles.

Rupert, four months younger than Charles, shared many of his enthusiasms, including a love of the sea. Though only nine years old, both boys were confident swimmers; indeed, they were too confident and tended to disregard warnings of danger. Caution was therefore far from their minds when, on the afternoon of 2 January 1907, they left the older boys on the sand and ran together into the surf at Bondi Beach. Within minutes, a strong undertow swept them out to sea. Charles fared the worst. According to one eye-witness, he was carried 'outside the farthest breaker' and for a while was 'completely lost to sight'.

Bondi was one of the few beaches to possess a surf-lifesaving team, and once the alarm was given, four lifesavers donned cork belts connected to lines and swam out to the children. Rupert was found quickly and reeled in. Charles, who by then was 'a considerable distance from the shore', took longer to be reached. By the time he was brought to shore, little hope was held for his life. Placing him stomach-down on the sand with his face turned to one side, a lifesaver pumped his back, trying to force air through his lungs, but not a flicker of life was seen. These efforts at resuscitation had virtually been abandoned when a nurse named Sadie Sweeney stepped out from among the bystanders. Pushing her way to the boy's body, she began to pump and pump until she saw signs of consciousness returning. Charles owed his life to her.

From Bondi Beach the boys were taken by the police to the Swallows' house, in Darlinghurst. There, stoical Catherine nursed Charles, 'sick and feverish', through the night. She must have been distressed, but she seems to have hidden her feelings, believing it would be harmful for him to dwell too much on his misfortune. Next day, in her diary, she dismissed the incident as a 'bad accident', and noted with satisfaction that 'Charlie' was 'none the worse for his terrible experience'.

But was he really none the worse? Some have thought that the terrible experience left him with a life-long fear of drowning, and they offer as evidence the anxiety he sometimes felt when he flew across stretches of open sea, such as the Bay of Bengal. His own activities, though, reveal no persisting fear of the water. After the Bondi incident, he returned to swimming and diving and boating with the same enthusiasm he had shown before.

However, another fear might have arisen from the experience. According to the diaries he kept during his flights, what worried him most was the prospect of a forced landing in a desert, or jungle, or ocean – a place so isolated that no rescuer could reach him. At Bondi, help had been close at hand, and the incident could be passed off as a 'bad accident'. In a jungle or desert or ocean, where no rescue was likely to be forthcoming, he would surely die. Perhaps this was the lesson he took with him from Bondi.

Seven months after the near-drowning, Charles and his mother returned to Vancouver, where William and Harold were running an expanding business. Emboldened by the now-booming economy, father and son had enlarged Smith Brothers, adding accounting, auditing, auctioneering and real-estate brokering to its activities. Trading under the new name of Kingsford Smith and Co., its imposing premises were located in downtown Granville Street.

The hopeful return to Canada quickly soured. Just as the family was moving into a new home at 1935 Comox Street, and Charles was starting a new school, race riots erupted across the city. Thousands of supporters of the Asiatic Exclusion League rallied outside City Hall to hear inflammatory speeches, and then rampaged through the Japanese and Chinese settlements. The riots were quashed, but the tension sim-mered on.

The following year, 1908, the harvest was lean and the Canadian economy slumped. William and Harold, ignoring the downturn, added shipping and manufacturing agencies, took on a partner and moved to even grander premises, at 167 Cordova Street. They also opened a further branch in North Vancouver, ignoring the fact that overexpansion in a time of falling prices had brought an end to William's business in Cairns.

At the end of 1908, they had little option but to wind up their company. Harold elected to migrate with his wife and daughter and son to California, where business opportunities seemed to abound.

William and his other children could think only of Australia. On 29 January 1909, he, Catherine and Charles boarded the *Aorangi* for Sydney, there to rejoin Winifred, Elsie, Leofric and Eric, who had already settled back into Australian life.

CHARLES
THE CHORISTER

Y ears later Winifred would describe her father, on his return to Sydney, as 'weary and disillusioned'. Certainly he had lost money, and, at the age of fifty-six, he possibly had difficulty in finding suitable work. However, the family story that he was forced to earn a living as a humble postman does not ring true. He seems to have taken a job as an accountant, and as such he was listed in the municipal rate books. Indeed, his finances were sufficient for him to rent an attractive villa called *Kintore*, which sat at the corner of Bydown and Yeo streets in Neutral Bay. The house still stands: in the Smiths' time it was number 68, now it is 18. It was here that Charles began to blossom in an unexpected role – as a chorister.

His aunt Caroline Swallow pointed the way. In her girlhood, Caroline had been a promising pianist, skilled enough to consider studying in Germany, and her children had inherited her musical talent. Her daughter Marjorie was a member of J.C. Williamson's Musical Comedy Company, and her son Rupert – Charles's close friend – was hoping to become a professional singer. Rupert was now a scholar and chorister at the St Andrew's Cathedral Choir School, where, along with a sound musical training, he was receiving the academic instruction and social polish that a leading boys' private school tried to impart. Dressed in his uniform with its Eton collar and straw boater, Rupert looked every inch the upper-class Sydney schoolboy.

The choir scholars paid fees of nine guineas a year, which was too high for William to afford, but knowing Charles's musical talent, Catherine hoped he might win a scholarship. In April 1909 she visited the school's headmaster, Rev. Edward Wilton, and was sufficiently persuasive to gain Charles an interview and an audition. On the day of the interview, his speech, appearance and interests must have worked against him. He was sporting an 'American Indian' haircut, shaved on one side and upswept on the other, and, in his strong Canadian voice, he talked volubly about baseball but knew nothing about cricket. Even so, he impressed the Rev. Wilton. When they tested his musical skill, the scholarship was his.

The Kingsford Smiths – adults and children – were regular attenders at Anglican church services, but it is doubtful that young Charles was prepared for the church-going that was now expected of him. He and his fellow choristers sang at four weekday and three Sunday services at St Andrew's Cathedral. They also gave sacred and patriotic concerts, joined in choir festivals, sang at neighbouring parishes and participated in special church ceremonies.

In both the choir and the classroom, the discipline was strict. Hard work, punctuality and obedience were demanded, and boys who broke the rules received a caning. Singing practice was conducted under the eagle eye of the organist and choirmaster, Joseph Massey, who, peering into his organ mirror, could easily spot a slacker, and was especially vigilant when church festivals were near. At Christmas the choristers performed Bach's *Christmas Oratorio* and Handel's *Messiah*, and at Easter they sang Bach's *St Matthew Passion*, Gounod's *Redemption* and Stainer's *Crucifixion*. It was challenging music demanding wholehearted concentration, not a quality always in evidence in Charles.

How, one wonders, did this cheeky, hyperactive twelve-year-old manage to apply himself successfully? His theatrical and musical talents were part of the answer, and so was his delight in catching the public eye.

Cathedral services were as much theatrical events as religious ceremonies, and to take a prominent part must have thrilled him, especially on those grand occasions when the congregation held some of Sydney's best-known citizens. Even so, there were times when mischief got the better of him. In 1910, a year after he joined the choir, he sang at a service to mark the death of King Edward VII. 'I am the resurrection and the life,' chanted the choristers as, white-robed and solemn, they processed down the aisle towards the choir stalls. Charles, passing his mother's seat, could not resist the opportunity. Catching her eye, he gave her a large wink.

All in all, there were surprisingly few lapses, and his marks were unfailingly satisfactory. In his first year he was placed second in his class. 'A clever lad,' wrote the Rev. Wilton, 'who has done extremely well': he 'has a good future before him, if he accepts his advantages and keeps them.' Perhaps one of his teachers had witnessed the wink. The following year his report was even more favourable. Charles's conduct was 'very good', and his marks were high in Scripture, Algebra, English and Geography. 'He is holding his own splendidly,' wrote the headmaster, 'and certainly approaches his tasks in the proper spirit. Charles will improve a great deal in the next half.'

In December 1910, Charles won a running race, and received an album as a prize. Dedicated to 'The Thoughts and Confessions of My Friends', the album contained fourteen questions. On 23 December, Charles decided to forget about his friends and respond to the questions himself. His answers give a glimpse into the serious side of his thinking.

What was his favourite line of poetry, asked the album. He answered: 'They also serve who only stand and wait' – a line from Milton's sonnet 'On His Blindness'. Who was his favourite poet? He replied, Henry Wadsworth Longfellow. Who were his favourite actor and actress? He chose the Australian-born Oscar Asche and his wife, Lily Brayton. A distinguished Shakespearian actor, Asche had just completed a theatrical tour of Australia, and Charles had been excited by the acting and the plays. When asked about his favourite song, he named the popular love song 'Because'. As for his favourite author, it was Ernest Thompson

Seton, a highly popular Canadian nature writer, and a founder of the American Boy Scouts.

On other topics Charles was equally illuminating. He declared that he was in favour of votes for women, but not in favour of women smoking. He believed a man should marry at twenty-five years of age, and that his wife should be twenty-one. He gave his favourite indoor pastime as tinkering with engines, and his favourite outdoor pastime as foot-running. As for his ambition in life, it was to be an engineer. Undeniably, these were the answers of a thoughtful, well-educated and intelligent thirteen-year-old.

Charles stayed at the St Andrew's Cathedral School until the end of 1911. By then he was almost fifteen, his voice was breaking and his time as a choir scholar was over. Thanks to the wisdom of his teachers, he had learned the satisfaction of mastering serious tasks and applying himself to serious study. Sensibly, his teachers did not try to stifle his exuberance, knowing it could be an asset as well as a fault, and he remained a joker and a daredevil.

There is no doubt that he was popular with his classmates. His sister Winfred was fond of recalling the numerous young friends who used to descend on their house at Neutral Bay. When his family was not at home, Charles would treat his friends to rowdy singalongs and illicit cigarettes. Winifred remembered returning unexpectedly one night to find the house reeking of tobacco, and Charles at the piano, pounding out popular songs. Turning a blind eye, she made cups of cocoa for the young smokers. 'I could not have been a spoilsport,' she remembered, 'they were all so happy together.'

During those early teenage years, Charles often indulged in schoolboy pranks and larks. He loved to walk on stilts, and the local shopkeepers came to dread his visits because he often toppled off his stilts and sent their merchandise crashing. On his bicycle he was an 'incorrigible speedster', swooping down footpaths and coasting 'no hands' down the steep hills of Sydney's North Shore. His worst prank was placing a lighted cracker in a tin, and tossing it over a sea-side cliff onto a couple beneath. The cracker incident was serious enough to warrant a visit from the police, but since no real harm had resulted, Charles refused to acknowledge his guilt. This refusal would be a life-long trait. If corrected gently for a prank, he would smile disarmingly but express no regret. If corrected angrily, he would set his jaw and remain stubbornly silent.

As a small child, Charles had been fascinated by trains and ships and in adolescence the street trams caught his fancy. Friends recalled how, when riding the tram to the Cathedral from the ferry at Circular Quay, he always rushed to sit near the driver, so that he could ask him questions about the engine. It was this passion for machinery that now dictated his future. Early in 1912 he entered Sydney Technical High School at Ultimo, with the ambition of becoming an engineering apprentice. His studies again gave him no trouble: he came twelfth in a class of thirty-one. And his popularity remained undimmed. His vocal imitations of a bee in a bottle, a bee in a lady's shoe, and a bee on a cat's tail kept his classmates in stitches. They called him 'Mouldy Tooth', 'Buccaneer', 'Pirate' and 'Mad Yank' – this last a reference to his Canadian accent, which had stubbornly evaded the Cathedral School's elocution lessons.

On Wednesday afternoons, when most at the school played sport, he would lead likeminded mates down to the docks at Woolloomooloo. At the head of a bunch of eager boys, he would confidently bluff his way on board a ship so that they could make a tour of the engine room. 'Smithy did not want sport,' wrote one of those mates, 'he wanted machinery', and he inspired his friends to think similarly.

Soon after his sixteenth birthday, Charles became an electrical apprentice at the Pyrmont workshop of the Colonial Sugar Refinery, a company with which his family had had commercial ties since their North Queensland years. The factory was a far cry from the refinement of the Cathedral, and some of the humdrum work assigned to the apprentices probably bored him – he seems to have spent hours winding the coils on armatures. But he had no trouble fitting in with his fellow workers, because he easily made friends from all walks of life. Even so, anyone who made fun of him or tried to pick a fight was in for a surprise, because he was skilled with his fists and gave as good as he got.

His spare time he spent on a battered motorcycle – his 'old bus' – which he rode recklessly on the hilly North Shore. A resident of Spofforth Street, Cremorne, would long remember 'the fair-haired youth who loved to open her out to 60 mph down the half mile hill of that street'. The bike's engine fascinated him. A friend would later remark that there was nothing inside or outside a motorbike that he did not understand. He thought of the machine as a living creature, almost in the same way as he thought of his pet dog. This communion between himself and his motorbike, he later extended to his relationship with aeroplanes.

For more-tranquil recreation, he sailed his family's boat around Sydney Harbour, or to beaches at the mouth of the Hawkesbury River. There, with parents, cousins, brothers, sisters and friends, he would camp during the long Christmas holiday. He would look back on those days spent sailing and swimming and camping and riding his motorbike as among the happiest of his life.

SAPPER SMITH

B y his late teens, Charles was fully grown, and rather dissatisfied with his appearance. Five feet and six and a half inches tall and ten stone in weight, he was, according to the statistics of the time, smaller than the average Australian boy of his age. Moreover, he lacked conventional good looks, even though his blue-grey eyes and sandy hair gave his face a boyish freshness and sweetness. His nose was too large and aquiline, his movements were almost too energetic, and his speech was rapid and clipped; some called it 'staccato'. The combination suggested dynamism rather than attractiveness. However, this changed when he smiled. Spontaneous and joyous, his smile lit up his whole being, and could be irresistible, especially if the recipient was a girl. He was already showing signs of the sexual magnetism that would characterise his years of fame.

Charles was certainly attracted to girls. He already had a crush on pretty Betty Tuckwell, a schoolgirl who lived nearby in Yeo Street. She came from a musical family and was a promising pianist, which added to her charm; and she was similarly smitten, and could not wait to open the romantic notes he sent to her. Keen to widen his social circle, he had recently joined a tennis club, less for the tennis than for socialising.

At any gathering, his smile and his bounding energy earned him female admirers, and he was beginning to realise that, when it came to girls, he could pick and choose.

This carefree adolescent life was about to receive a jolt. In the middle of 1914, rumours of an approaching European war filled the newspapers. Like most Australian boys of his age, Charles had been obliged by law to undergo part-time training in a branch of the militia known as the Military Cadets and, having played at being a soldier for three years, the thought of a real war was exciting rather than frightening. In prospect, it seemed like a grand adventure which would take him to far-off lands he might never otherwise see.

In August 1914, German troops invaded Belgium and France, and Britain and Russia declared war on Germany. Patriotic sentiment that had been simmering for months now rose to boiling point, and fit young men aged between eighteen and thirty found themselves expected to enlist. If they did not, they were likely to receive a white feather, symbolising cowardice. The previous year, Charles's brother, Eric, had joined the Royal Australian Navy, and was soon on HMAS *Sydney*, escorting troop ships through the Indian Ocean. When in November Eric's ship sank the German cruiser, the *Emden*, off Cocos Island, he and his shipmates were hailed as national heroes, and the Kingsford Smiths received many congratulations.

Charles was determined to enlist, and would have done so if his mother and father, who seldom denied him anything, had granted the parental consent required for those under eighteen. According to his mother's recollection, Charles and his father had a blinding row. When tempers cooled, he extracted the promise that his parents would grant their blessing once he reached his eighteenth birthday.

On 10 February 1915, the day after his birthday, Charles hurried to the Army recruiting office, where, although physically fit, he almost failed the medical examination. A minimum height of five feet six inches was mandatory for soldiers, and he passed by a bare half-inch. Eight days later, at the military camp in Liverpool, on the outskirts of Sydney, Private Kingsford Smith, number 1017, swore his oath of allegiance to

King George, and was posted to an artillery unit of the 19th Battery. Three days later, at the Engineer Depot in the Sydney suburb of Moore Park, his posting was altered. Since he had testified on his enlistment form that he could ride a motorcycle and a horse, and had trained to be an electrical engineer, he was reassigned to the newly formed 4th Signal Troop, which was attached to the 4th Brigade of the Australian Light Horse Signals. He was to become a motorcycle dispatch rider. Since wireless was in its infancy, and even field telephones were unreliable, dispatch riding and running continued to be important forms of communication in the military.

Soon his cousin Raymond Kingsford – who was not yet seventeen but had lied about his age – joined him at the Signals training camp at Broadmeadow, near Newcastle. Their Officer Commanding, a respected Boer War veteran named John Fraser, later recalled that the younger, shyer lad relied on his cheeky and self-assured cousin to take the lead. And cheeky and self-assured certainly described Charles's behaviour. In a letter to his family he told his 'Dearest Mater' that he had been promoted to corporal, and consequently was 'badly in need of a little spare cash until pay day', because 'the living of a corporal has to be on a better class than a private'.

It was an impudent ruse because according to Army records, he would not become a corporal until the following year. It is doubtful that his shrewd mother was taken in, although she seems to have sent the money, and also a generous hamper – he had earlier begged for condensed milk and a bottle of Worcestershire sauce. Family tradition relates that when he received his new Army motorcycle, he took it home and tried to entice his mother to ride on the pillion. Since dispatch riders on duty were permitted to travel at breakneck speeds, and were given precedence over other vehicles, he presented this as an intoxicating prospect, which his 57-year-old mother declined. She said she preferred to ride on the ferry.

But not all was fun at the Broadmeadow camp. With overseas embarkation approaching, the training intensified, and the two cousins applied themselves to transmitting messages by means of heliographs and field telephones. Wrote Charles to his mother:

I tell you Mummy, it is hard graft at that for your budding 7th. But I don't mind that as I have already set myself to do it or die in the effort. We have to do in a month what in other circumstances would take six months. You see, the best part of the troop are old hands, and we are absolute beginners. Anyway I am making excellent progress, I am told.

In the Signals – which was considered a branch of the Engineers – privates were called Sappers, a term that dated from the seventeenth century, when military engineers dug tunnels known as 'saps'. Sapper Smith was already well known in the camp as a daredevil and a practical joker, but now, to everyone's surprise, he began earning a reputation as a hard worker.

In May 1915, the lists of deaths and casualties began arriving from the landing at Gallipoli. Charles appeared undismayed, although the grim news may have lent an urgency to a private wish that he was keen to see fulfilled. Like many Australian theatre-goers, he ardently worshipped the Australian actress Nellie Stewart, and he had the good fortune to have relatives who knew her. His sister Elsie and cousin Marjorie Swallow were members of her theatrical company, and Marjorie had recently acted with her in Belasco's play, *Madame du Barry*, at the Theatre Royal in Sydney. Charles had seen the play, and longed for a signed photograph of Nellie to take to war as a good-luck charm.

When Nellie Stewart learned of his wish, she was quick to respond. Selecting a photograph, she inscribed it with words that she had composed especially for young soldiers going to war. 'May you live,' she wrote, 'for those who love you; for the work God has assigned you; and the good you can do. God bless you all.' Charles would treasure the photograph and its inscription for the rest of his life. Indeed, he was said to have been so overwhelmed by Nellie's kindness that he rushed to thank her in person, and by doing so greatly embarrassed his family. For reasons now obscure, he arrived with his head and eyebrows shaven.

On the last day of May, the 4th Light Horse Signal Troop sailed off to war. Charles and his companions watched with interest as carpenters erected ramps so that the horses could travel with them aboard the *Ajana*. They could not foresee that the horses would turn the voyage into a nightmare. As the ship crossed the Indian Ocean, the stench from their stables in the hold became unbearable, and the animals began to sicken and die. 'The poor horses are suffering terribly,' Charles wrote to his mother. 'They are dying every day and it is a gruesome sight to see them being quartered and slung over the side.' The soldiers on the *Ajana* did not fare much better. Rough seas brought nausea, and over-crowding fostered infections; Charles caught influenza. Deep was the general relief when the ship stopped unexpectedly at Bombay to allow the horses and the men to stretch their legs.

One month after sailing from Sydney, the *Ajana* docked at Suez, where the men and the surviving horses were conveyed to a camp at Heliopolis, close to Cairo. Here the Army had requisitioned a casino, palatial hotels and a racecourse, all recently built by a Belgian entrepreneur. The grand buildings were now transformed into a barracks for the troops and a military hospital with 3000 beds, ready to receive the wounded from the nearby battlefields.

To their surprise – for they had assumed that they were bound for France – Charles and his comrades now found that it was not the Germans but their allies, the Turks, that they were about to confront. The battle at Gallipoli, on Turkish soil, had been waged for more than three months and now – like the battle in France – it was almost dead-locked. The original aim of the British and French had been overly ambitious. They had hoped to capture the rugged Gallipoli Peninsula, seize control of the adjacent strait that linked the Mediterranean and the Black Sea, capture the city of Constantinople, and employ that new sea route to send munitions and equipment to the aid of the Russian Army, then the largest in the world.

The losses at Gallipoli had been heavy, and Charles watched the wounded streaming back to the hospital at Heliopolis, and heard stories of the hellhole that the Peninsula had become. Although the ethos of the time demanded a show of patriotism and a stiff upper lip, the sights he saw and the tales he heard certainly distressed him. A letter to his 'Dearest Mum and Dad' betrays that distress. He saw lads of his own age with arms and legs blown off. One had lost most of his face, while another – a dispatch rider like himself – was totally blind, although he had suffered no physical wound. The cause was said to be concussion, and was called 'shell blindness', he told his parents. Although their wounds shocked him, their courage appears to have reassured him. He expressed no regret that he was on his way to war.

Notwithstanding the distressing sights and the heat – for, according to his letters, the temperature in their tents sometimes soared to 125 degrees Fahrenheit – Charles enjoyed his days and nights in Cairo. The 4th Signals Troop had been reformed as the 2nd Divisional Signals Company, and he found his new work engrossing. He rose at four in the morning and worked from five to nine am, in order to escape the worst of the heat. Then he enjoyed free time until four in the afternoon, when he resumed work for another two hours. Although the curfew was at ten pm, he seems to have treated the night as his own, exploring the city with his mates: drinking and playing pranks, and prowling through the bazaar with its quaint, cramped shops. Wherever they went, touts and children followed, offering various services and pestering them for money. Cairo provided 'vice of the worst sort', according to the doctor who lectured the men sternly on the perils of venereal disease.

At the other end of the scale, there were visits to 'the swell place of the town', Shepheard's Hotel, where English tea was correctly served. And there were sightseeing trips to the wonders of the ancient world. Charles wrote his mother a long and sensitive letter about the interior of the Great Pyramid:

> When the candles went out, we stood there and smelt the faint dead sort of smell and felt little cold winds, and realized that we were surrounded by 200 feet of stone on every side, it's enough to give you the horrors. It seemed as if lots of people, unseen presences, were there, who didn't have any business there.

Most thrilling were the rides on his dispatch motorcycle. The eight-mile stretch between Brigade Head Quarters and Cairo was the perfect speedway – 'absolutely like a table' – and he once covered the distance in the record time of seven minutes and forty seconds. His 'hair-raising stunts along the Mena Road' were the talk of the camp.

In August, Charles learned that pride went before a fall. He was tossed off his bike and damaged his knees and thigh and hip. The pain was bad and the diagnosis was sciatica or rheumatism, maybe the result of the fall, or maybe related to a rotten tooth, which was summarily extracted. Charles realised that his mother was probably worrying, so he quickly put her mind to rest. Their officer in charge, or OC, John Fraser, was a second father to him, he wrote reassuringly: 'a splendid man' who 'looks after us as if we were his own children and pays us out of his own pocket if we are short. Also, he gets us the best of everything, and works hard to do so. We are fearfully proud of him.'

By the first week of September, Charles was mending fast, and driving colonels around in a 'jolly fine Sunbeam car'. By now his comrades were preparing for the train trip to Alexandria, where they would board their troopship for Gallipoli. His cousin Ray, with whom he had been inseparable all these months, was preparing to go: 'lucky young Ray,' Charles wrote enviously. But it was 'heart-breaking' to watch 'the old troop' dispersing. Still, he was well enough now to return to his motorcycle, and he knew that his own embarkation orders would soon arrive.

The day they came, Charles set out on his bike for a last frenetic 'flutter'. But as he reached a corner, with the speed dial clocking 50 miles per hour, an Egyptian pedestrian loomed in sight. The 'poor devil' was 'chucked fully 12 yards into the sand'. Defiant rather than repentant, Charles confessed the details to his mother by letter. The unfortunate 'Gypo', he told her, 'was a brainless fool' who 'hadn't enough gumption to get out of my way'. 'I got him to hospital,' he continued. 'I think he will recover all right, as concussion and a broken leg is all he got.' As a type of patriotic afterthought, he added, 'I hope we get the bikes over pretty soon, and perhaps I will be able to do for the Turks in the same way.'

His letter was written on 22 September, aboard the troopship *Melville*, which was bound for the firing line. It seems likely Charles's defiance over the accident was superficial; really he was sorry and scared, and craving maternal forgiveness and reassurance. He addressed his mother as 'Dearest Mummy', and signed himself in the way he had done since childhood: 'Your loving boy, Chilla'. One doubts that either of them was reassured.

GALLIPOLI:
NOTHING TO RAVE ABOUT

Arriving at Anzac Cove on the Gallipoli Peninsula on the early morning of 25 September 1915, Charles saw a narrow, open beach, a straggling pier and jostling troops, all within range of enemy guns. In the high cliffs that rose from the beach, and in the deep valleys and high ridges behind them, trenches, dugouts and tunnels stretched for miles, housing or protecting exhausted soldiers from both sides. Locked in a deadly stalemate, the Allies were unable to advance far inland, and the Turks were unable to drive them back into the sea.

Had Charles arrived the previous month, he would likely have been wounded or killed. At the very least, he would have witnessed carnage so horrible that it might have haunted him for the rest of his days. By the end of September, life on the quietened battlefield had become a mixture of misery, fear and boredom. In their dugout homes – holes gouged from the hillside – the soldiers lived almost humdrum lives between the bouts of warfare. Hours were spent each day cooking, cleaning their kits and scavenging firewood. Other daily tasks included fetching water, which was severely rationed, and delousing themselves and their clothes. For recreation, they played cards, sang songs, composed poems and gossiped about happier times, trying somehow to 'forget the present in the pleasure of the past'. Early on, Charles acquired a portable gramophone, which soon made him popular. 'I was in the dugout next to him,' remembered one of his mates, 'and many a merry time we had in his dugout.' He was always 'a good sort' and 'a hard case'; meaning, in the slang of the time, that he was good mate and an incorrigible joker.

Charles's dugout lay high on a ridge close to the heavily fortified Plugges Plateau. In his first letter home, on 6 October, he tried, unsuccessfully, to describe his life in reassuring words. 'We have been here about ten days and under fire practically the whole time,' he told his mother, 'but nothing very heavy. Our dugout is pretty comfortable, with its stove and kind of sandbag balcony and I am dispatch carrying.' (By 'carrying', he meant running on foot with dispatches, because the rugged terrain precluded the use of motorcycles.) He confessed that it was perilous work: 'Snipers are pretty bad at the foot of our gully and get our chaps fairly often. One has to do a sprint or get a bullet after him.'

In another letter, he recounted a close encounter with a bullet. 'When coming back,' he wrote, 'I got into the wrong sap [trench] and found myself in an open sand patch near the beach, and within range of the Turkish machine guns. I hurriedly sought shelter in the sap, but not before a bullet frayed the edge of my cap.' John Fraser later recalled a breathless Sapper Smith coming at dusk to his dugout in search of reassurance. He admitted to having taken a short cut over open country on his way to visit friends in the 7th Brigade. Fraser made him promise not to leave the relative safety of the communication trench again. The young daredevil who so enjoyed playing with danger was beginning to discover what it was like to seriously court death.

By late October, winter was approaching, heralded by heavy storms that flooded the trenches and dugouts. Drainage gutters were hastily cut and dugouts were roofed, but there was no remedy for the sudden drop in temperature. Still in their summer clothing, the soldiers shivered, and none more than Charles. In response to the cold, his rheumatism returned in force: he could not walk 'without twinges of actual pain'. 'I have to chuck up dispatching for a while, and become a cook's mate until I can move about more,' he informed his parents. 'I am afraid if it gets much worse I will have to go back to Egypt until the motor cycles come over.'

The storms also brought winds so fierce that food and water could not be unloaded, and rations were cut to a minimum. The wild weather wrecked the mail boats, which vexed Charles, who had asked 'Dearest Mummy' to 'forward some chocs but label them sox if you want to be certain'. He had also begged for a warm sheepskin vest, but that too failed to arrive. Fraser, concerned about Charles's condition, advised him to apply for repatriation, and he must have been sorely tempted because the chronic pain was dragging him down. But pride would not let him yield. Stoically, he replied that he would 'stick it out unless it absolutely cripples me'. Fortunately, Fraser was able to send him as a clerk to the Signals office, where conditions were less harsh. 'I live jolly well, considering the circumstances,' Charles wrote home more cheerfully. 'Out of our issue of biscuits, I grind sufficient to make porridge. It is A1 stuff. Also I make pancakes and rissoles and am quite an expert cook.'

In November came a blizzard of snow, followed by three days of freezing wind, reportedly the worst blizzard on the Peninsula for forty years. To those Australians who had never seen snow, the first fall seemed magical and provoked jolly snowball fights behind the lines. But soon the soldiers were suffering frostbite, some even freezing to death. To Charles, who had spent winters in Vancouver, the snow was no novelty and the intense cold greatly increased his misery: the 'pain at night', he wrote to his father, 'is pretty solid'. Huddled in his dugout, he longed for warmer days. When the warmth came, so did the thaw: trenches and dugouts flooded, and corpses of buried soldiers came tumbling down the swollen creeks. So severe was nature's bombardment that rumours of evacuation, which had been circulating for weeks, suddenly gained credence. 'I don't think this war is going to last much longer,' Charles wrote to his mother, 'and when it does stop I will be the happiest in existence, as I didn't realize what home was until I left it.'

The evacuation came in December, having been planned with amazing stealth and speed. The Signals, who in four strenuous days laid many miles of telephone cable, ensured the quick and accurate coordination of the evacuating groups. During the nights of 17, 18 and 19 December, 80,000 men, 2000 vehicles, 200 guns and 5000 horses and mules were shipped to safety from the Gallipoli Peninsula. They left in darkness, and in such a way that when the sun came up, everything appeared unchanged: 'the Turks,' wrote Charles, 'were absolutely bluffed.' Delighted to be able to post home cheerful news, Charles wrote a lengthy account of the evacuation:

> Everyone behaved splendidly and all was an absolute success. The casualties amounted to one or two men, which is unsurpassed in history ... No doubt there will be mixed feelings at home about this great retreat, but in everyone's opinion it was the only thing to be done as advance was impossible and to continue there only meant the loss of valuable lives every day which could be ill spared.

Charles wrote his account on 30 December 1915 on the island of Lemnos, his first landfall after leaving Anzac Cove. Five days later he sailed for Egypt. In camp at Heliopolis he caught up with Rupert Swallow, who had contracted mumps and was deeply disappointed that he had not managed to reach Gallipoli. In response to his cousin's lament, Charles 'set the record straight'. 'I told him it wasn't anything to rave about,' he wrote firmly. 'I can really say I'm glad to see the last of it.'

Back in the warmth of Egypt, Charles's rheumatism improved, and he was reunited with his beloved motorcycle. 'I am doing a lot of work on my bike,' he wrote to his mother, more contented than he could say to be tinkering again with its engine. Early in March his former OC, John Fraser, took command of a new signals unit, the 4th Division Signals Company, and Charles and his cousin Ray Kingsford soon found themselves transferred to it.

For months there had been fears that the Turks would advance through the Sinai Desert and capture the Suez Canal. Accordingly, while some of the Anzac troops sailed directly to France, others – the 4th Signals among them – remained behind to defend the Canal Zone, and were housed at a large tent camp on the edge of the desert to the north-east of Cairo. Here, at Tel el Kebir, Charles was promoted to provisional corporal. On 27 March 1916, he and his fellow Signallers were ordered to march with the 12th Infantry Brigade to Serapeum, a dusty little port on the bank of the Suez Canal. It was a distance of some 40 miles through the desert, and each soldier was required to carry full battle kit and ammunition, in all about 40 kilograms. Three days were allotted for the journey, with a stop on the second night at the Army camp at Moascar. The heat was intense and the laden men suffered cruelly from heat exhaustion. By the time they staggered into the tents at Moascar, many were dehydrated and required medical attention. So affected was one of the brigades that there were rumours of deaths, and the brigadier in charge was relieved of his command.

Charles not only survived the march, he did well, and he was confirmed as a full corporal when he reached Serapeum. 'By Jove, it's great having meals in the NCOs' mess,' he wrote to his family. 'No more washing up and good tucker in abundance.' His weekly pay was increased to ten shillings, of which 'Dearest Mummy' was to receive half. Knowing his spendthrift ways, his mother had insisted on this since his enlistment, and was putting her half-share towards a block of land for him, as a nest egg for his future.

Best of all, on those afternoons when the temperature rose to 120 degrees, Charles and his cousin Ray were permitted to swim in the Suez Canal. When 'steamers come along', he wrote, 'we swim out to them and are occasionally lucky enough to have a tin of cigarettes or some other luxury thrown to us'.

The urgent topic of conversation at the camp was their next destination. This was likely to be France, where the fighting was intense and the casualties were high. At the beginning of June, the 4th Division Signals were ordered to Alexandria, to sail on the troopship *Kinfauns Castle*. Voyaging across a submarine-infested Mediterranean to Marseilles, Charles confided a fear that would gain importance as his life progressed. 'I don't mind being in action on land,' he informed his parents, 'but the thought of being stuck up miles from nowhere at sea with mighty little chance of rescue gives one a nasty taste.' Was he remembering that day at Bondi Beach when he was plucked from the surf by the lifesavers? In the Mediterranean there were no lifesavers, and consequently no reassurances. If his ship sank, he stood little chance of survival.

Thankfully, the voyage was accomplished safely and the Signallers were unloaded in Marseilles. From there they were transferred to Calais, bypassing Paris, and from Calais they moved north to Bailleul, in Flanders. Their final destination was the Ypres Salient, the war zone adjacent to the Belgian border. Fought over since October 1914 by the entrenched armies of Germany and the Allies, its situation on the road to the northern channel ports made it a place of immense strategic value.

On 14 June, in the small town of Merris, Charles was promoted to sergeant and placed in charge of the motorcycle section of the 4th Signals. 'Good luck, wasn't it?' he wrote modestly to his parents. Luck was on his mind that day because he had just survived his first bombardment, infinitely 'heavier and more plentiful than on Gallipoli'. On the first night 'sleeping in a shallow open trench', he 'woke up to see whiz bangs unpleasantly close. At about 4.30 am stones and earth almost buried me.' A short while later, on his way back from the supply column, a nine-inch shell burst on the road about thirty yards ahead of him. 'The concussion,' he wrote, 'sent me skidding up the footpath and back across the road and into a ditch. I was not hurt but I got a nasty fright.'

For the next three months, his life was a succession of nasty frights, as thousands more shells burst around him, killing many of his comrades and blasting the flat French fields into a scene from hell. In July he exchanged Flanders for the Somme, which was now experiencing the heaviest fighting. With the 4th Signals he travelled to Rubempré, Canaples, Vignacourt and Doullens, towns in the vicinity of Amiens. There, amid the shells and bombs and aerial grenades, with the ground swaying and rocking from the concussion, the Signallers extended and buried the telephone cables connecting the trenches; and Charles, on his motorcycle, dodged from headquarters to headquarters delivering dispatches. 'Some of the sights out there are really sickening,' he wrote to his mother. 'But one gets hardened quickly.'

He had already developed a self-protective shell; otherwise, he could not have survived emotionally. He saw many who did not, and who sobbed like children, 'their nerves completely gone'. But no amount of self-protection could fully shield him from the sights and noise and stench. To escape to the peace of the YMCA hut was a solace, even though he felt 'horribly homesick' as he listened to a pianist play 'Night and the Stars Are Gleaming', a sentimental ballad that his family had often sung at home.

His other solace was girls. 'My word, Mummy,' he wrote on 19 July, 'there are some bosker girls in this country. I don't know whether I am susceptible, but some of them would turn the head of a statue.' There seems little doubt that he was indeed susceptible, and that thereafter there were many sexual encounters from which he derived much-needed comfort. That he wrote like this to his mother suggests that he knew she would not approve, but hoped, given the circumstances, that she would understand. And since he continued to write in this way, the indications are that she did understand, and this also gave him comfort. One of the more respectable girls invited him to dine with her family. 'Although she can't speak a word of English,' he reported, 'and my French is far from fluent, we get along fine.' This easy empathy with women was one of his gifts.

Late in August, while he was at Rubempré, rumours began to circulate that the Royal Flying Corps was about to seek recruits from the AIF. Formed in 1912 and devoted at first to reconnaissance, this airborne branch of the British Army had by now become a formidable fighting force. Its current pilots flew planes equipped with machine guns, studied the tactics of aerial combat, and were vying with the Germans for supremacy in the air. Usually British boys were recruited, however a few Australians had performed so well that the War Office now proposed to admit 200 soldiers from the AIF. They could be of any rank, provided they were deemed suitable, and would take a 'Special Reserve Officers' Course'.

John Fraser, recognising officer-like qualities in Charles, seems to have alerted him to the offer. He also seems to have pointed out the opportunities that such training might bring when the war was over. Fraser's encouraging words fell on receptive ears because only the previous year, Charles had thought of joining the 'Aviation Corps'. Then he had viewed it as an exciting way to reach the battlefield more speedily; now he saw it as an honourable escape from the horrors and privations of trench warfare.

On 16 September 1916, still on the Somme, Charles made his application, and Fraser endorsed it. In his written reference, Fraser praised Charles's engineering and signalling skills, his reliability and his 'good moral character'. Charles himself declared on the application form that he had a sound knowledge of internal combustion engines, had been trained as an electrical engineer and had excellent speed with Morse code: he could read twelve words a minute and send eighteen words a minute.

Before Charles could be selected, he was obliged to appear at an interview. A fellow Signaller who applied at the same time has left samples of the questions asked: 'What was your school? Have you attended a university? Have you yachting experience? Or played polo? What is the extent to your musical knowledge?' Many Australians taking the interview were from humble backgrounds, and they found the questions objectionably elitist. But they were shrewd enough, and cheeky enough, to lie unblushingly.

Charles did not need to lie. Although he had not been to university or learned to play polo, he was well able to ride a horse and manage a yacht, and – thanks to St Andrew's Cathedral School – had had an excellent general, musical and social education. He was able to answer the questions truthfully.

Back in Flanders at Reningelst, Charles wrote home, overjoyed at having been accepted:

I have thought this matter over very seriously for a long time, and I think you will agree I am doing something to my own advantage. I hope so, for I would not do it if you would rather not. My intentions for the future are to take up the game permanently after the war in Australia. At home there will be big possibility for our services; and it is an honourable and interesting career for a young man as well as a splendidly paid one. Of course all this depends on my being able to earn my commission, which I feel capable of doing in a few months.

In November, he excitedly cabled his parents: 'Address now One RFC Cadet Battalion, Denham, England.' He had started on his new life, and nothing would be the same again.

ONE OF THE VERY
BEST FIGHTERS

T he village of Denham in Buckinghamshire was the home of
the Royal Flying Corps Special Reserve Cadet Battalion,
devoted to the training of the AIF recruits. Charles, like
most Australians of his time, had a strong emotional attachment to
the 'Mother Country', and to have exchanged the horrific trenches for
this picturesque English village made his spirits soar. All aspects of his
training thrilled him. The uniform was 'really fine', he wrote to his parents: 'a double breasted tunic and a Glengarry cap'. And the schedule
was really demanding: they rose at six, paraded from nine to twelve-
fifteen, paraded again from two to four, then it was tea and work until
nine pm. 'We keep going like the very dickens,' he reported proudly.
'I was only 11 stone when we came over here but I now weigh about 12
stone, and am as fit as anything. I am nearly 40 inches round the chest
which is fairly hefty.'

In the classroom the trainee officers studied military law, interior
economy, hygiene, topography and infantry training, this last subject
enabling them to pass the standard examination that was mandatory for
all officers in the British Army. Charles wrote home excitedly:

By Jove I wish we were already flying, but I don't suppose it will seem very long to wait. Of course they will shove us straight out to France on completion of the course, to do our bit in the air, for which calling, though perhaps not too safe, I am quite convinced I was cut out for.

His social life was equally absorbing, and such were its demands that he was obliged to cable home for £10 to buy clothes; 'one has to get a start with some good clothes,' he explained to his mother. Remembering what it was like to be young, Catherine sent the money. She may well have suspected that the clothes were to impress a girl, and for that she had sympathy. And of course she was right – he was keen on a local girl. On most nights they were out together until well past his curfew. Discovered as he sneaked back to camp, he would be obliged next day to pound the parade ground as punishment, carrying his heavy pack. Charles's flourishing sex life was a source of wonder and envy to his comrades.

Of wonder also was his penchant for poaching, for, like many Australians, he did not accept that wild game could be privately owned. Moreover, he had discovered that wartime shortages of food had produced a more tolerant attitude towards poachers. Outwitting the gamekeepers at a nearby estate, he and a few friends brought back regular pheasant suppers. With a cheery fire to eat them by, and musical offerings from the host and guests, Charles's hut became one of the most convivial in the camp. In this era of nicknames, some called him Kingy, others called him Smithy, while he remained Charlie to old friends and always Chilla to his family: he invariably signed his letters home 'your loving boy, Chilla.' But Smithy was the name he now most answered to, and the name by which he would be widely known in the years ahead.

That December, Smithy – as he shall now be called – received leave to travel to Scotland to see Eric, whose ship, the *Sydney*, was with the Grand Fleet in the Firth of Forth. Living aboard the cruiser for a day or so, with a cabin to himself and the use of a servant, he and his paymaster brother had much to tell one another. Smithy reported the visit excitedly to his parents.

Less than a month later, he hitched a ride to the aerodrome at Hendon and went up in an aeroplane for the first time. It was a moment he would never forget. 'We went up to 8,000 feet,' he reported, 'and everything looks like a map. You can't see people at all, and the roads and streams are just wee streaks of light.' Since childhood he had disliked heights, and had sometimes felt apprehension at the thought of flying high. But what struck him most as he rose into the air was the 'feeling of absolute security'. Why, being in the open cockpit was not much different from being in a racing car, with the wind rushing past his face. He could not wait to begin his instruction.

Having passed his Denham exams, he was transferred on 26 January to the RFC number 3 Military Aeronautical School at Oxford, where he was billeted at Exeter College. 'My word, it is an improvement on Denham,' he wrote home. 'We are put three to four in a room – the same rooms that varsity students occupy in normal times – and treated just like officers.' But again, the workload was heavy. Divided into squads of ten, the recruits were up at six am and paraded and drilled; after breakfast they marched to the Museum for workshops and lectures until late afternoon. Their studies were 'all about internal combustion engines, wireless, morse code, construction of planes, etc' – subjects that, to some extent, he already understood.

Falling ill in February, Smithy spent a week in hospital in Oxford, but returned in time to study for his exams. 'Commissioned,' he cabled home happily on 16 March. He was a second lieutenant, and the prospect intoxicated him. Now, at last, it was time for flying lessons, and he was buoyed up by the advice he had had from his instructors. 'They tell us,' he wrote, 'that a motor cyclist always makes a good flyer.' The prospect reassured him as he travelled to Number 8 Reserve Squadron, at Netheravon, Wiltshire. He needed confidence because a bleaker place could scarcely be imagined. Situated on the edge of windy Salisbury Plain, about ten miles from the nearest railway station, it was a rambling village of huts and hangars, and old and rickety planes.

Smithy began his training on a Maurice Farman Shorthorn: a 'strange sort of biplane' made of wood and canvas, with 'an astonishing confusion of wires holding the wings together'. The pilot and instructor sat in two bucket seats, one behind the other, and 'right behind the pilot was a hefty rotary motor, ready to land on the back of his neck if, as was more than likely, he ever crashed'. Early in the war, the Maurice Farman Shorthorn had been a valuable reconnaissance aircraft, but by 1917 it was judged fit only for training. To Smithy, however, it was the most desirable plane in the world. High was his excitement when, at six-thirty am on 27 March, he climbed into the pilot's seat and listened to his instructor's orders, shouted over the roar of the engine. The lesson lasted twenty minutes.

The following morning his second lesson lasted thirty-five minutes, and he was allowed a brief time at the dual controls. From then on he flew daily with his instructor, and practised straight flying, turning and landing, and was in charge of the dual controls for increasingly long periods. On 15 April at seven-fifteen am, he flew solo for the first time, and before the week was over he had flown two hours and fifty-five minutes solo. 'By Jove it was grand having the "old bus" up in the air by myself. I am "some" pilot now', he cockily informed his parents; soon, he boasted, he would be stunting and looping with the best of them.

His cockiness was premature. Four days later, while landing solo, he smashed the plane's undercarriage. 'I was going to land', he told his parents,

had judged things nicely and my wheels had just touched the ground, when a large ditch which was hidden from me by grass, struck my wheels and bounced the machine violently up to about 50 feet ... Of course, I fell with the bus like a ton of bricks. The machine was wrecked, but I being marvellously lucky, only got shaken and a bit bruised.

The crash dinted his confidence and earned him a reprimand; even so, he was judged sufficiently competent to take the advanced course at the RFC's Central Flying School at Upavon. Here the landscape was equally bleak, and the school was just as dismal, but at least he was nearer to the town of Devizes, where he could go to dances and meet lively girls. Best of all, the planes were of the latest design.

The star was the SPAD S.VII, a single-seater biplane equipped with one machine gun (sometimes two) that was synchronised to fire through the gaps in the propeller. Able to rise to 20,000 feet and fly at 130 miles per hour, it was an excellent climber and a wonderful diver, ideal for scouting, aerial combat and ground attacks. Such planes were designed to meet the recent revolution in aerial warfare. In the early days of the war, lone flyers had stalked their prey as though engaged in airborne hunting trips. Now – largely through the initiative of the German air ace Manfred von Richthofen – aerial fighting had become a team game, and planes hunted in packs and operated to a plan.

At the start of June 1917, Smithy wrote home jubilantly: 'At last I have my wings.' He had been airborne for forty-six hours and fifteen minutes, of which thirty-six hours and forty minutes was solo flying. He was also stunting, and proud of it. 'This morning,' he told his parents, 'I looped nine times, flew upside down for 30 seconds; I also did a few things known as stalls, side loops, spinning nose dives. It's grand to be able to chuck a machine about, any old way.' Since loops and rolls and dives and spins were vital manoeuvres in aerial warfare, his pride was well-founded. His instructors were delighted with him. He was beginning to show that rare quality of being reckless enough to try anything, and skilful enough to rescue himself when his recklessness was misplaced. His training was now virtually over, and he was ready to join a squadron.

His first posting, a temporary one, was to Saint-Omer, in French Flanders. 'Things look very bright over here,' he told his mother in a long letter home. He had been practising formation flying and had engaged in a dogfight, but whether or not he downed the Hun he did not know. He had also visited his old Signals unit. John Fraser, now Major Fraser DSO, was delighted to see him 'swanking it in an officer's garb', and his Kingsford cousins, Jack and Ray, were hoping to join him in the Flying Corps. Late in June, an attack of German measles grounded him, but by July he was well enough to join Squadron 23 at La Lovie, on the fringe of the Ypres battlefield. Equipped with SPAD VIIs, and specialising in patrols, the squadron was preparing for the third Ypres offensive, to be launched by the Allies on the last day of July.

To Smithy, La Lovie was familiar territory, since he had been stationed at nearby Steenvorde and Reningelst the previous year. He knew the good and bad points of the place. On the good side, it was hops country and the beer was famous. Just two miles away was the town Poperinge – known as Pop to thousands of soldiers. There he could find recreation at a lively officers' club called Skindles, or at a cafe called La Poupee, where the proprietor's daughter – nicknamed Ginger – flirted enthusiastically with British officers. Or he could drink a cup of respectable English tea at Talbot House – known in Signals' jargon as Toc H – run by the Rev. Tubby Clayton.

Writing to his parents on 10 July, he emphasised the good points. He assured them that he was very comfortable, 'in spite of the fact that the Hun sometimes shells us with long-range guns'. 'The squadron people are a tip-top crowd,' he added reassuringly, 'and I get on very well with them.' He was adept by now at painting a rosy picture, whatever the true circumstances. He did this partly to spare his family's feelings, and partly to evade the scissors of the military censor.

What he dared not tell was the shocking death rate. A pilot was lucky to last a couple of weeks, and all the pilots knew it. On his arrival, the OC, Major Wilkinson, was brutally frank with him. 'Do you know,' he told Smithy, 'that we are losing three men a day from this outfit? And every one of them are young fools like you. The ones who live are the ones who obey orders ... Obey your patrol leader always.' As Smithy looked around the mess, he saw that his young comrades had old faces. 'There were many vacant places,' he remembered, and alcohol was their consolation. 'We drank a lot because there was nothing else to do.'

On 10 July he began to fly patrols. Describing them years later, he wrote:

Sometimes our squadrons would sweep the sky in bands twenty strong, looking for trouble in the shape of Hun machines, and generally finding it. We flew low over enemy aerodromes and trenches, ground strafing and attacking anything in sight, with our drums of Lewis fire. At other times we flew high, waiting at 15,000 feet to pounce on our enemies.

On 14 July he 'got his first Hun'. 'Eight of us attacked about twenty Huns,' he informed his parents, 'and had the dickens of a fight ... Altogether we downed about five. I had bad luck just after bagging my bird. My gun jammed and I had to leave the scrap and tootle off home.' The day before, he had been in another scrap which turned into a 'narrow squeak'. His gun jammed early and he had to head for home with three Huns on his tail. When he finally reached La Lovie there were 'holes all over the machine and one burst of a dozen alongside my ear'. Afterwards he found a hole in the collar of his flying jacket.

Two days later, Smithy thought he had downed a German plane that was about to attack an Allied balloon. 'I gave him 200 rounds,' he reported excitedly, 'while he was running billy-o.' But the following week an emotional reaction set in. Exhausted and despondent, he fretted over his lack of letters from home, fumed over his deferred pay, and developed a boil on the back of his neck so painful that he was obliged to visit the hospital.

By August he was in better health and spirits, which was just as well because the third Ypres offensive was gathering pace and patrols were long and frequent. He often flew for six hours a day, which was extraordinarily tiring since he needed to stay constantly alert. Nevertheless, it was now that he did his most effective fighting. On 7 August he spied a farmyard in Beselare, crowded with men and lorries. He swooped down, guns firing, and saw the men fall. On 9 August he forced an enemy machine down over Ten-Brielen. On 10 August he had his greatest success. Writing to his parents, he described the scene:

We were out at 4.30 am looking for something to strafe. I saw this chap flying very low, just beyond the Hun trenches. So I dived and fired at him, driving him back into Hunland, and lower down all the time until he hit the ground and turned right over. So I came at him again and had the satisfaction of seeing him catch fire. Later on, I killed a lot of Hun troops and set fire to some wooden huts in Hunland. So you see I had a good morning; and have been mentioned in the Army Commander's report as having done bold and valuable service.

According to the record books of Squadron 23, the 'Hun plane' went down over Gheluvelt, and the 'Hun troops' were about forty German infantrymen, resting on the Langemaarck to Poelkapelle road. Recounting the incident later, he recalled that the road was sunken, lined on each side with shattered poplar stumps. He remembered that his mind was 'completely occupied with one unearthly desire – I must sneak down on them', which he did with blazing guns. As his bullets ripped along the road, he could see the soldiers scrambling and falling. He recalled being filled 'with quite unholy joy' and screaming at the top of his voice. When he reached the aerodrome he realised what he had done. 'I leaned against the fuselage,' he said, 'and vomited.'

Four days later, on 14 August, Smithy's was one of eight planes that set out from La Lovie on an 'offensive patrol'. The cloud was heavy and visibility poor, so after half an hour the leading pilot signalled 'Go home'. Six planes obeyed but two did not, and one of those was Smithy's. Instead, he turned east in search of prey, continuing on through cloud until he found himself over the German stronghold in Houthulst Forest. He was about to turn back when he spotted 'two Hun 2-seaters' far below him. His own account, written soon after the event, takes up the story:

So I thought I'd have a hurry-up scrap and tear off home as it's not too healthy at low altitudes 10 miles over and by oneself. I proceeded to turn the old bus bang on her nose and dived at the nearest Hun at speed of 220mph. That was one of my last coherent recollections. I sort of recollect a fearful clatter just in my ear and a horrid bash on my foot which made me think the whole leg had gone. Then I fainted. It turned out the bullet had busted numbers of nerves and the shock sent me off. When I came to, about 30 or 40 seconds after, I was spin-ning nose first, down to Hunland. The little holes and chips of wood etc were suddenly appearing all around me where Hun bullets were chewing up things. I tried to turn round to scrap. But my whole leg was paralysed, and I could only fly straight ahead as fast and steadily as I could for our lines and pray the Hun wasn't a very good shot.

It so happened that a small German scout plane had been hiding in the clouds, and seeing Smithy attacked and vulnerable, it had rushed in to finish him off. Fortunately, they were now approaching Allied lines, so the assailant soon fell back.

'I was feeling fairly groggy then,' Smithy remembered. 'Blood was gradually filling my flying boot past the knee. I wondered if I could get back to the aerodrome. I chanced it, and more by instinct than anything else made a moderately good landing, and then crawled out of the bus and fairly collapsed.'

A BIG THING
TO HOPE FOR

From his hospital bed, Smithy wrote to his parents: 'It was a remarkable escape, considering there were about 180 bullet holes in the machine and dozens round my head within inches.' A kindly fate had spared his life, but the wound in his left foot was so severe that it had needed urgent attention. Carried first to the nearest casualty clearing station, he was given emergency surgery before making the long train journey to the military hospital at Abbeville. Two of his toes were amputated, 'leaving a long and deep V indentation between the big toe and the fourth toe'. Although he felt 'rather blue' at losing his toes, and was often in pain, he made light of his injury; 'they say I'll be able to walk alright in three or four months,' he told his family reassuringly. He also presented a brave face to his visitors, among whom were Major Fraser and Major Wilkinson. Wilkinson had informed Smithy's parents that their son was 'one of the very best fighters I have had, full of grit and a splendid war pilot'.

As soon as he was well enough, Smithy was shipped to England and taken to Sir John Ellerman's Hospital for disabled and paralysed officers. Run by the Red Cross, this new nursing home was housed in St John's Lodge, a splendid Regency villa set in five acres of garden in the Inner Circle of London's Regent's Park. He was glad it was such a pretty place because he expected to be there for many months, since several operations were scheduled for his wounded foot. Time passed slowly, and he was restless and frequently homesick. The tantalising knowledge that a convalescent officer was sometimes awarded home leave was almost more than he could bear. 'To have the chance, slight though it is, stuck before me is driving me frantic,' he told his mother in a letter.

In September his surgeon cancelled the operations. The indentation was to remain unsewn, and his foot was to be left to heal unassisted. Smithy tried to joke about it. 'It will give me a certain Satanic aspect – a cloven hoof as it were,' he wrote sardonically. Even so, it was a somewhat depressing moment, because he had been hoping that surgery might at least partially repair the injury. Anxious questions flooded his mind concerning his fitness to fly. Would his foot be strong enough to work the controls? Would his nerves have recovered? 'I'd like to go up for a flip to see how my nerves are,' he wrote to his mother. 'We can't properly tell until we get in the air what state a wound has left them in.'

Late in September, to his great surprise he received a telegram that completely changed his mood. Major Wilkinson's report had impressed his superiors, and Smithy was to receive the Military Cross for 'conspicuous gallantry and devotion to duty'. The citation was glowing:

He has brought down four machines, and has done most valuable work in attacking ground targets and hostile balloons. Of the latter he forced at least nine to be hauled down by his persistent attacks, during which he was repeatedly attacked himself by large hostile formations, and his efforts undoubtedly stopped all hostile balloon observation during a critical period.

Feeling 'frightfully bucked', Smithy began to imagine himself returning to Sydney 'covered with stars, wings, Military Crosses and large grins'. But happy imaginings failed to banish all his anxieties, and he was struck by the irony of it. He had lived for so long refusing to be afraid in case fear captured him, and now that he was safe and could look danger in the eye, it seemed to be overwhelming him. On 5 November he wrote to his parents: 'My nerves have gone to the pack. I'm afraid I'm in for breakdown if they get worse.' Then, suddenly, another telegram bolstered him: he was to attend Buckingham Palace at ten am on 12 November to receive his Military Cross.

His investiture became one of the greatest days of his life, and never would he forget the kindness of George V. At the sight of this wounded young airman limping towards him on crutches, the King's formality melted and he became gentle and fatherly. He spoke to Smithy for nearly five minutes, asking where he came from and where he had fought. As he pinned the cross to Smithy's breast, the King declared warmly: 'Your mother will be proud of you today.'

Etiquette required Smithy to back away from the royal presence but, overcome by emotion, he tripped over his crutches. Officials rushed to rescue him, but the King reached him first. He helped Smithy to his feet, gave him his arm while the crutches were repositioned and said how sorry he was for the mishap. Smithy was allowed to leave in the normal way, and by the nearest exit. Later, he would observe that he was one of the very few who was ever invited to turn his back on the King.

On 20 December Smithy was summoned to Caxton Hall to appear before a medical board. He had been dreading this summons, because his home leave lay in the board's hands. Fortunately, its decision was quick and favourable: he was deemed unfit for service until 5 May 1918, and there was no objection to his returning home, although it was not judged a medical necessity. Smithy had been hoping that the Air Ministry would pay his fare home, and he knew its decision depended on the medical board's verdict. Smithy now became a little devious. In applying for the money, he claimed that a sea voyage had been recommended as part of his cure. Whether this was believed one does not know, but his application was accepted and his fare paid. He was to sail via America, as this was considered the safest wartime route. 'I'm frightfully excited and can hardly believe it,' he wrote to his parents triumphantly. 'It will mean a month or six weeks at home.'

Smithy sailed from Liverpool on the *Star of Lapland*, arriving in New York on 28 January 1918. It was just twelve days before his twenty-first birthday, but he confessed to feeling far older. When he reached home, his family and friends at once noticed the change in him. The Smithy they had known was exuberant, boastful, at times bumptious. Now, when asked about Gallipoli or France, he was reluctant to speak; and when he did speak, he said very little. No words could describe the horror he had witnessed – and even if he did try, how could anyone in peaceful Sydney possibly understand it? Moreover, if he told the truth, they might condemn him for a lack of patriotism. It was far safer – when obliged to speak – to adopt a patriotic tone, say as little as possible and echo the official propaganda. 'The Huns are well licked, and they know it,' he told a reporter who sought an interview. 'The German is a good pilot, but when it comes to actual fighting in the air he has to play second fiddle to our men.'

Smithy was reluctant to wear his Royal Flying Corps uniform. Having watched so many of his comrades die in it, the uniform was sacred to him, but to his fellow citizens, who had never seen it before, it looked comic, even effeminate. Onlookers whispered about the jaunty Glengarry cap, and a few sniggered. While travelling across Sydney Harbour one day on the ferry – his parents now lived in Shadforth Street, Mosman – two youths made fun of his cap. Silently he bashed their heads together before returning to sit beside his brother, Leofric, as though nothing had happened. Thereafter he avoided wearing his uniform in public, until a sympathetic girl, whom he was hoping to impress, cajoled him into donning it again.

The warmth of his family began to heal his emotional wounds. What a relief it was to be back at last with his 'dearest Mother and Father', to whom he had addressed so many letters. A photograph shows him sitting with his brother Wilfrid's family beneath a large fig tree; there would be many more of these happy family gatherings in the months he was at home. Physically, however, he was not well. He still needed a stick to walk and his foot continued to give him pain. The injury had left him with chronic neuritis, for which there was no cure. Smithy applied to the Air Ministry for an extension of leave; on 23 April he went before a medical board in Sydney and his request was granted. It was June 1918 before he sailed for England on the *Orontes*.

The *Orontes* was serving as a troopship, but Smithy was permitted to travel as an independent passenger, and he made the most of it. His spirits now were high, and at the ship's concerts he was 'the hit of the show', singing amusing verses about his fellow passengers, and appearing as a scantily dressed showgirl, revealing 'two yards of hairy leg to a highly delighted audience'. On 12 August – almost a year to the day since his wounding – he was back in England.

In London, he discovered that the Royal Flying Corps had been reorganised and renamed, and he was now a lieutenant in the Royal Air Force, with a smart sky-blue uniform that was 'altogether topping'. He also found that he must appear before another medical board, which judged him still unfit for active service. As an alternative, he was sent as an instructor to the 204 Training Depot Station in Eastchurch, on the marshy little Isle of Sheppey in the Thames Estuary. But was he actually still capable of flying? The question tormented him.

An answer was soon forthcoming. On his first day aloft in a small Scout plane, he found that he was still able to operate the rudder, although his skills were 'quite rusty'. 'I was horrible at stunting,' he reported, 'did all sort of weird antics I never meant to.' But this 'being as I expected, I'll be quite all right in a day or so'. In fact, being 'quite all right' took rather longer: three weeks later he was still feeling unconfident. A two-week course at Shoreham, where he qualified as an instructor in all types of aircraft, helped to restore some of his authority. Nevertheless, when asked how he felt, he often replied, 'A bit nervy, as usual.'

Although he told his mother that teaching was 'not a bad job' and his pupils were 'turning out A1', Smithy continued to see teaching as a fifth-rate alternative to fighter-piloting. In no way could it supply those aerial thrills to which he had grown accustomed, and for which his nature now craved. He started to manufacture his own excitement, sweeping his fellow officers along with him. He devised aerial pranks and jousts. He led aerial poaching parties, wherein pheasants were shot from his plane and served at boozy dinners at a nearby hotel – until a local landowner threatened to prosecute. At night, in the mess, he sang satirical songs of his own composing, which amused some but annoyed others. And every girl in sight seemed to end up in his bed. According to a fellow officer, he 'just seemed to hypnotise them'. The same officer also recalled that Smithy's companions at Eastchurch had a nickname for him – 'King Dick'. The recollection was perhaps tinged with envy, but it captured the life that Smithy was leading.

The war was now in its final days, and Smithy was jubilantly watching its progress. The armistice was signed on 11 November 1918, and had he been capable, he would undoubtedly have rejoiced. As it happened, by that day he was fighting for his life. The previous day he had been admitted to the Royal Naval Hospital at Chatham, gravely ill with the dreaded Spanish influenza. Classed as one of the worst pandemics in history and the killer of 20 million people worldwide, this flu seemed to target strong young men, and in that pre-antibiotic age, pneumonia and death often followed quickly. Smithy did develop pneumonia, and his family was notified that he might not live. But by some miracle he survived. On 22 November he wrote a shaky note to his parents. He admitted to being 'horribly weak', but said he was 'getting on'. By Christmas he was himself again.

Smithy by now was in debt, and urgently needed to earn money. Early in 1919, while waiting to be discharged from the RAF, he secured an interview with A.V. Roe and Company, the makers of the Avro 504K aircraft, on which he had been training his pupils. Roe seemed interested in the notion of selling planes in Australia, and Smithy declared he would be 'tickled to death' to demonstrate a plane there on Roe's behalf. By early March, having received no firm offer, he began to negotiate a similar deal with the Central Aircraft Company.

Meanwhile, he had teamed up with a pilot from Perth named Cyril Maddocks, who, like Smithy, had been in the Australian Army before transferring to the Royal Flying Corps. Maddocks had fractured his skull in a crash and been rendered unconscious for ten days.

Surprisingly he had recovered, and he now superintended, in a rather eccentric fashion, the disposal of unwanted planes at the RAF depot in Henlow. The planes at the depot were 'perfectly good', Smithy assured his mother, and 'by pulling strings' he and Maddocks could obtain three or four of them 'for about 50 pounds each'. They were hoping to ship the planes to Sydney and use them in a flying school or for short passenger flights. Payment would be covered by their war gratuities of £300 each; if these failed to arrive in time, Smithy would try to raise a loan from his brother Harold, who had prospered in business in California. 'The scheme has been gone into in every detail by the both of us', he informed his parents, 'and so far seems perfectly sound.'

Quite suddenly news from Australia knocked their plans. In mid-March 1919, the Australian government offered the substantial sum of £10,000 for the first successful flight from England to Australia. The main condition was that the journey must be performed by an Australian crew in a British plane, and take no more than thirty days. Smithy's excited mind grasped the challenge. To be the first person to fly across half the world and connect his native land with its motherland would be a wonderful act of daring and patriotism. 'Should we pull the job through it will mean that we're made for life', he wrote on 17 April. 'It is a big thing to hope for.'

Fortunately, he and Maddocks had the prospect of a sponsor. At Brough Aerodrome, near Hull, Robert Blackburn, founder of the Blackburn Motor and Aeroplane Company of Leeds, was converting war planes for commercial use, and had shown interest in the England to Australia air race. He was already at work on a plane that he was proposing to enter for it. Aptly named the *Kangaroo*, because it had originally been fitted with a pouch for carrying a machine gun, this large, twin-engine biplane had been built as a wartime bomber but had excelled in anti-submarine patrols in the North Sea. To pilot the plane, Blackburn had tentatively engaged two other Australians: Valdemar Rendle from Brisbane, and a South Australian named Booker. But Blackburn was easily won over by Smithy's enthusiasm, and when Booker dropped out, he and Maddocks were included in the team.

Over the following weeks the *Kangaroo* was carefully prepared. Rolls-Royce engines were fitted, wireless communications were installed, and dual controls were set up between two of the three cockpits, so that Rendle could relieve Smithy as pilot. Maddocks was nominated as the engineer, and Rendle took on the extra task of navigator. Care was put into selecting stopping places, arranging fuel deliveries and receiving weather forecasts. The itinerary of some twenty days was tentatively chosen as Turin, Brindisi, Crete, Alexandria, Suez, Damascus, Baghdad, Basra, Delhi, Calcutta, Rangoon, Singapore, Sumatra and Java, with the terminus at Darwin. All this was confided to the reporters who flocked eagerly to Brough to talk to the crew, and to interview Smithy in particular. Of Smithy, one reporter wrote:

> I judge him venturesome, but not rash; efficient but not ultra-methodical; with the full dash of twenty-two and stern coldness in his determination to get through. He is fair-haired, blue-eyed, a slim and wiry Sydney type of youth, with undoubtedly a brilliant mastery over the science of flight. Though young, a trusty leader of an exploration. His crew are older than he – Rendle about twenty-four, Maddocks twenty-eight.

During the first fortnight of June, Smithy and his friends were champing to leave, but always there were objections. The chief objectors were from the Fédération Aéronautique Internationale in Paris – the world governing body in air sport – and the Royal Aero Club, which the Australian government had nominated as organiser. One cause for concern was the early start of the monsoon season; another was the entrants' lack of navigational training. Nevertheless, many influential people lent their encouragement, General John Monash among them.

In mid-July the Australian prime minister, W.M. 'Billy' Hughes, arrived in London, having just attended the Paris Peace Conference. He was an enthusiast for aviation and conferred with the secretary of the Royal Aero Club and other officials over the problems of the air race. Sensibly, they agreed to postpone the departure until after 8 September, by which time the monsoon season would be almost over and the entrants would have had more chance to prepare. Maddocks was in London, and seems to have had a meeting with Hughes: it was he who relayed news of the delay to Smithy at Brough.

Although disappointed, Smithy and Maddocks were determined to make good use of the delay. They formed a company named Kingsford Smith Maddocks Aeros Ltd, and grandly offered passenger flights 'anywhere in the British isles'. Most of their business, however, lay in exhibitions of aerobatics and in short pleasure flights known as joy rides or joy flights, popular at fairs and carnivals. Often called 'barnstorming', this type of aerial entertainment was gaining followers in England and America, and Smithy took to it wholeheartedly.

It was a rough activity for planes, and frequently led to mishaps, but neither Smithy nor Maddocks seemed much concerned. Replacement planes were cheap and easy to obtain at the aircraft disposals' depot; and since the damaged planes had been insured for somewhat above their real price, a modest profit was made, which helped to keep the company afloat. Smithy seems to have had no pangs of conscience about over-insuring, and to have regarded it as an amusing prank. But there were many, his parents included, who would have seen it as downright dishonesty.

One who regarded it as dishonest was Robert Blackburn. Rumours of the over-insuring came to his ears, and he shared his concern with Lieutenant Richard Williams, the Australian Liaison Officer for the air race. Blackburn and Williams agreed that, as well as cheating the insurers, Smithy was undermining 'civil aviation control' and damaging the development of aviation insurance. Soon after his meeting with Williams, Blackburn fired Smithy and Maddocks from the crew of the *Kangaroo* and installed Valdemar Rendle as the chief pilot.

In letters home, Smithy concealed the true reason for his leaving the air race, blaming his 'withdrawal' on lack of funds. At the time this seemed a convenient explanation. However, when he spoke of it some months later, he recognised that poverty was a tame excuse – and of course he dared not give the real reason. So he devised a more colourful version of the story. He said that the Billy Hughes had considered him too young – and far too inexperienced as a navigator – to compete in so dangerous a venture. When he attempted to enter, Billy Hughes 'absolutely forbade it'. Smithy had no real evidence that Hughes had stopped him, but this did not prevent him telling the story. And in time, it seems, he came to believe it. In later years, whenever he heard Hughes' name, Smithy would refer to the England to Australia air race and exclaim, 'That bloody man!'

THERE'S A LOT OF FIGHT LEFT IN ME

'Luck seems to be persistently against poor old me, doesn't it?' Smithy wrote privately. He bemoaned the 'black weeks I've had with this darned Australian flight falling through'. By the end of October 1919, newspaper readers in Britain and Australia knew that he was out of the air race, and those in aviation circles probably knew why. Smithy deflected the blame as best he could, and complained to his family about 'how badly I've been let down'. But in his heart he knew that he had disgraced himself, which added to his humiliation.

To increase his distress, the barnstorming company was collapsing, and he was 'stony broke'. The obvious solution was to return home, but to slink back to Sydney seemed like admitting defeat. Perhaps it would be wiser to visit his brother Harold in California: with Harold's help, he might reverse his luck. Moreover, the fare to California by ship and train was less than the fare to Sydney. Having somehow scraped the money together, Smithy sailed for New York in late November.

Worn out by his inner struggles, Smithy's vitality was low, and it was not surprising that he fell ill. Arriving in New York with a sore throat, a fever and a rash, he failed to pass the medical test for disembarking passengers, which was especially strict in this time of the Spanish flu.

He was consequently bundled off to the Alien Immigration Hospital on Ellis Island with what was diagnosed as scarlet fever – a serious illness at the time, sometimes fatal, and often leaving behind serious heart defects in those who recovered. It was testimony to his basically sound health that he recovered so quickly and completely. Discharged after little more than a week, he took the train across America. His diet was mostly oranges, a cheap food, because he was obliged to watch every cent.

Harold was living in some style near San Francisco with his wife, Elsie, and their three children. And with them, on an extended visit from Sydney, were Smithy's sister Elsie and brother Wilfrid. All were alarmed by Smithy's gaunt appearance, and the two Elsies set about nursing him back to health. Rest and healthy food soon restored his body, but loving care failed to raise his spirits. While in New York he had received the depressing news that the South Australian brothers Ross and Keith Smith – competitors in the air race – had reached Darwin on 10 December 1919, completing the flight in just under twenty-eight days. He also learned that the *Kangaroo* had crashed in Crete. None of this altered his belief that if he himself had been permitted to compete, he would have been the winner.

It seemed to him that the cure for his disgrace was to achieve a resounding success, and now, miraculously, an opportunity for success awaited him. A prize of US$50,000 – or £10,000 sterling – had been offered in the British and American press for a flight across the Pacific. The donor was the Californian movie tycoon Thomas Harper Ince, remembered today as the father of the American Western. The prize was for the first aviator to cross the Pacific Ocean, and the flight had to be made in the six months from 1 September 1919 to 28 February 1920. 'If only I could manage to do that job,' he wrote to his mother from California, 'I would be able to justify myself in the eyes of the Australian people with a vengeance.'

The time limit on the prize was close to expiring, and although several Australians had shown interest, so far no one had entered the contest. The scarcity of stopping places – for it entailed crossing vast expanses of ocean where refuelling was impossible – made it a formidable, if not suicidal, challenge. But Smithy refused to be disheartened. 'I am once more full of optimism', he informed his parents:

> Dears, there's lots of fight left in me, and I have every intention of coming home to you by air ... I think I can get people over here sufficiently interested to back me. Hal's influence has already helped me and I am getting close to several big cinema concerns.

In search of a plane and a sponsor, Smithy sent persuasive letters to the Curtiss Aeroplane and Motor Company, to the Los Angeles Chamber of Commerce, and to Charlie Chaplin, who, along with his half-brother Sydney, owned the first domestic airline in America. Writing on paper that was headed 'Lieutenant C.E. Kingsford Smith MC (late AIF, RFC, RAF)', Smithy included a list of possible stopping places and estimates of costs and probable profits. The stopping places nominated were San Francisco, Honolulu, Fanning Island, Enderby Island, Pago Pago, Levuka, Noumea and Brisbane. The expenses came to $22,000, of which the main costs were the plane, estimated at $16,500, and a life insurance policy costing $4000, payable to the sponsor if Smithy died in the attempt. In the profits column were movie rights, prize monies and lecture tours. Smithy optimistically estimated that he and his sponsor could make a profit of $92,500, which they might share equally. Smithy's dreams and calculations were too late. February ended and Ince's offer expired. An entreating letter was sent to Ince, but the tycoon remained firm: the offer was closed. 'He is a rotten four-flusher,' wrote Smithy indignantly, 'and was only out for cheap advertisement.'

Smithy did not give up his dream. His letter-writing continued over the following eight months. He lobbied the Manufacturers Aircraft Association, the British firm Vickers, the Yolo Fliers Club and the US Navy. He also approached a small Californian firm called Davis Douglas Company, the general manager of which, Donald W. Douglas – not yet famous in aviation circles – replied encouragingly. As most pointed out, the insuperable problem lay in the flight to Hawaii. Being over sea the whole way, there was nowhere for the plane to stop to refuel.

Smithy's sister Elsie, a professional typist, helped him with his letters. 'Poor boy,' she wrote to their parents in Sydney, 'I don't think he has a chance in the world of raising the money', even though 'there is plenty of money over here and any amount of millionaires'. She felt acute sympathy for him in his dilemma: 'However I dare say it will all be for the best in the long run, and he will get over his disappointment in time.'

Meanwhile, Smithy was obliged to earn his keep. 'I can't live on poor Harold indefinitely and don't intend to borrow money from him,' he told his parents. His first impulse was to study 'wireless', which would be useful in the air and, if no aerial avenue opened to him, might also secure him a 'job on a boat'. He expected Harold's help in finding him a job on a boat, because Harold worked for the American-Hawaiian Steamship Company. Neither he nor Harold seemed to understand that to work as a wireless operator on a US ship – or to fly the US mails, another job he was seeking – the applicant had to be a US citizen. Not that Smithy wanted to work on a ship. As Elsie explained to their mother:

> His whole heart is in the flying game and nothing else seems to interest him – except a pretty girl and the banjo – and as the former requires a certain amount of dollars to be entertained, he has to fall back on the good old banjo and spends all his spare time practising.

Eventually, Smithy found work as a pilot. According to his own testimony, one of his first jobs was to fly for the silent-movie stuntmen, and in particular the famous Ormer Locklear, who could change planes midair and climb from an airborne plane into a speeding car and back again. Locklear's life is well documented: it is known that, because of the extreme danger of his stunts, he worked almost exclusively with two flyers, Shirley J. Short and Milton 'Skeets' Elliott. Elliott was Locklear's pilot on the night of 2 August 1920, when, while working on a film for William Fox called *The Skywayman*, they plunged to a fiery death over De Mille Field in Hollywood. Smithy claimed to have witnessed their death-dive, but this could not be, because – as his sister's letters reveal – he was in Nevada and Arizona in July and early August of 1920.

Although Smithy is unlikely to have flown for Locklear, he certainly flew for Edward Moffett of the Moffett–Starkey 'Aero Circus', joining the troupe in March 1920 and leaving it the following September. Billed as 'Chas Kingsford Smith, RAF Ace', he performed awe-inspiring spins and rolls and loops that were highly popular with rural audiences. He also piloted joy rides and charmed the girls at the Aviation Dance which invariably rounded off the day. It was his young flying partner George Bernoudy – billed as the 'Cowboy of the Air' – who performed the death-defying stunts that the American public demanded of its barnstormers. Cowboy George – an 'awfully nice' boy – hung from the undercarriage of the plane and walked along its wing. In England it was illegal to unduly risk one's life in a public performance, but in America it was believed that no one would watch unless there was a fair chance of somebody dying.

Smithy later claimed that he himself also performed stunts, and he used to describe how he had hung by his calves from the undercarriage of a plane and then hauled himself upright against the force of the slipstream. The secret of the trick, he used to say, were two stay wires attached to the undercarriage. By tucking his toes around the wires he could achieve the necessary stability. A photograph of a man hanging thus has often been reproduced, and although the face in the photo is too distant to be recognised, the man has been said to be Smithy. But what of his maimed, neuritis-prone left foot? Would it have been strong enough to grip a strut, even in a strong boot? Surely this is open to conjecture, although nobody seems to question it.

At first the Moffett Circus confined its performances to California, but in July 1920 Moffett bought an Avro 504K similar to those Smithy had used in Eastchurch. It was a treat for Smithy to pilot such a powerful plane again, and soon he was across the high Sierras and performing with the troupe in Lovelock, Nevada, a small town proud of its unrestricted gambling and three licensed brothels. In those far-off regions, 'air thrills' were still a novelty and the circus earned big money. In August Smithy returned to California, boasting that he had made $3000 for Moffett in Nevada and Arizona in little over a month.

Smithy adored the appreciative audiences, but the constant moving from town to town, coupled with the high risks involved in the aerobatics, were sapping his strength. He returned thankfully to his sister Elsie, arriving at her door with an escort of local children, who proudly carried his banjo and his bags. 'Charles being an aviator is an object of deepest veneration to the youngsters around', explained Elsie. However, she went on to add that her brother had had 'a bad time with rheumatism or sciatica and is also very nervy. All that exhibition stuff is frightfully strenuous.'

At first, Moffett seemed sympathetic to the idea of a trans-Pacific flight, and promised to pay Smithy generous commissions. But then Smithy learned that Moffett was 'absolutely broke', and that he would be lucky to receive the $500 still owed to him in wages. Hoping to recoup some of his losses, he took a job with Moffett's partner – a man named Hunt – in the town of Willows, in northern California. Engaged on the 'aerial rice patrol', he chased wild ducks off the rice fields in a popular American plane, a Curtiss JN-4D, known as a 'Jenny'. He rose at five am and flew until eight, then did another three hours at dusk. Even when he was among the rice fields, the idea of a Pacific flight obsessed him.

'Hunt's people are worth all kinds of dough,' he told Elsie, 'and if we make a success of the patrol, they will back him for anything at all, and I have got him all enthused about the idea.' Elsie, however, was worried. She wished her brother would 'get a settled job and quit rushing around the country with sundry weird gangs'. At times he may have thought similarly, because one day he said to her: 'I'll be glad when all this knocking about is over and the flight accomplished, so I can settle down in Australia to a good steady job. 'And get yourself a wife,' added Elsie, at which he only grinned.

Elsie was correct: at heart, Smithy longed for home. A cable from his brother Wilfrid, now back in Sydney, brought him closer to a decision. Wilfrid was now running a successful import and export business, and had been investigating opportunities in Australia for funding the Pacific flight. His cable was encouraging: 'Can you leave on Nov. 23 Sonoma. Prospects could not be more satisfactory.' Smithy had no money for a berth aboard the *Sonoma*, so he booked a cheap fare on the mail steamer *Tahiti*, and paid for it by painting signs for Shell Oil at $15 a week. 'He was sick of his paint brush by the end of his third week,' Elsie remembered, 'but he was young and strong and took life as a joke.'

Elsie sailed with him, and so did a group of demobilised Australian soldiers, who, like Smithy, had been touring America. The soldiers were full of fun: one of them, Albert Pike, so charmed Elsie that she eventually married him. But the life and soul of the ship was Smithy. He conducted community 'sing-songs', told jokes, sang comic songs and was in his element in fancy dress. At the ship's costume ball, Elsie went as a Quakeress, and Smithy, remembering his success on the *Orontes*, went as a sexy girl, 'showing an expanse of leg'. Needless to say, he won first prize.

However, it was not all playing the fool. As he steamed towards Sydney in the new year of 1921, he soberly took stock of his life. Compared to his brothers, he had achieved very little. Harold had done well with the shipping company; Eric was a lieutenant commander and acting paymaster on board HMAS *Melbourne*; Wilfrid was running his own import-export business; Leofric was a successful draughtsman and builder – and all had wives and children. Even his sister Winifred had a job, running a tea room in Sydney's Angel Place. They had all made good use of their time, and he had squandered his in dreaming and barnstorming. If he were to regain his self-respect, he must very soon fly across the Pacific.

HARUM SCARUM ANTICS

S mithy returned to his parents' home. They were now living in Longueville, in a 'little, brown weatherboard bungalow' on a steep slope in Arabella Street, with a splendid view of Sydney Harbour. The 'bungalow' – which was number 73 – was like an old-time Queensland house, built high off the ground with lattice-enclosed rooms beneath and a wide verandah at the front and sides. It was rented from a Miss Baynes, of Brisbane, and presumably she had named it, because it carried the Queensland name *Kuranda*. It was often said that the Kingsford Smiths enjoyed living in close proximity to one another, and this was certainly true in Longueville. Next door to *Kuranda*, fronting Arabella Street but on the other side of Dunois Street, was a simple weatherboard cottage named *Werona*. Here Leofric and his wife, Elfreda, had been living since their marriage in 1917. Further up the road, at number 93, Wilfrid and his wife, Ernestine, and their children would live between 1923 and 1925.

Smithy's first days at home were happy ones, enlivened by family jokes and chatter, and singsongs. But the happiness did not last and was succeeded by gloom, for he soon discovered that there were no satisfactory prospects; Wilfrid's cable had been a well-meaning tactic to lure him home. In America, the habitat of so many millionaires, success had seemed almost within arm's reach; in Australia it looked unattainable. Even finding a job seemed impossible, because the only thing he cared about was flying and the aviation industry in Australia was in its infancy. To make matters worse, he had no money, and was existing on humiliating handouts from his family. Above all, he was bored, and for him that was a dangerous state.

Sometimes Smithy was angry and resentful, and sometimes he was depressed. Sometimes he thought about the war, and many incidents he would rather have forgotten rose to the surface of his mind. One in particular seems to have rooted itself in his memory, and on rare occasions he spoke of it to his closest friends. This was the slaughter of the enemy soldiers on the sunken road near Ypres. He remembered the 'unearthly joy' he had felt as he fired his gun. He remembered leaning against his plane after he landed and vomiting his heart out. He, of course, was not alone in such recollections. All over the world, young men who had fought in the war were similarly afflicted. The memories were war wounds, as real as the gap in Smithy's foot. But whereas his foot had, for the most part, healed, the demons in his memory would haunt him for the rest of his days.

Smithy sat in his room and brooded, which greatly disturbed his parents. This might have gone on for months but for a fortunate meeting with an old acquaintance in a Sydney hotel. This was Lionel Lee, a former Royal Flying Corps pilot who had trained with Smithy at Denham and Oxford. Recently, Lee and several comrades had pooled their war gratuities and set up the Diggers Co-operative Aviation Company. Their long-term ambition was to set up a regular air service between Sydney and central New South Wales, but their short-term aim was to conduct taxi flights and barnstorm. Hearing from Smithy that he was an experienced barnstormer and a former Hollywood stuntman, Lee realised his value. With little hesitation, Lee offered him a job at £12 a week, and since Smithy had no war gratuity to invest, he waived the rule that employees must be investors. Lee came from the small country town of Wellington, about 250 miles by rail north-west of Sydney. The Diggers

Aviation secretary, Gordon (Dick) Wilkins, lived there too, and so did Lee's father, the proprietor of the local newspaper. Wellington was therefore chosen as the company's base, and Smithy was invited to move there.

On 9 March 1921, Smithy set out from Richmond Aerodrome, north-west of Sydney, as a passenger in an Avro 504K, ready to take up his duties at Wellington. The pilot of the plane was another member of Diggers Aviation, Lieutenant G.A. Doust, a native of Oberon, a township on the western edge of the Blue Mountains. Doust was determined to be the first pilot to fly into Oberon, so he took it upon himself to stop there on his way to Wellington. No sooner had the smart silver Avro – catchily named the *Silver Streak* – circled the town than the residents rushed out to greet it. A reporter on the local paper exclaimed that he 'never knew that the residents of Oberon could climb fences like that – and especially the women!'

That night the aviators were feted at the Oberon Town Hall. According to the local newspaper, there were speeches and dancing and three deafening cheers. Doust and Smithy, and their mechanic, a Sergeant Kerr, made suitable replies, promising public joy rides the next day, which were duly performed. Nevertheless, rumours began to circulate that the plane had been damaged while landing, and secretly – and shoddily – repaired. The rumours were firmly denied by Doust, who made a reassuring statement to the press when, two days later, he and his crew flew on to Wellington. There was 'not a vestige of truth in them', he told a reporter; 'the machine has been constantly flying in Oberon since its arrival there and only left this morning'.

This quashed the rumours for the time being, but a few years later they gained fresh life. In the revived version, the accident was attributed to Smithy, who by that time was world-famous. One still hears tales that, while Smithy was landing, the aircraft turned over and was severely damaged. People have forgotten that, on that first visit to Oberon, it was Doust, not Smithy, who flew the plane – which raises the problem of how much faith one should put in rumour and anecdote. They permeate much of what has been written about Smithy's time at Diggers Aviation, and separating the fact from the fiction is a tricky task. It seems safe to say that there appears to be a core of truth in most of the stories, although the more colourful details often prove untrustworthy.

This same question worried Norman Ellison, Smithy's early biographer, when, soon after Smithy's death, he completed his chapter on Diggers Aviation. Unsure of how to proceed, he sought help from Smithy's close friend and nephew-in-law, John Stannage. Were the crazier of the joy riding incidents true, asked Ellison, and would their inclusion upset Smithy's family? The 'joy riding incidents are all true', replied Stannage, and he advised Ellison to retain the incidents in his book. In Stannage's view, Smithy was all the 'more lovable to those who matter because of his harum scarum antics'.

It was certainly true that, in central New South Wales, aviators were celebrities. On reaching Wellington, Doust and his companions were mobbed by admirers. However, they could not linger, because Dubbo was hosting its annual railway picnic, and the event was drawing big crowds. The Avro, normally a two-seater, had been converted to a three-seater by making the rear cockpit wider, and was thus able to accommodate two passengers. A long line of railway workers and their girlfriends were queuing up for rides, many of which Smithy was to pilot.

Late in the afternoon, Smithy prepared to fly a young local couple, Oliver Cook, a railway clerk, and Dulcie Offner, a farmer's daughter. Almost seventy years later, Smithy's biographer, Ian Mackersey, tracked them down to see if they remembered that afternoon. And they did remember: Oliver recalled that Smithy had asked, 'Would you like some loops?' and Oliver said yes, because he had paid the large sum of £2 for the flight and wanted his money's worth. Cuddling together in the rear cockpit – without seatbelts, which were then virtually unknown – Oliver and Dulcie enjoyed the first loop, but the second was a disaster. As they rose steeply they heard a loud crack: a wing had split. As Oliver described it: 'We hit the ground like a thousand bricks.' Smithy called out to his passengers, 'Are you alright?', to which Oliver replied, 'Yes, but jeez, what a rough landing.' The landing broke the propeller and demolished the undercarriage, but miraculously no one was hurt.

In response to further questioning, Oliver recalled that Smithy 'was shaking and in a hell of a nervous state. He'd obviously been drinking.' Nor was Oliver the only one to come to that conclusion. Lionel Lee made a similar observation when, in later life, he was interviewed by a number of Smithy's fans. Like all good storytellers, he embellished his memories and so must be read with caution. But when one remembers the daredevil pranks of Smithy's past – in adolescence, in the Royal Flying Corps and in his English and American barnstorming days – much of what Lee tells seems credible. The overwhelming impression given in his recollections is that from March to July of 1921, Smithy was often out of control.

There were prolonged bouts of drinking, and boozy all-night parties in which Lee, by his own admission, was Smithy's partner. There were drunken party tricks, such as Smithy standing on his head while drinking a glass of beer. There was a drunken fight, which ended with Smithy spending a night in the Coonamble lockup with a broken nose, and a ten shillings' fine. There were also wild displays of aerobatics. One time Smithy is said to have buzzed the local hospital, where he fancied one of the nurses. Another time he skimmed along the river and tried to fly under a low bridge, barely missing the telegraph wires.

The aerobatics were perhaps an attempt to sustain his reputation as a former Hollywood stuntman. An intriguing advertisement, running in the local newspaper, had recently announced that Smithy had 'just fulfilled a 12 months engagement in America, trick flying and wing walking in the air'. Requests had then poured in from readers eager to see Smithy 'wing walk'. But the chairman of Diggers Aviation is said to have forbidden it: wing walking in Wellington was a step too far.

Smithy's fondness for alcohol had begun during the war. It had made bearable the prospect of almost certain death. In peacetime England he continued to drink because it made the humdrum life of an instructor or barnstormer more palatable. When he reached America in 1920, prohibition was in force and the sale of alcohol was illegal, although, needless to say, he was able to buy it. In letters home, Smithy's sister, Elsie, admitted that Smithy sometimes drank bootlegger whisky, which made him 'deathly sick' or 'half blind and dizzy', and that he sometimes ended up in drunken fights. But as illegal spirits were expensive and Smithy was poor, his consumption was restrained.

In Wellington he had money and leisure, and the alcohol flowed legally and freely. Supervision of the pilots was minimal and government regulations were few. Moreover, this was an age of dangerous stunts, which today would be publicly condemned, but then were publicly admired. If Smithy felt like drinking and flying recklessly, he did it with surprisingly little criticism, and often with a surprising degree of amused or affectionate tolerance. After all, he was only twenty-four and had lived through a fearful war, and, like many young ex-servicemen, he was finding it hard to adjust to the blandness of peace. Many felt that his youthful high spirits should be indulged – aviators were known to be a little crazy – and, as Stannage told Ellison, some thought more highly of Smithy for 'his harum scarum antics'.

The climax came in the middle of July. In preceding weeks, Smithy and the *Silver Streak* had barnstormed towns in central New South Wales – Bathurst, Orange, Lithgow and Cumnock, to name just a few – and while there were rumours of unrestrained behaviour, nothing untoward seems to have reached the newspapers. In Cowra he was in such tearing spirits that he and a passenger named Ken Richards flew under the local traffic bridge that crossed the Lachlan River. Not content, Smithy is said to have repeated the dive two or three times, some say under the same bridge, and some say under, or just over, the railway bridge. A farmer and his heavily pregnant wife were on the bridge in a sulky, along with a wagon and team of horses. The diving plane and the bolting horses brought on the wife's contractions. The baby was born there and then, or so the story goes. Folklore insists that the local paper brought out a special edition headed 'Fate was tempted today by Charles Kingsford Smith'. But a close study of the files of the *Cowra Free Press* for July 1921 has failed to discover the article or the special edition.

Soon after, Smithy blundered again. On 18 July he flew a passenger to a christening party at the homestead of Riverslea Station, near Cowra. According to a Sydney newspaper, he landed safely in a paddock but burst a tyre when he took off. Lionel Lee's account claimed that he landed on the drive, burst a tyre and, impatient to join the celebrations, forgot about the tyre until, hours later, he took to the air again. 'What with the slope, the trees, a flat tyre, and judgement not at its best,' wrote Lee, 'the plane swerved off the drive and plunged into a deep hole.' George Campbell, the passenger, received a few cuts, and Smithy's head was gashed and perhaps two ribs were broken. The plane was a wreck.

The company's insurers had heard about Smithy's drinking, and their subsequent inquiries cost Diggers Aviation its insurance claim. This insurer employed a former Royal Flying Corps pilot named Lee Murray as its assessor, and he was not easily fooled. He is said to have discovered that Smithy had a secret arrangement with the local barmaids, who, when he ordered a ginger beer, gave him a toxic cocktail of half ginger and half gin. Murray refused the claim, which placed Diggers Aviation in such bad debt that it folded a few months later. Smithy was sacked, but not without regrets on both sides, for his ebullient conviviality had won him many friends.

Back in Wellington he reviewed his situation. He was worn out by the war and subsequent disappointment; he was frustrated in his attempts to fly across the Pacific; he was haunted by wartime memories and bored by peacetime life; and his efforts to cure his ills with alcohol and daredevilry had backfired. Even he realised his plight. At the start of August, Smithy told to his mother that if a 'decent' job turned up, he would take it. Perhaps he would even try for a permanent place. 'If Dad or Wilf know anyone in the Air Force,' he wrote to his mother, 'let me know as soon as possible. Meantime I will ring off now. Your loving son, Chilla.'

HAPPY AND
CONTENTED

S mithy returned to Sydney in August 1921. While his desire
to cross the Pacific was as strong as ever, he knew that, in the
meantime, he must find congenial work, which was likely to
prove difficult in a city where many were unemployed. Moreover, by
'congenial' he meant 'flying', which further reduced his chances. He was
greatly relieved when a promising advertisement caught his eye.

Earlier in the year, the Commonwealth government had agreed to
subsidise a regular airmail service in Australia's north-west, the first of
its kind in the country. Following the Australian coast for 1200 miles –
which made it the longest scheduled air route in the world – it would
serve the isolated coastal ports of Geraldton, Carnarvon, Onslow,
Roebourne, Port Hedland, Broome and Derby. The contract had been
awarded to Western Australian Airways Ltd, a small company which
was now calling for five pilots. Those applicants deemed suitable were to
be tested at the Central Flying School, run by the Royal Australian Air
Force (RAAF) at Point Cook, near Melbourne.

Smithy applied at once, submitting a glowing reference from his
old OC at Eastchurch, listing the twenty-seven types of aircraft he had
flown. He also enclosed a generous letter from his mates at Diggers

Aviation, praising his initiative and resourcefulness – the very qualities that the airline was seeking. Taken together, these references constituted a persuasive testimonial, and almost by return of post he was offered a job, conditional on his passing the interview and the practical tests. The salary was high: £500 a year, with possible bonuses.

At the flying school at Point Cook, Smithy was happier than he had been for months. It was like 'a repetition of old service days', with the same discipline and the same living arrangements: everything seemed 'quite natural'. He performed his tests excellently. He told his mother he was top of the class and, what was more, he had made a good impression on his new boss, Major Norman Brearley, 'an awfully nice chap and an excellent pilot'. Flying with his new pilot in an Avro 504K, Brearley ordered Smithy 'to demonstrate his advanced flying'. The result, wrote Brearley, was 'quite outstanding'. Brearley had taught instructors at the Gosport School of Special Flying in Britain, and after that had spent two years as a barnstormer in West Australia, so he was no novice at aerobatics. Nevertheless, he discovered that Smithy could outshine him. Unlike Smithy, Brearley was no daredevil. He placed emphasis on technical expertise, efficiency and safety. Even so, he was happy to accept Smithy's occasional outbursts of daredevilry, provided they were accompanied by a genuine humility. Asked by Brearley how he had won his Military Cross, Smithy replied, 'For various acts of foolishness.' Brearley, who had himself won a DSO and an MC, thoroughly approved of the answer.

In November 1921, Smithy sailed on a coastal steamer to Western Australia. When he took time from his new duties to write home, his happiness overflowed onto the page. 'I have quite fallen in love with Perth,' he told his mother; 'everybody is so good and anxious to help, and we have been royally treated.' He was bidden to more civic receptions, and asked to make more speeches than he could possibly have imagined. And at the grand opening of the new airline, on 3 December 1921, it was he and Major Brearley who were invited to give exhibitions of flying. The planes they flew were selected from those to be used by their airline: Bristol Tourers, simple biplanes, adapted from wartime fighters and seating a pilot and two passengers.

Just two days later, one of their planes crashed at the Murchison House landing field, north of Geraldton. The pilot and mechanic were killed. The cause of the crash was the roughness of the field, and the incident shattered public confidence; a survey of the landing sites between Geraldton and Derby became a matter of urgency. At the end of 1921, Brearley and Smithy and their mechanic, Peter Hansen, set out on a tour of inspection.

In this pioneering, long-distance flying, the pilot kept a meticulous logbook. At eight-thirty am they left Perth and were in Geraldton in three hours and ten minutes; a further fifty minutes brought them to Murchison House. Then they spent two hours and ten minutes covering the 200 miles to Carnarvon, and then two hours and fifty minutes to fly the 250 miles to Onslow, where the summer temperature rose to 110 degrees Fahrenheit in the shade. What a shock it was to come 'down from the delightful coolness of about 8000 feet', Smithy told his parents, 'to the sweltering heat of 3000 feet and below'.

From Onslow they flew in two hours to Roebourne – where everybody came out to greet them – and then 110 miles to Port Hedland, that journey taking one and a half hours. They slept the night at Port Hedland aboard the coastal steamer *Bambra*, and at sunrise flew 320 miles – in almost four hours – to the pearling port of Broome. By now the aviators were dog-tired, but the town was in celebration and they had yet another official reception to attend. A quick flip of 120 miles – in just over an hour – took them to Derby to collect the mail, and then back to Broome to officially conclude the flight. Few Australians, reading an account of the trip in the newspapers, could believe the speed with which they had travelled. What would have taken many weeks on land was accomplished with ease by air in a couple of days.

Now it was time to return to Perth and, to avoid the heat of the day, they set out at two in the morning. To their dismay, as they flew over the ocean, the engine suffered an oil leak. Desperately searching for a suitable place to land, they came down on the Eighty Mile Beach near Cape Latouche. Of course they carried no wireless – wireless was in its infancy – so Hansen set out on foot for the nearest telegraph lines, fourteen miles away, to call for help. Meanwhile, Brearley and Smithy watched the tide come in. As the sea swept close to the plane, they made brave efforts to drag it above the high-tide mark, but in saving it from the sea, they pushed it further into the sand, where its wheels sank deep. When Hansen at last returned, there were futile attempts to free the plane. Their only course was to sail away in the local launch that had come to rescue them, and leave the plane for the *Bambra* to carry as cargo to Perth.

That remarkable flight, with its searing heat, its lonely townships, its vast distances and its dust and insects and other hazards, gave Smithy a glimpse of the weekly challenges that lay ahead. It also brought him closer to Brearley, whom he increasingly admired. Seven years older than himself, Brearley had a wife and son and a stern sense of responsibility. Here was the mentor and role model Smithy had long required.

The regular mail services were not to commence until the numerous landing fields were ready, but while the pilots waited, a call came that could not be ignored. On 7 February 1922, two days before Smithy's twenty-fifth birthday, a call came for a Perth doctor to be flown to a critically ill man in the port town of Carnarvon. Since night-flying was considered unsafe, Smithy and the doctor left Perth in daylight, and spent the first night of their journey at Geraldton. The next morning they landed in Carnarvon at nine-thirty, the patient was operated on at once, and the doctor was able to return that night to Geraldton. The following morning – his birthday morning – Smithy flew the doctor back to Perth in a record time of two and three-quarter hours. This seems to have been to be the first time in Australian history that a doctor was flown to the bedside of a patient: a significant event, and a precursor to the famous Royal Flying Doctor Service.

By April, Smithy was flying regularly on the far-northern route, and transforming the lives of those who lived in Carnarvon and Port Hedland and Broome and Derby. On each flight he carried mail and often a couple of passengers – and sometimes medical aid, for it became almost commonplace to deliver a patient to a doctor or a doctor to a patient. Among the mailbags there would be unusual items, which never failed to interest Smithy. There might be a sealed packet of pre-cious pearls for the European jewellers, a botanical sample for analysis in a government laboratory, reels of film to be shown in northern cin-emas, spare parts for cars or boats or engines, even a batch of newly hatched chickens.

Although a strict timetable was enforced, the work was not ardu-ous. Each pilot flew only two sectors of the route a week, so there was ample time, as Smithy found, to create a satisfying life unrelated to flying. Brearley declared that he had selected pilots of 'initiative and resource', who were capable of acting as 'pioneers in the West', and his men consequently stood high in public regard. Smithy welcomed the respect shown to him and did his best to live up to it. 'I have had excel-lent reports of you from all and sundry', wrote Brearley on 23 May. 'I am very pleased indeed with the way you are carrying on. I know that you are all you appear to be, and that is saying a lot.'

In the simple country hotels which he now called home, Smithy was admired, whether drinking beer with his mates – and his drinking now was moderate – or playing his banjo or ukulele or the broken-down pub piano. Kath Blizard, of nearby Yarrie Station, remembered Smithy bouncing into the bar at the Gascoyne Hotel in Carnarvon, wearing half-mast khaki trousers with his shirt hanging out, and a pair of large and grubby sandshoes because his left foot was hurting. With dis-arming confidence he would break into his favourite song, 'Every Little Breeze Seems to Whisper Louise', and his energy would carry his listen-ers along with him. He was, Kath said, 'one of the most happy-go-lucky people I have ever known'. In those months he was happier than he had been for years.

Needless to say, there were girls in his life. June Dupre, a shy teen-ager, regarded Smithy with 'utter awe'. A war hero, he was 'most people's favourite among Brearley's pilots'. He would fly low over her house when landing or taking off, and her mother once exclaimed, 'that boy's going to be embarrassed one morning, when he lands with my bloomers on his wheels'. When Smithy heard of this remark from June, he quickly replied, 'Tell your Mum I'd be delighted to use them as a windsock.'

In the late summer of 1922, Smithy and two of his fellow pilots went down with the dreaded mosquito-borne disease, dengue fever. Smithy was in the pearling port of Broome when he fell ill, and he remained there until he recovered. It was during those weeks that he met Verona Grave, born Maria Caroline Veronica Rodriguez, the daughter of the captain of a pearling lugger, who was said to be the leading pearl diver on the northern coast. Although there is no mention of her in his letters home, Smithy did describe his visit to a lugger and his descent to the bottom of the oyster beds in a cumbersome diving suit. It seems likely that the lugger belonged to Verona's father.

The biographer Ian Mackersey, who interviewed Verona's son, believed that she was a rich widow who was visiting her father, in company with her two small daughters. Smithy is said to have joined a search to find her when her car broke down south of Broome. This may well have been their first meeting, but Verona was not yet a widow.

She was the wife of Frederick Grave, who did not die until 1924, and their home was in the Perth suburb of Claremont. Grave was thirty-seven years old and wealthy, being co-proprietor of a firm that owned the Ford franchise in Western Australia, and was reputedly the most successful motor business in the state. He had met Verona when she was in her teens: a bright and pretty girl, the winner at her Loreto convent school of prizes in music, English and Catholic doctrine. She had gone on to become a schoolteacher, but teaching had failed to satisfy her. She craved excitement and romance, and this brought her notoriety.

In 1917, when Verona was twenty-one, she had been thrust into the spotlight when she appeared as the 'other woman' in a sensational divorce action brought by Grave's first wife. Verona's ardent love letters to Grave were read aloud to the court, from where they quickly found their way into newspapers across Australia. Verona and her letters caught the public imagination. She was gossiped about and caricatured and became the butt of satirical verses and jokes.

Not that she was unused to courtrooms and publicity. Five months previously, she had found herself pregnant with Grave's child. Desperate to hide her lover's identity, and at the same time to save her reputation,

Verona had accused a Mr John O'Mara of Adelaide of having given her reason to expect he would marry her. She had sued him – unsuccessfully – for breach of promise. The Chief Justice of the Supreme Court of Western Australia, Sir Robert McMillan, who presided over both cases, described her breach-of-promise action as 'the most impudent claim he had heard in any court of law'.

Passionate, volatile and reckless, Verona stirred all Smithy's senses. Small wonder that they became lovers. Spice was no doubt added by their need to outwit her rich and influential husband. At the end of June, when he had largely recovered from his illness, Smithy told his parents that he was still feeling 'rocky', and was hoping to convalesce for 'a spell in Perth'. Verona's son believed that the affair had been carried on 'far from discreetly' in Perth as well as in Broome, and that they wrote to each other for some time thereafter.

Much as Smithy must have wanted to remain in Perth, his convalescence could not be prolonged indefinitely. At Brearley's insistence, he returned to his regular flights, and in the following months he was involved in two near-crashes which forced him down in remote country south of Broome. Given his fears of being marooned 'miles from nowhere', he must have found these incidents unnerving.

Mercifully, help was at hand. After the first accident, he was able to tap into the overland telegraph line with the 'portable instrument' that he always carried with him. Seven hours later, his fellow pilot Len Taplin arrived, and so by chance did a RAAF survey plane. Smithy found it piquant to see 'three machines together there out in the Never Never'. His second forced landing came as he was crossing Roebuck Bay. He managed to land on a mudflat but, as he landed, his plane sank and tipped over. Thirty hours later he and his passenger were rescued. It is worth noting that, in contrast to his barnstorming days, in none of these incidents did the plane sustain real damage. Brearley expected careful flying from his pilots, and Smithy gave it.

The forced landings, while finding their way into his correspondence, were pushed aside in his letters by matters of finance. Before his departure for the west, his parents had initiated stern measures intended to curb his extravagance. His brother Leofric, nicknamed by Smithy 'The Old Dragon', was made the guardian of his money, and Smithy was required to send home part of his salary and deliver a regular account of his spending. He was now earning £600 a year, three times the basic wage.

Originally, Smithy had insisted that his savings should go towards a block of land for himself, the same scheme that his mother had instituted during the war. As the months passed, he changed his mind. On 25 June 1922 he asked that his savings be used to help buy their family home in Arabella Street. When his mother spoke of arranging a loan to buy the house, and taking in boarders to repay it, Smithy would not hear of it. 'I would ever so much rather you would take more of my money and have the house to yourselves,' he told her. A few months later he forgot his generous resolve to help his mother and bought a motorbike, being obliged to borrow from his employer to pay for it. Then he began complaining that his income tax, unpaid for the past two years, would soon be due, and he hoped Leofric would find the money for him. In between times, he was buying expensive presents for his sisters and begging 'Mummie dear' to 'take a few pounds of my dough and buy yourself and Dad a little treat'. Nobody at home was much surprised: his carelessness with money was all too well known to them.

Maybe it was the aftermath of his dengue fever, maybe it was the stress of his volcanic affair with Verona, but by August 1922 Smithy's letters were revealing dissatisfaction and weariness. He was apprehensive of spending another summer in these steamy northern ports. The high humidity caused his feet to swell, and his left foot was painful. Moreover, the boredom of flying the same coastal routes and keeping to the same strict schedules seemed to be sapping his energy. Writing to his mother, he confided his dismay: 'It's a ghastly coast after one has been up and down it a few dozen times; and to finish one's run with the only prospect of going "home" to "a bush pub" – Gawd!'

It was not surprising, therefore, that he rejected the bush pub in favour of Mrs Mousher's homely boarding house when he stayed in Port Hedland during the last week of August. He was lucky to find a room because it was Race Week, when families from the outlying properties poured into town. For the next seven days there were horseraces, balls and dances, a cricket match and a concert. There was also an infestation of 'stickfast' fleas, and a shortage of water so acute that no flea-bitten visitor could take a bath. Nevertheless, an air of merriment prevailed.

Staying also at Mrs Mousher's boarding house were a mother and daughter. The mother was Mrs Gertrude McKenna, whose husband, Maurice, was the manager and part-leaseholder of a cattle station of 700,000 acres about 200 miles inland. The daughter was Thelma Corboy, a buxom, dark-haired girl of twenty-one, the eldest child of her mother's first marriage. Thelma told Ian Mackersey that she never forgot her first sight of Smithy. She and her mother were inspecting dress material in Mrs Moseley's drapery shop when Smithy and a group of chattering friends walked by. 'The Kingsford Smith crowd,' exclaimed Mrs Moseley, and pointed to their leader: 'a slight-looking chap with a big nose'. Then she took Mrs McKenna aside and whispered knowingly, and Thelma assumed that this meant that the slight-looking chap with the big nose had a bad reputation with women. Her assumption was confirmed when, on leaving the shop, she asked what had been said. 'You don't need to know,' was all her mother would say. Thelma was intrigued rather than deterred by this cryptic statement. When she finally met Smithy at the Race Ball, she was immediately drawn to him – and so, she believed, was he to her.

Smithy was indeed drawn to her, and it was possibly her wholesomeness that attracted him most. She was an uncomplicated country girl, the antithesis of Verona, who, for all her allure, seems to have presented more risk than he could handle. Moreover, Thelma was just the type of girl of whom his parents and Major Brearley would approve.

There was no time to pursue the relationship. Race Week was over, and since Smithy was the airways' senior pilot – 'our best pilot', according to Brearley – he had work to do. About 150 miles north of Port Hedland lay the tiny coastal settlement of Wallal, and already a line of ships stood several miles offshore, unable to come closer because of the enormous difference between low and high tide. A phalanx of distinguished scientists was arriving from America and Canada and India. They brought with them costly scientific equipment, which was carefully unloaded into small boats, rowed ashore, then hauled by donkey teams to a site close to the beach. Here labourers had set up giant cameras and a wireless station. On 22 September 1922 there was to be a total eclipse of the sun, and Wallal was calculated to be one of the finest places on earth from which to observe it. If the day was clear and the position of certain stars near the sun could be measured accurately, the world might at last have proof that light is bent by gravity, a theory recently put forward by Albert Einstein in his general theory of relativity.

Smithy made many trips to the vicinity of the scientists' camp, carrying mail and stores and scientific instruments. He examined the sky through the great telescopes, and, as a keen photographer, took particular interest in the giant cameras. Listening to the voices beamed by radio from Bordeaux, Lyons, London and Rome, he exclaimed that it was 'really marvellous to hear those messages from thousands of miles away'. Though wireless communication was still in its infancy, he could see its potential, in the air as well as on the ground.

Alas, on the day of the eclipse, he and Brearley were in Port Hedland, which was not quite as perfect an observatory. Nevertheless, they watched the weird black shadow pass across the sun and were amused by the way the 'fowls went to bed and generally made goats of themselves'. At Wallal the scientific work went smoothly. Einstein's theory was validated, and the relevant photographs and data were flown to Perth. Tiny Wallal had played a vital role in furthering the march of science.

With the eclipse off his agenda, Smithy turned to Thelma Corboy. His desire to see her had not abated. Using the ingenious excuse that he was searching for emergency landing fields, he set out for her stepfather's cattle station, a difficult place to reach. He was obliged to travel over a hundred miles by train from Port Hedland to the remote goldmining town of Marble Bar – celebrated as the hottest town in the continent – and then ride a horse for another sixty miles before reaching the homestead. It was 'about 107 in the shade (of which there ain't any)', he told his mother, and the land was arid and largely uninhabited. He could 'easily understand men doing a "perish" in such country', but his welcome at

Meentheena made up for the discomfort. Though simply built of timber slabs and corrugated iron, the house had twelve well-appointed rooms, a cellar and a wide verandah, and the unusual luxury of running water, which was piped from a high tank to the house and garden.

Smithy proved an ideal companion at Meentheena, whether he was mustering with the stockmen or riding over the hills with Thelma. The station seemed to cast a spell on him, Thelma recalled: 'he was just so relaxing to be with'. At the same time, he made no attempt to hide his forcefulness, and that also won him friends. 'Though such a little man, he had this distinct aura – the air of someone of consequence,' remembered Thelma. 'I think it had a lot to do with the very brisk, authoritative way in which he always spoke.'

A young part-Aboriginal station employee named Bill Dunn also remembered him fondly. 'Smithy was a stockman alright,' he recalled, 'and no fool. He had a good hand on a horse.' And he was 'one of the best with a banjo. He kept us going round the campfire, singing and telling stories.'

It was testimony to Smithy's social versatility that he moved with ease between the workmen's camp and the homestead parlour. He was at pains to win approval all round, but of course he particularly wanted to please Thelma's parents. Motherly Gertrude McKenna liked him at once, but the rough, tough Maurice McKenna took longer. An Irish battler with a hot temper and not much regard for the law, McKenna was difficult to please, but fortunately he had a generous side. He had given shelter to Gertrude and her five children when Corby deserted them, and after her divorce in 1918 he had married Gertrude and provided for her family. Recognising that Smithy was courting his stepdaughter, McKenna was prepared to tolerate the young airman.

Often Smithy played the banjo and piano and sang for the family. Thanks to a parcel of sheet music sent over by his mother, his repertoire had enlarged wonderfully. To the usual popular songs and traditional ballads he now added American jazz and minstrel songs. Thelma was an efficient pianist and singer, and liked to perform with him. Her manifold talents often surprised him. He discovered that, like his sister Elsie, she was a competent stenographer, with a diploma from the City

Commercial College in Perth. In his letters home, Smithy began to hint that he was in love. 'Mrs McKenna and Thelma gave me a splendid time,' he wrote his mother, 'and I enjoyed the break immensely.' He added a postscript: 'Thelma Corboy (Mrs McKenna's daughter) heap nice girl. Am very interested!'

By the end of October, Smithy was back in Port Hedland, and wondering when he might see Thelma again. He did not wonder long because, in mid-December, she appeared in the town for Christmas week, during which he seems to have elicited another invitation to Meentheena. Writing to his parents on New Year's Eve from Broome – where he may still have met Verona – the hints were broader than ever. 'Next weekend I go to Meentheena for a few days' spell,' he wrote. 'Guess I'll end up a family man all right. Can't find anyone I fancy better and I'm tired of pub life. What say you?' He knew that his parents deplored his footloose lifestyle, and he hoped they would endorse his decision.

When he returned from Meentheena on 7 February, he wrote again in the same strain: 'I will jolly well have to get spliced. Not that I relish losing my erstwhile freedom, but I must have a home at this job, other- wise I'll have to chuck it.' It was an offhand way to describe his feelings. Did he really see Thelma primarily as a housekeeper, and did he choose her mainly because he could find no superior alternative? Perhaps his words were designed to reassure his family – and maybe also himself – that he was making a rational decision and not simply an emotional one.

According to Thelma's recollection, he did not propose to her that January: he did not even hint at marriage. However, his two long visits to Meentheena had aroused gossip in Port Hedland. An intriguing item appeared in the *Pilbarra Goldfield News* on 20 February 1923. A reporter writing under the pen-name of 'Ida No' asked jokingly: 'Is it true that pilot KS is anxious to ascertain the date the residents of the Port have fixed for his wedding and where his honeymoon is to be held?' If Thelma read the newspaper – and it is likely she did – whatever did she think? Did she see this as referring to herself? Or did she suppose he had another girl in mind? After all, she knew his reputation as a philanderer.

Perhaps he had tried to propose, but Thelma avoided the question. It so happened that there was something in her past that may have made her wary. Three years previously, when she was only eighteen, she had become engaged to Leslie Edgecliffe Tilney, the immature nineteen-year-old son of a distinguished soldier, Colonel L.E. Tilney. An impulsive engagement entered into while the boy was visiting his uncle at Mount Edgar Station, it had lasted only a few months. The troubled young man then returned to Perth, where, less than a year later, he married another young girl and deserted her soon after. With these events in mind, Thelma may have wished to avoid making another foolish mistake.

In the last days of May 1923, Smithy began his third journey towards Meentheena. The annual Race Ball, the highlight of Marble Bar's social season, was fast approaching, and he intended to attend it as Thelma's escort. Bill Dunn remembered how, a day or so before the ball, McKenna drove his family into town in his Ford Model T. The station hands rode behind, bringing a strong horse for towing in case the car became bogged in a sandy creek. From this point on, the stories of the witnesses differ. Thelma's recollection was that she and Smithy met by chance at the hotel in Marble Bar, that he proposed that evening and that they were married three days later. Bill Dunn recalled that the wedding took place the day after he and the McKennas arrived in town.

Smithy's version, written home on 9 June, is probably the more accurate, although he may have slightly fudged the facts because he felt guilty at not having notified his parents of his intentions in advance:

Some days ago I went out to Meentheena to bring Thelma into the Marble Bar Ball, which was held last Thursday. While we were in the Bar we sort of came to conclusion that after all why wait – anything might happen in the interim and we would be losing the best time of our lives. When I told Thel that it might prove unfair to her to take on a partner whose future might prove somewhat uncertain she said that if we couldn't face the downs of life as cheerily as the ups with me we shouldn't get married at all. Anyway we definitely made up our minds at about lunchtime and were married by special licence at 3pm.

They were married by Warden Kelly in the stone courthouse in Marble Bar on the afternoon of 6 June. Bill Dunn and others waited outside and cheered and clapped as they emerged and jumped into the Ford, under which, in the custom of the time, 'some big empty kerosene tins had been tied'. They drove off, amid rattling, to send an urgent wedding telegram to Sydney. Then they made for the Ironclad Hotel, so named because its walls and roof were built of corrugated iron.

Since it was Race Week, the hotel was filled to overflowing, and the wedding turned into a celebration which lasted well into next day. Smithy, to his credit, was mindful of the young Aboriginal stockmen who lingered outside, and brought them bags of oranges and apples. These were the first apples and oranges most of them had seen, let alone tasted. 'It was the best part of the fun,' Bill Dunn remembered.

The bride and groom spent their wedding night at Meentheena. The next morning, returning to Marble Bar, they were farewelled at the little railway station. There was no train scheduled that day but there was a motor trolley: a simple platform on wheels with a small motor attached, designed to run along the rails. Sitting on the trolley with their legs dangling over the side, they chugged along at a snail's pace, and took many hours to cover the 112 miles to Port Hedland. That night Smithy wrote to his parents:

My Darling Mum and Dad,

Of course you got the shock of your lives to learn of Thelma's and my marriage. As a matter of fact we are rather surprised ourselves, but very happy and absolutely content and satisfied that we have done the right thing … You will both love the kid and admire my taste when you see her. We haven't rushed into this in any rash way at all, we do understand each other and can be and are very happy and contented.

A LAST FLUTTER
AT THE FLYING GAME

S mithy's mind over the past months had been largely focused on work and courtship, but the dream of crossing the Pacific Ocean had never left him. If only, he wrote to his mother, 'I could aviate across the Pacific or do some damn thing, but even that is fading into "things impossible" tho' longed for'.

In the middle of 1922 his fading hopes received a boost. A tall, shy young pilot named Keith Vincent Anderson joined Western Australian Airways. A year younger than Smithy and, like him, a former fighter pilot, Anderson's childhood in Western Australia had been secure and happy until his tenth birthday. In that year his father sailed to Ceylon to manage a rubber plantation, where, a couple of years later, his wife joined him. The adolescent boy found himself packed off to cousins in South Africa and lodged in a lonely boarding school.

At the age of nineteen, Anderson joined the Royal Flying Corps, and was sent for eight weeks to the Western Front. There he shot down five enemy planes, a finer war record than Smithy's, although he received no medal. The stress of those few weeks of combat plunged him into a severe mental breakdown, from which he was rescued by his now-widowed mother. She sent him for two years to a farm in Portuguese

West Africa, where the calm country life soothed his nerves. When sufficiently recovered, he returned to Western Australia, where Brearley, reading of his war record, quickly recruited him. Anderson proved to be, in Brearley's words, a skilled and reliable pilot who 'never thrust himself forward as a newsworthy man'.

In temperament Anderson was Smithy's opposite: quiet and hesitant, a follower, not a leader. These qualities, so at variance with his brilliant fighting record, stood alongside one attribute that the war had not quenched. Anderson had a capacity to dream. He dreamed of flying across a wide ocean: admittedly, it was the Indian and not the Pacific, but it stirred Smithy's enthusiasm. He began to think of Anderson as a possible co-pilot for his own ocean-crossing scheme.

Smithy insisted that they tackle the Pacific Ocean before the Indian Ocean, and Anderson accepted this without argument. Indeed, he was so overwhelmed by Smithy's friendship that he would probably have accepted any ocean Smithy cared to name. By now he was conversant with Smithy's plans, and agreed that the next step was to secure a rich backer. Piloting brought them into contact with wealthy landowners, and one had started to show interest. He was young Keith Mackay, who, on the recent death of his father, had inherited Mundabullangana Station, on the mouth of the Yule River. Mackay was looking for excitement and novelty, and was prepared to sell off part of his inheritance to pay for it. And what higher excitement and novelty could there be than to buy a plane and join Smithy and Anderson on their Pacific adventure!

By now it was July 1923, and time for Smithy and Thelma to visit Sydney and stay with Smithy's parents at Arabella Street. It proved an uneasy visit because the family was still shocked by the suddenness of the wedding, and although they put on welcoming faces, the atmosphere was often tense. Wilfrid's eldest son, John, who was twelve at the time, remembered Thelma as 'a very pretty, well-built woman, extremely well spoken'. Smithy, he recalled, 'made a great show of affection for her', and fondly called her by the pet name of 'Beeb', a reference to a star of the silent screen named Bebe Daniels, to whom she bore a slight resemblance.

Once Smithy had introduced his bride to his family, he pursued his own concerns. 'I was just left with these elderly parents,' Thelma recalled, 'old enough to be my grandparents. We kept running out of things to talk about.' Thelma believed that Smithy was out 'drinking with old mates in bars', and no doubt he was, but at the same time he was laying plans for the Pacific flight. Now that he had Anderson as an assistant and had almost secured Mackay as a sponsor, he needed to buy a plane. Since Mackay insisted on flying with them, they would require an aircraft with three seats, and since their route lay almost entirely over water, he was considering buying a flying boat.

Smithy confided his needs to Leofric, who had a brainwave. They should contact Lebbeus Hordern, a son of the late Samuel Hordern, of Hordern's department store: a playboy in his early thirties, so spectacularly rich that his annual income was reputed to be about £100,000. Hordern loved flying boats and seaplanes. In June 1914, while a pupil of the visiting French aviator Maurice Guillaux, he had flown a Maurice Farman seaplane around Sydney Harbour. Two years later he presented his seaplane to the Australian government and sailed to England to enlist in the Royal Flying Corps, where, ironically, he failed an eyesight test. Undaunted, he joined an artillery regiment, was speedily wounded and repatriated to Sydney, where he co-founded an air-charter company.

Leofric arranged an introduction, and Smithy accompanied Hordern to a hangar at Botany Bay to inspect a Short Brothers Felixstowe F.3 flying boat, which was capable of carrying ten passengers. Hordern's asking price was £2000 – a bargain – but there were many drawbacks. The flying boat's fuel range was limited, and it would need additional tanks. Worse, it had never been completely assembled, and parts were still mouldering in their original packaging.

Schemes jostled one another in Smithy's mind during that visit to Sydney. He felt increasingly stifled by the boring regularity of the air-mail route in Western Australia and longed for excitement and action. Above all, he wanted to devote himself solely to the Pacific crossing. This would, of course, require an independent and substantial income, for even on Brearley's generous salary he was usually in debt. How to achieve his independence he had no idea.

Curiously, it was Thelma's stepfather, Maurice McKenna, who seemed to provide an answer. In the following months, he urged Smithy to take over a crown lease of 63,000 acres on Meentheena's eastern boundary. He argued that it would give Thelma and Smithy a secure home and income. As a further inducement, he offered to manage the livestock so that Smithy could continue flying. Poor grazing land, a mixture of scrub and spinifex, it was incapable of yielding the regular income that McKenna seemed to be proclaiming, but Smithy agreed. He bought the lease – number 3421 – and registered it in his own name.

Smithy's next idea was to turn his land into a returned soldiers' commune, run by himself and a few close relatives. His sister Elsie was briefly recruited at the end of 1923, along with her husband, Albert Pike, and her cousin Phil Kingsford, both of whom were returned soldiers. So excited was Smithy by the prospect of the commune that he considered taking leave from the airways. This 'station proposition looks so remarkably good', he told his mother, that maybe it would 'be worthwhile for me to spend some months helping to get it started'. But in February 1924 the bank foreclosed on Maurice McKenna, and the station was put up for sale. Smithy retained his lease but abandoned his idea of communal living.

He and Thelma were now living in Port Hedland, where Thelma often found herself alone, since Smithy's schedule obliged him to be elsewhere on five days out of seven. Fortunately, she was seldom lonely, for Port Hedland was a sociable town with fewer than 200 inhabitants, many of whom she had known since childhood. In Race Week her mother and sister joined her, and Smithy put on aerobatic displays and joy rides at two guineas a time: Mrs McKenna was one of his first customers. This way of life suited Thelma, so she was dismayed when, as the heat increased, Smithy decided they should move down the coast to the larger and less friendly town of Carnarvon. It was cooler there, he said, and had fewer sandflies. More to the point, it had fewer creditors. 'From the very first week of our marriage,' Thelma remembered, 'we were pressed by creditors.'

For most of his time in Brearley's employ, Smithy behaved sensibly. 'He was a really first class pilot of the type needed for overcoming the hazards that faced us in 1921 and 22 period,' Brearley would later say. 'He had to be "tamed" for our course, and he submitted when he realized this.' Taming, however, did not come easily to Smithy, and occasionally he ran wild. He was certainly guilty of 'buzzing' a coastal steamer; and he often rode on the lower wing of one of the Bristol Tourers when he was travelling as a passenger and the cabin was full. Understanding his nature, Brearley made allowance for these brief lapses. But now Smithy allowed his restlessness to boil over into a protest.

As it happened, Smithy was not alone in his restlessness. The pressure had been so intense over the previous eighteen months that three of Brearley's six pilots were feeling the same. What was more, knowing that their highly skilled services were essential, they decided to demand higher pay. Whether Smithy was the ringleader one does not know. Brearley received a strong letter from 'the majority of the pilot team saying that they considered the time had come for their pay and conditions to be substantially upgraded'. Otherwise, they would behave 'like Arabs' and fold their tents 'and silently steal away'.

Brearley and his fellow directors refused to meet this ultimatum. Two pilots were retained and the others were sacked, Smithy and Anderson among them. The parting cannot have been bitter, because Smithy was rehired temporarily some months later as an emergency pilot. That he was happy to leave was made clear in a letter to his parents on 15 February 1924. He claimed – in highly coloured and somewhat mendacious terms – that he had been instrumental in winning an increase in pay for himself and his colleagues, but thereafter the bosses demanded he 'climb down in various ways', which he refused to do. A deadlock followed, which he broke by resigning. 'The pilots,' he continued, 'are all for me and would have come out too should I ask, but my principles never included Bolshevik methods of strikes so I won't hear of it. It leaves my good name intact which is a fine asset.' Thelma, he added, was 'very philosophical about it all. She is a great little kid, too, God bless her, and stands by me in everything.' Recalling the incident years later, Thelma is said to have remarked: 'Philosophical was not exactly how I would have put it.'

Smithy had already found another job. He and Anderson had joined a local garage owner named Tom Carlin to transport goods to and from the sheep and cattle stations inland of Carnarvon. Previously the cartage had been done by donkey or camel teams, and it had taken weeks to carry wool to the port; with motor trucks the time could be considerably shortened. Of course, Smithy and Anderson needed money to buy into the business, and both began to badger their families for loans.

Smithy also invited Elsie and Bert Pike to Carnarvon to help him run the business. Bert, wary by now of his brother-in law's schemes, refused to come, but motherly Elsie hurried to his aid. Over the years she had tried to supervise many of his business negotiations, and she knew his strengths and weaknesses. His grasp of complicated propositions was admirable, and his ability to make them seem simple and acceptable was little short of brilliant, but his management of everyday dealings left much to be desired. Years later she would comment:

> I never knew anyone who could write a better business letter than Charles and put matter more clearly and lucidly. But in ordinary business routine he was hopeless – his office methods were haphazard in the extreme, he was hopelessly unpunctual as far as office hours were concerned (though never in his technical work) and just harassed and bewildered by the usual financial adjustments and worries always connected with running one's own business.

He was also hopeless about collecting payment.

Although Smithy and Anderson had several competitors, new carrying contracts rolled in and their prospects seemed rosy. In June 1924 Tom Carlin was bought out. The business was renamed the Gascoyne Transport Company, additional drivers were hired and two new International Speed Trucks with specially constructed bodies were purchased in Perth and driven to Carnarvon by the two proprietors. Landowners were intrigued to see former Western Australian Airways pilots driving trucks and carting freight. The personal goodwill Smithy had built up in the north-west over the past two years now paid off handsomely.

Since most trips were 300 to 400 miles long, it was usual for Smithy to be absent from home for a month at a time. There were no roads, only clay tracks that in wet weather became bogs into which the heavily laden trucks sank. When it rained, the usually dry rivers also flooded; Smithy, ever resourceful, once floated his truck across on empty petrol drums. Breakdowns were frequent. If near enough to a homestead, the driver would walk there under the hot sun, but if not he would simply wait for a search party to find him. Smithy became adept at digging loaded trucks out of bogs and wrangling angry rams onto the top of loads. But he thrived on challenges. In June 1924 he wrote to his mother: 'We really seem to be on the right road at last.'

With the business well established, Smithy turned his mind to the Pacific Ocean, although he did so with some trepidation. Thelma, disturbed by the loss of Meentheena, was not adjusting well to his absences. Worse, she was becoming increasingly unsympathetic to his plans, which were beginning to seem like the height of folly. Lying in bed in the Gascoyne Hotel in the early hours of the morning, listening to his voice floating up from the bar where he drank and talked and played the banjo, she began to wonder if he would ever grow up. Smithy tried to shrug off her opposition. 'Damn it all,' he wrote to his parents defiantly, 'one must have a last flutter at the flying game before one quits.'

Smithy's hope of crossing the Pacific still rested with Keith Mackay, who had agreed to buy Lebbeus Hordern's flying boat for £2300 – just as soon as Hordern made it airworthy. Mackay was keen; indeed, he was taking flying lessons so that he could act as a relief pilot on the Pacific crossing. Notwithstanding the difficulties with Thelma, Smithy had good reason to believe that everything was turning out well.

On 16 July, Mackay, as yet unlicensed, hired a plane and a pilot to fly him from Port Hedland to his station. By a terrible mischance the plane crashed into a creek: it was shallow, but Mackay was drowned. Drowned also were Smithy's hopes of acquiring the flying boat, for he had no money to pay for it. Grieving alike for his friend and his hopes, Smithy felt unable to do the simplest tasks. He was so distressed that he delegated to Leofric the task of telling Hordern.

The shock of Mackay's death affected Smithy's health. He had always been thin, and now lost weight so rapidly that he seemed nothing but skin and bone. Realising how badly her brother needed assistance, Elsie prevailed upon her husband to join her in Carnarvon, and urged him to take over the management of the trucking company. At first Bert was reluctant, but once he had assured himself that the business was sound, he took control with a firm hand. And sound indeed the business must be, joked Smithy to his mother, because Bert was 'never unnecessarily optimistic'!

On the marriage front, there were no signs of improvement. Living in the Gascoyne Hotel, with insufficient money and too many creditors, Thelma felt angry and lonely, and scarcely able to tolerate Smithy's absences. She especially resented his occasional trips to Perth, where he bought or leased trucks which he would drive back to use in the business. One wonders whether, during those stays in Perth, which became increasingly frequent, he saw Verona Grave. Her husband had died in March 1924, and she was now living the life of a rich widow. With his marriage falling apart, Smithy may well have rekindled their affair.

It was true that Thelma had Elsie's company in Carnarvon, but the Kingsford Smith women had never liked Thelma, even though she had done her best to please them. Their united opinion was that the marriage was a mistake, and local details, gleaned by Elsie, more than confirmed their belief that the McKennas were 'common'. The knowledge that Maurice McKenna had lived with Aboriginal women and fathered a half-Aboriginal son was enough to damn Thelma and her relatives in the Kingsford Smiths' eyes. In those times, there were many who would have supported them in such beliefs.

The family's disapproval reached new heights when, in February 1925, Maurice McKenna was committed for trial at the Marble Bar Police Court. He was charged with stealing cattle from the Warrawagine Station, which shared a boundary with Meentheena. It had been alleged by an Aboriginal stockman that for some years Warrawagine cattle had been falsely branded with a Meentheena brand. The allegations were readily believed: McKenna had been convicted of a similar offence twenty years before.

Being the leaseholder of land close to both Warrawagine and Meentheena stations, Smithy was interviewed by Detective Sergeant Harry Manning from Perth, and he was shocked to learn that he and his mother-in-law were suspected of complicity. According to Sergeant Manning, the brand that Smithy had been persuaded to register – OKQ – bore a suspicious similarity to the OIO of the Warrawagine brand – as did QIO, the brand registered for Gertrude McKenna. A few deft touches of a branding iron and the I could become a K and the O could become a Q, and a Warrawagine beast could be passed off as Smithy's or Mrs McKenna's.

Fortunately, Smithy had spent so little time on his pastoral land that he was easily able to establish his innocence. But there seemed no escaping the fact that, by urging Smithy to take up the lease, McKenna had an ulterior motive, and had used Smithy and Thelma as pawns in his game. Whether Thelma defended or condemned her stepfather is not known, but either way, the forthcoming trial must have caused them both distress, and further damaged their fragile marriage. It was greatly to Thelma's credit that she remained strong and calm.

Wounded by her stepfather's duplicity, harassed by debt and hurt by Smithy's neglect, Thelma entered bravely into the life of Carnarvon, becoming a popular figure at social gatherings. She played the piano and sang at concerts, and acted in the local dramatic society. It was while acting in a play that she became friendly with Henry Doyle Moseley, a cultured young man from a prominent family of Perth lawyers. He had arrived in 1924 with his wife and eight-year-old son to fill the post of police magistrate, but within weeks of their arrival his wife had dropped dead in the Gascoyne Hotel, leaving him a grief-stricken widower. In his bereavement, he seems to have turned to Thelma, who understood loneliness only too well. And he in turn may have provided comfort and wise counsel at the time of her stepfather's arrest.

Smithy later recalled that in the early months of 1925 he had noted a coolness in his wife. He even asked her was there 'anybody else'. She replied, 'Yes.' They talked the matter over and he asked her to go with him to see 'the man'. 'I went and saw the man first,' said Smithy.

'I accused him of having alienated my wife's affections. The man admitted that he loved my wife, but he assured me that there was nothing improper between them. And I am satisfied,' Smithy added, 'that there was not.' It would seem that Moseley, with a position of trust to uphold, had sensibly kept the relationship within the bounds of propriety.

At the end of March 1925, Thelma and Smithy travelled to Perth, she on holiday, he to take delivery of a truck. Thelma refused to travel back with him; she stayed on for a further week and returned by coastal steamer. Alarmed by what he now glimpsed as the end of their marriage, Smithy offered to do what he could to mend matters. She replied that she 'no longer cared for him', and intended to leave. Early in June he went on another trucking trip, and when he returned she was gone. He presumed she was with her family, who were now living in Nullagine, where, before his arrest, McKenna had been working a goldmine. But more likely she was in Perth, supporting her mother through McKenna's trial, which commenced on 5 June.

Fortunately, Gertrude McKenna was quickly cleared of suspicion, but the evidence against Maurice McKenna was so conclusive that, after a trial lasting less than a week, he was sentenced to two years' imprisonment. Chief Justice Sir Robert McMillan said he was 'sorry to see a man of the accused's type in the position McKenna occupied', and had sentenced him 'as leniently as possible'.

Elsie, deeply concerned about her young brother, sent regular bulletins to their parents. On learning that the marriage was over, Catherine Kingsford Smith wasted no time in arranging for her boy to come home. 'Thanks muchly for your letter and understanding of my troubles,' Smithy replied on 26 July. 'They are pretty heavy on me and will take some getting over, but all will be OK some day.' Thelma's rejection of him was undoubtedly a blow to his pride. Yet even in this moment of pain, his thoughts were more firmly focused on his Pacific dream than on Thelma. 'I do hope there really is a good chance for this Pacific flight,' he told his mother. 'I need a change tremendously, and if there is something doing I will be able to throw myself into it with some energy.'

A STEP NEARER
HIS DREAM

T helma's departure caused Smithy pain, but it also brought relief. Now he could pursue his Pacific obsession without her opposition. On 25 August 1925, a week before setting out for Sydney, he sent a letter to the federal member for Perth, Edward Mann, to be passed on to Stanley Bruce, the prime minister, outlining plans for the flight and begging for an interview. When Bruce replied through Mann that Smithy had given insufficient thought to the difficulties, in particular to the problems of refuelling, Smithy broke his journey in Melbourne to explain in person. He did not succeed in speaking to the prime minister but he did manage to speak to Mann, who, while undoubtedly sympathetic, was a realist. He explained that to devise a plan with a 'reasonable chance of success' was not enough. Smithy must propose a plan that promised 'a more than reasonable chance of success'. If he could do that, Mann assured him, the prime minister would be prepared to listen.

Thanks to his visit to Mann, Smithy arrived in Sydney far happier than anybody could have expected. To have official recognition seemed a momentous gain, and put Thelma's departure somewhat in the shade. Of course he had loved Thelma, and at the beginning had held sincere hopes of settling down with her. Likewise, he had held sincere hopes of raising a family, partly because it was expected of him, and partly because he liked children and family life. But marriage could never engage his whole self. Excitement and adventure were always beckoning him – as, for that matter, were attractive women. This had been the pattern of his life for ten years, and he was not strong enough to change it. Maybe when he was older and had achieved his aeronautical goals he might settle down – but not yet. Meanwhile, he must put aside his feelings of loss and concentrate on his plans for the Pacific.

Unfortunately, Lebbeus Hordern was still asking £1000 pounds for his flying boat. On Mann's advice, Smithy sought the opinion of the RAAF's chief technical officer, Lawrence Wackett, who told him it would take at least that sum to make the flying boat airworthy; also, he doubted if the flying boat could carry enough fuel to cross the longest stretch of sea. Far wiser, said Wackett, to buy a single-engine, amphibian biplane designed by himself and called the Widgeon. In November 1925 Smithy applied to Mann for a loan to buy the Widgeon, only to cancel the loan a few weeks later. He had learned that neither machine had adequate fuel tanks.

Instead, Smithy turned his attention to the latest British aircraft. He also canvassed influential people to take up his cause. One was the minister for defence, Neville Howes. Another was Billy Hughes, a curious choice, when one remembers Smithy's belief that Hughes had denied him the opportunity to compete in the 1919 air race. On this occasion, Hughes showed Smithy conspicuous goodwill, writing to Bruce on his behalf.

As part of his submission, Smithy also drew up a provisional itinerary. The first stop was to be New Zealand, followed by New Caledonia, Fiji, Samoa, the Phoenix Islands, Fanning Island, Hawaii and San Francisco. He assured the government that the Vacuum Oil Company would supply fuel at these stopping places and, even more importantly, would supply a ship to refuel a flying boat in mid-ocean between Hawaii and San Francisco. He needed such a ship because there was no suitable island stopping place, and in 1925 no aircraft could cross so long a stretch without refuelling. Although it was risky to refuel a flying boat from a ship in an open sea, it appeared to be the only solution.

When he was not communicating with politicians, Smithy attempted to cope with the practicalities of his broken marriage. By Christmas 1925 he had had no contact with Thelma for six months, and on Boxing Day he wrote to her from his parents' house:

> If you are willing to return, I can offer you at least a home although you didn't seem over anxious to share one with me prior to your leaving me when I was in the west. Unfortunately, dear, I cannot tell you truthfully that I have anything financially attractive to offer you, as the longed for Pacific flight seems as far off as ever and I have almost decided to give up the flying game for something steady, if uninteresting, in the way of ground jobs over here or back in Carnarvon. However, Thel, this place is really my home, and if your affection for me is what it should be, you will I know return to me.

The letter ended peremptorily: 'So there it is in nutshell, Thelma, and I'm asking you to come back to me, and if you won't do that, do at least let me know. Your affectionate husband, Charles.'

In truth, Smithy did not want Thelma back, and he certainly did not want a humdrum job in Sydney or Carnarvon. The letter was written at his solicitor's direction, and was the first step towards seeking a divorce on the grounds of desertion. He no longer mourned the loss of Thelma because he had a new love. This was Lyal Hilliard, usually known as Bon, the daughter of a well-connected Sydney family. Her father, Arthur Hilliard, was a successful solicitor, and her socialite mother, Maude, seems to have been a friend of Catherine Kingsford Smith.

Unlike Thelma, Bon was neither young nor inexperienced. A year older than Smithy, she was almost thirty, and although still unmarried had had plenty of admirers, to one of whom – Blakey Laycock, the son of a rich pastoralist – she had been engaged for a time. Described as warm, sympathetic and very pretty, she possessed, from all accounts, a refined but powerful sexual appeal. She also had a lovely speaking voice;

she would later become a radio broadcaster. But her most endearing quality was her kind heart. She seems to have enjoyed helping the emotionally troubled, and Smithy, in his present mood, needed her.

It was April 1926 before Smithy could tear himself away from Bon and return to Western Australia. Elsie, Bert, Phil Kingsford and Keith Anderson had run the business expertly during his absence, and there were now six trucks, all paid for, and a fully equipped garage and machinery workshop. But none of this made him happy. What he craved, above all, was to be back in the air, and back in Sydney. He was deeply disappointed when he failed to gain an instructor's job at the Aero Club of New South Wales.

Fortunately for Smithy, Anderson was also anxious to be free of the business. In September 1926 they advertised the Gascoyne Transport Company for sale, and although they asked more than £2000, they quickly found a buyer. Half the money they spent on two of the elderly Bristol Tourers, which Western Australian Airways was discarding. They were hoping to enter the charter-flight business, and thereby raise money for their Pacific venture.

While purchasing the Tourers, Kingsford Smith and Anderson had renewed their friendship with Henry Smith Hitchcock, known as Bob, an aircraft mechanic somewhat older than themselves and a veteran of Gallipoli. Despite poor health, little money, a broken marriage and almost no education, he had battled his way to a secure position with Western Australian Airways, but such was his admiration of Smithy and Anderson that he was prepared to abandon his job and join them in the charter-flight business. Whether or not he was to cross the Pacific as part of their crew Smithy and Anderson seem to have left unclear. However, there was no doubt in Hitchcock's mind: he was counting on being part of the Pacific team.

Hitchcock also dreamed of launching a charter business in New Guinea to connect the thriving inland goldmines with the coast, and he infected Smithy and Anderson with his enthusiasm. They decided to fly the Tourers to New Guinea, but first they needed to convey them to Sydney to be refitted. At this point, Smithy had a brainwave. What a fine publicity stunt it would be if they could break the Perth-to-Sydney record of twenty-one and a half flying hours – and since a Tourer could accommodate three people, they could carry a couple of fare-paying passengers. Taking the idea a step further, Smithy persuaded John Marshall, the Perth representative of Smith's Newspapers Ltd, to buy fares for himself and his wife. Mrs Marshall would thus be the first female passenger to make an aerial crossing of the continent, which would please Mr Marshall and provide colourful copy for the Smith newspapers.

Smithy also enlisted help from Norman Brearley, who gave him free use of the Maylands Aerodrome in Perth, and from the manager of the Vacuum Oil Company, who organised their refuelling along the way. A small crowd of friends watched the planes take off on 28 January 1927. The Marshalls and Smithy flew in the first plane, with Hitchcock, Anderson and the luggage in the second. Their first stage was to the goldfields town of Boulder, and that was easy, but afterwards they encountered problems. A buffeting headwind made a record time unlikely. Worse, Anderson's sore throat, over the course of the journey, would develop into tonsillitis.

That evening, a failing engine prevented them reaching their overnight stop at Cook, on the transcontinental railway. Instead they landed at the tiny township of Naretha, where Mrs Marshall would long remember the kindness of their hosts. 'The wife of a railway employee came across and invited all our party to her place to dinner,' she recalled. 'We went and dined excellently, and next morning she gave us a fine breakfast.' After leaving, the Marshalls were distressed to learn that they had consumed the railway family's rations for the following two days. In Mrs Marshall's words, it was a fine example of 'that wonderful warm heartedness that only the people of the great Never Never can extend'.

Their next overnight stop was at Wirraminna, a remote cattle station far to the north of Adelaide. 'I was the only woman,' Mrs Marshall remembered, 'within a radius of 100 miles, and there were 25 men. But they were grand.' The following day they lunched at Broken Hill, and then flew on to spend the night in Parkes. 'As we descended,' recalled Mrs Marshall, 'we could see hundreds of motor cars hurrying to the landing ground.' The next morning, while crossing the Blue Mountains, they almost froze and the chilling wind did nothing for poor Anderson's tonsillitis.

The reception at Mascot Aerodrome made up for these discomforts. A large crowd was waiting, and the smiling faces of the pilots and their passengers appeared the next day in newspapers across Australia. Having taken thirty flying hours to complete the journey, they had missed the record, but the publicity was elating.

Smithy's high mood did not last. A few days later, on 9 February 1927, he turned thirty. Soberly reviewing his life, he could feel little satisfaction. Where had his youth gone, he asked himself; had 'the locusts eaten it'? And why did all his plans come to nothing? The previous year he had written to Lieutenant Colonel Horace Brinsmead, the Controller of Aviation in Australia, asking if a slur had been cast on the 'once-good name' of Kingsford Smith. Brinsmead, a veteran of Gallipoli and the Royal Flying Corps, was no stranger to Smithy's recklessness, but he had a soft spot for young air aces, and assured him tactfully that there was no slur. Smithy did not entirely believe him.

To make matters worse, their plans for New Guinea were receding. Studying aerial photographs of the rugged countryside, Smithy saw how difficult the construction of landing grounds would be. Hitchcock disagreed but was overruled, and Smithy and Anderson resolved to forget New Guinea and run charter flights from Sydney instead. Renting an office in Eldon Chambers, at 92B Pitt Street, they appointed Bert Pike as manager of their brave new company. Called Interstate Flying Services, it offered charter flights at thirty minutes' notice to anywhere in Australia – and at longer notice to anywhere in the world!

Now that he was back living with his parents, Smithy renewed his relationship with Bon, which had languished for nearly twelve months. Anderson also came to know Bon, and at first Smithy was pleased that his best mate and his best girl got on so well. As the weeks passed Smithy grew less pleased. Bon's generous sympathy, which had formerly been showered on himself, was now being showered on Anderson. The blow fell in April, when Bon and Anderson announced their engagement.

Bon's change of heart was not hard to fathom. The previous year Smithy had been grieving and in need of her; this year he was jaunty and independent, and less appealing. Shy, trusting Anderson, on the other hand, needed her, and was ready to marry her. Like most women of her time, she wanted a husband and children, and at almost thirty-one she was conscious that her chances of marriage were diminishing. She would have been foolish to refuse him.

Smithy was in no position to marry anybody. His letter to Thelma in December 1925 had eventually been answered. 'I have no intention whatsoever of returning,' she wrote in April 1926, 'and absolutely refuse to live with you again.' While her letter paved the way for a divorce on the grounds of desertion, no action, according to the law in New South Wales, could be initiated until Thelma had been absent for three years. Not until 1929 would Smithy be free to marry again.

There was no real breach between the partners. Anderson went to live with Bon and her family, but he remained on close terms with the Kingsford Smiths. Nevertheless, for Smithy, who prided himself on his prowess with women, Bon's rejection of him must have been painful. As the victor was Anderson – a man he felt was his inferior – Smithy's pain must have been all the keener.

Fortunately, a new friend was about to transform Smithy's life. Business had been almost at a standstill at Interstate Flying Services when a letter came from Cherry Willis, a well-known solicitor, intro-ducing one of his clients. The client was Charles Ulm, and Smithy dimly remembered meeting him at Mascot Aerodrome while flying for Diggers Aviation. The following day, Ulm presented himself at their office and was greeted by Smithy, Anderson and Pike. Nearly six feet tall and ruggedly built, with dark good looks and an impeccable suit, Ulm exuded competence and worldliness. His presence filled the tiny room and made a deep impression on all three.

Although some months younger than Smithy, Ulm 'looked and acted considerably older than his years' and easily dominated the inter-view. He explained that he was not a licensed pilot – his war wounds at Gallipoli and the Western Front precluded this – but he believed passionately in the future of civil aviation, and the scheme he was about to outline called for experienced pilots. He wished to apply for a government contract to operate an air service between Adelaide and Perth, and he believed he stood a stronger chance if he allied himself with their company. He whipped out a sheet of paper. 'There are the whole of my figures,' he told them; 'you fellows chew them over.' He promised to return the next day for their answer.

The incisiveness of Ulm's thinking and the forcefulness of his manner were too much for Anderson and Pike, who were instinctively wary of him. But Smithy was afire with excitement, sensing a kindred spirit. In their first conversations, the Pacific flight was never mentioned, but it was certainly in Smithy's mind. And in no time it would be in Ulm's mind as well.

THE BIG FEAT

There was much about Ulm that was not yet known to Smithy. Behind the impeccable façade was a man with no job, almost no money, a string of failed businesses and a failed marriage. Nevertheless, some of Ulm's special qualities were already apparent: he was ambitious, enjoyed danger, and, as he himself put it, was 'inoculated with the aeronautical germ'.

Born on 18 October 1898 in the Melbourne suburb of Caulfield, and reared in the seaside suburb of Albert Park, Charles Thomas Philippe Ulm was the third child in a family of five. His mother, Ada Emma Ulm, was Australian. His father, Emil Gustav Ulm – notwithstanding his German surname – was French-born and a photographer by profession, being employed by the distinguished firm of Talma, which owned studios in Melbourne and Sydney. In 1909 he would move to Talma's Sydney studio, and later would become the well-known head photographer at the Cruden Studios in Pitt Street.

At the Albert Park State School, Charles Ulm was recognised as something of a loner, 'quick of temper, and inclined to obstinacy'. He was also inquisitive, shrewd and intolerant of rivals. He claimed he wanted to be a businessman or a lawyer, but he also had a passion for physical danger.

In childhood he rode his bicycle at breakneck speed – just like Smithy.

A few months before his sixteenth birthday he enlisted in the AIF, using the assumed name of Charles Jackson. 'I perceived,' he wrote later, 'that here was an escape from the commonplace and an opportunity to express certain things which were in me.' When he was wounded at Gallipoli in 1915, it was discovered that he was only sixteen. He was sent home. He re-enlisted under his own name in February 1917, fought on the Western Front and, while taking an officer's training course in England, fell in love with flying. A friend gave him secret lessons and helped him to make his first solo flight. 'I subconsciously realized,' Ulm wrote, 'that I had illegally discovered something which appealed to my entire heart and brain.' So fired was his imagination that, after the armistice, he determined to enter commercial aviation. 'Make no mistake,' he told his friends, 'the whole world will be on wings soon, and I'm going to put it there.'

In 1920 he set up Aviation Services Pty Ltd with a wartime comrade, Captain 'Billy' Wilson, who flew a single-engine Dove between Sydney and Bathurst. 'The route was a complete failure,' Ulm recalled. In desperation, he and Wilson then tried 'the airman's last resort', barnstorming, and it too 'was a complete flop'. The following year he set up another company, Commercial Air Transport, to tender for an airmail service between Brisbane, Sydney, Melbourne and Adelaide. When the federal government rejected the tender, the company collapsed. In 1923, in a flash of inspiration, he decided to photograph two Royal Navy battleships that were about to visit Sydney. He hired a plane, pilot and photographer to follow the battleships as they made their way along the coast towards Sydney Heads. The photos he sold to Herbert Campbell Jones, chief executive of the Sydney *Sun*, and a special edition of the newspaper was on the streets just as the ships entered the harbour. The *Sun* had a coup, and Ulm had his first taste of success. One year later he submitted a tender to the federal government for an Adelaide–Perth airline. His tender was rejected. It was chiefly for this reason that now – three years later – he proposed resubmitting the tender under Smithy's and Anderson's names.

Ulm's new tender – with its careful costings and detailed schedules for three different types of planes on three different routes – was so superior to anything Smithy and Anderson could have produced that they accepted it almost without hesitation. It now required the support of a syndicate of responsible men. Ulm volunteered to find a suitable chairman, and to Smithy's and Anderson's amazement, he nominated no less than Sir Charles Rosenthal KCB, CMG, DSO, a five-times-wounded major general with a known enthusiasm for planes. As well as submitting a summary of their tender to Rosenthal, Ulm included a résumé of Smithy's flying career. He claimed that Smithy had been the first pilot on the US Transcontinental Airmail Service, and the personal pilot of Ormer Locklear. Both these were untrue, but neither he nor Smithy allowed this to stand in their way.

From the very beginning, Smithy and Ulm 'got on like a house on fire'. Anderson, on the other hand, was initially distrustful of Ulm, but while travelling with him to Melbourne to submit their tender, he seems to have revised his opinion. On their return to Sydney, he put forward a scheme that was almost certainly placed in his mind by Ulm. He proposed that Ulm should replace Bert Pike as their business manager, citing the business contacts and money-making potential that Ulm could bring to the failing company. It was a sad situation for faithful Bert, who knew that he was outclassed but, being Smithy's brother-in-law and long-term supporter, felt entitled to the job and did not give in easily. The deadlock continued for some days before Bert resigned. Ulm was then installed as business manager on a retainer of thirty shillings a week, with generous commissions on the business he attracted.

Ulm tried, some years later, to dissociate himself from the charge that he had ousted Pike. He claimed that at first he had refused the manager's job, preferring to wait until Smithy and Anderson 'had made an arrangement with Pike': he had no wish to be seen as 'an interloper pushing Pike out'. Even so, he believed that the superiority of his own qualifications fully justified his replacing him as manager. Unlike himself, Pike had had 'very little aviation experience'. 'I often used to give him advice as to people with whom he might make payable contracts,' Ulm added condescendingly.

It was now the first week of May 1927 and the Duke of York was in Australia, ready to open federal parliament in its new building in Canberra. A crowd of visitors was expected for the opening, and Smithy and Anderson flew their planes to Canberra and set up camp in a tent near the airfield. Ulm had arranged for the trip to be sponsored by Fays Limited, a large Sydney shoe store, and over the following week Smithy and Anderson made advertising flights for their sponsor and offered joy rides to the public. In the confines of the tent, formality fell away and – in Ulm's words – 'in a day or two we were calling each other Smithy and Andy and Charles'.

In Sydney a disappointment awaited them. The Adelaide–Perth contract was awarded to Western Australian Airways. Ulm, however, was unabashed, believing that his position remained secure. Smithy was – according to Ulm – 'a wizard pilot who had little more idea of organising big scale finance than he had of conducting a Sunday School'. And Keith Anderson 'was a simple soul with a broad grin and intensely parochial outlook'. They needed him quite as much as he needed them.

By now Ulm was conversant with Smithy's dream of crossing the Pacific. He also dreamed of performing aerial exploits, but he recognised that he lacked the flying skills. His strength lay in organising, and his best chance of achieving his dream was to enable Smithy to achieve it for him. Smithy had the same thoughts in reverse: he believed that he had a remarkable flying talent, but lacked organising ability. To discover a 'twin soul' whose needs and talents so nearly dovetailed with one's own seemed an answer to both their prayers. Later Smithy would write in his autobiography:

> It was at this time that I met Charles Ulm who for the next few years was destined to be my close associate and friend. Ulm had similar ideas to mine. He was ambitious; he wanted to do something which would make the world sit up; he had a good business head; in fact he was a born organiser. Why should not we two capitalise our combined

assets. We began to talk of some Big Feat which would bring us what we wanted, fame, money, status.

Smithy was right in believing that his new soulmate was ambitious. Ulm would later say of himself: 'I am an ambitious man, determined to push aside all obstacles in my path and if such obstacles cannot be pushed aside, then some form of dynamite must be used.' Having found his twin soul, he was determined that no obstacle should come between them. He had disposed of Pike, and now he looked for a way to sideline Hitchcock and Anderson. Hitchcock, who was not essential to the venture, could easily be moved aside, but Anderson was Smithy's legal partner, and had been his faithful friend and follower for the past five years. Even so, Ulm was determined to push him out, and over the following weeks the pushing gained momentum.

What did Smithy think of Ulm's behaviour? He had been brought up by his parents to believe in loyalty and fairness and the obligations of friendship. And since his teens – and especially during the war – he had come to know the value of mateship. By any calculation, Anderson was a mate, and yet Smithy now countenanced Ulm's tactics with scarcely a protest. The truth was that he had come to rely so heavily on Ulm that he dared not go against him. Moreover, he seems to have believed that he had some justification for what he was doing. Anderson had taken Bon from him, and so had forfeited some of the right to his loyalty.

After work of an evening, Smithy would drive Ulm back to his house in the North Shore suburb of Lavender Bay. Situated high above the harbour at 4 Cliff Street, it soon became Smithy's second home. Ulm was sharing the house with two sisters, Josephine and Amy Callaghan, and Smithy soon learned much about Ulm's personal life. Josephine, dark-haired and vivacious with a dimple in her chin, was a schoolteacher and Ulm's fiancée. He was planning to marry her as soon as his divorce became final. His marriage to Isabel Winter had ended in separation, and their son, John, lived with his mother and scarcely knew him.

On the morning of Monday, 23 May 1927 – two weeks after their trip to Canberra – Smithy opened his newspaper and received a shock.

The newspaper bore an astonishing headline: 'Air pilot Lindbergh, flying alone, flew from New York to Paris between Friday morning and Saturday night.' Millions of readers rejoiced at Lindbergh's flight, but Smithy and Ulm and Anderson responded with mixed feelings. Until that moment they had believed themselves safe from competitors, because no pilot as yet had been able to carry sufficient fuel to cross the widest stretches of either the Atlantic or the Pacific. By flying nonstop in his Ryan monoplane for thirty-three hours, Lindbergh had shown that such crossings were possible, and soon any number of aviators would be copying him. Nor were they the only readers to come to this conclusion. A few days later, James D. Dole, the American pineapple king, offered a prize of $25,000 for the first pilot to fly successfully between California and Hawaii.

Smithy's first thought was to somehow raise the money to buy a Ryan monoplane so that he and Anderson could set out immediately. However, during their nightly discussions at Cliff Street, Ulm came up with a bolder plan. The Ryan had room only for two, and as he and Smithy discussed this defect, Ulm put forward a proposal that he would later describe under oath in a court of law.

To Smithy he said, 'If I get the finance will you come along with me?'

'You bet your life,' Smithy answered. 'I am all for it.'

'If it does come off, you will have a problem with your partner Anderson,' Ulm then replied. 'He will naturally think you are dropping him if you join me.'

To this, Smithy is said to have answered: 'We will take care of that difficulty when it arises.'

Now that each knew where the other stood, they discussed the question of finance. According to Ulm, the best way to attract money was to perform a record-breaking flight, and the obvious choice was a flight around Australia. It had already been done twice, most recently in 1924 in a plane piloted by Captain E.J. Jones and carrying the Controller of Aviation, Lieutenant Colonel Brinsmead. That flight – a distance of over 7000 miles – had been completed in twenty-three days. Smithy believed he could fly it in twelve days. When Brinsmead heard the news, he said generously, 'They are taking on a tough proposition, but if any man in Australia can do it, it should be Captain Smith.'

Distance, as they would soon discover, was not their only problem. Their two Bristol Tourers were seven years old, of obsolete design and limited power, and equipped with only the simplest of navigation aids. They carried no wireless and no instruments to aid night flying, and so could only fly safely by day. And at each stop they would need refuelling with petrol, which had to be carted in advance to the remote landing fields. Smithy's jaunty statement that he would circumnavigate the continent in twelve days seemed a fantasy.

Not deterred, Ulm sought sponsors for the flight. He struck a deal with Campbell Jones of the Sydney *Sun* that their progress reports and interviews, in return for finance, would be exclusively with the Melbourne *Herald*, the Sydney *Sun* and the Brisbane *Daily Mail*. An arrangement was also made with Vacuum Oil to supply fuel at each stopping place. It now remained only to choose the pilot and passengers. As a business partner and a co-owner of the Bristol Tourers, Anderson had naturally assumed that he and Smithy would fly together in one of them. As a Tourer could carry three people, there was also room in it for Ulm, but the weight of a third person would reduce their speed. Also, some stages were so long that space had to be kept for a dozen or so petrol cans, which could be emptied into the tank while the plane was airborne. Anderson could see no earthly reason to invite Ulm anyway, for he was not a trained pilot and could not take over the controls while Smithy rested. Moreover, Ulm was a pushy interloper, and Anderson had ceased to trust him. What Anderson failed to recognise was that Ulm was determined to fly around Australia with Smithy, and Smithy was determined that Ulm should have his way.

Anderson was justifiably angry, and a 'devil of a row' erupted. In response, Smithy and Ulm offered a concession. Perhaps both planes might make the journey, with Smithy flying one and Anderson the other. This proposal cheered Anderson, until he discovered that additional conditions were to be imposed. Instead of the planes flying together as a team, Ulm decreed that Anderson should not start until Smithy had finished the course. He also insisted that Anderson should take two passengers: one was Bob Hitchcock, and the other Charles Vivian, the advertising manager of George Bond and Company, a hosiery manufacturer which had just begun to make flying suits. The Bond company had agreed to back the flight, and was entitled to reap the publicity. While the new arrangement was sensible commercially, it was unfair, and Anderson knew it. But being reluctant to sever the partnership, he stifled his resentment and accepted it.

Selecting the faster of the two Tourers, Smithy and Ulm took off from Mascot on 18 June, but a faltering engine forced them down near Newcastle. They promptly travelled in the train to Sydney and commandeered the second Tourer, which Anderson and Hitchcock were preparing for their flight; needless to say, this did nothing to improve their relationship with Anderson and Hitchcock. The following day, Smithy and Ulm set out again from Mascot in the second Tourer and reached Brisbane, where a large crowd welcomed them. Then they were off to Longreach and Camooweal, and then to faraway Darwin. On the Camooweal–Darwin stage an exhaust valve faltered when they were just short of their destination. It was testimony to Smithy's skill that he managed to coax the Tourer to Darwin.

No spare valve could be found in Darwin, so Smithy and Ulm now faced the prospect of abandoning the trip or nursing the ailing plane on the long flight westward to Broome. This entailed flying over an inhospitable tract of country nearly 200 miles from the nearest telegraph line, and equally far from the curving coastline. A mishap in such remoteness could easily have cost them their lives, but the weather remained calm, the engine was reliable and they landed safely in Broome.

There the valve was replaced by Brearley's mechanic, and Smithy and Ulm set out for Carnarvon, where Smithy was planning to meet old friends. But as they neared their destination, rain so pounded the plane that they were forced to take refuge at Minilya Station, and their eventual stop in Carnarvon was disappointingly brief. The contrary weather continued between Carnarvon and Perth. Fog blinded them, but by barely skimming the treetops and following the telegraph wires, they safely reached Perth's Maylands Aerodrome. Here Norman Brearley was waiting to greet them. He was holding a Geraldton waxflower, a gift for Smithy's father to plant in the garden at Arabella Street.

Perth was the turning point of the journey. The weather brightened, and Smithy and Ulm began to believe that that they could complete the journey in less than twelve days. Aided by a brisk tailwind, they raced across the Nullarbor Plain to Adelaide, and from there to Melbourne. That night in Melbourne, in 'talks and conferences' with reporters and businessmen, they openly publicised their proposal for the trans-Pacific flight. Anderson's name never was mentioned.

On 29 June 1927 they landed safely at Mascot, having flown around Australia in the astonishing time of ten days and five hours. In the excited crowd stood the New South Wales premier, Jack Lang, and Smithy's favourite actress, Nellie Stewart. In his speech of welcome, the premier claimed that their flight was as hazardous as Lindbergh's, while Nellie Stewart embraced Smithy and Ulm in turn, draping giant wreaths around their necks. Smithy's feelings for Nellie were as warm as ever, and her photograph, which had gone to war with him, still accompanied him as a talisman on his flights. As she kissed him in front of the 'closely clustered crowd', he blushed like a schoolboy.

For Ulm it was an even happier day. His divorce had become final while he was in the air, and he planned to marry Josephine Callaghan on 2 July. But their welcome at Mascot was so exciting that his imagination jumped ahead of him: what if he were on a ship sailing to America on 2 July? He begged Josephine to marry him immediately, and that evening at ten pm they were married in the Congregational church in North Sydney.

But what of Anderson? Four days previously, on 25 June, just as Smithy and Ulm were approaching Perth, Anderson ignored the injunction to wait and flew off with Hitchcock and Vivian for Brisbane. There seems little doubt that he was determined to break Smithy's latest record. Indeed, at first it appeared he might succeed, because his plane reached Brisbane forty minutes ahead of Smithy's time.

Thereafter, many things went astray. By mischance, on leaving Longreach, Anderson flew many miles off course and was obliged to spend the night in a remote paddock. Two days later, he tried to make up lost time by flying from Darwin to Derby via Wyndham – a dangerous but quicker option – and this improved his position. In Perth he actually arrived a day earlier than expected. But after reaching Melbourne, where bad weather grounded him for two days, he knew he had no chance of beating Smithy's time.

On their final night in Melbourne, he and Hitchcock and Vivian were guests of honour at a Bond Company dinner at the Menzies Hotel. Just prior to the dinner, Anderson opened the afternoon paper, the Melbourne *Herald*, and read the report of a Sydney luncheon to celebrate Smithy's record. Running his eye down the column, he was shocked to read that Smithy and Ulm would be sailing for San Francisco on 14 July. Once there, they would be purchasing a plane like Lindbergh's for £5000, which had been guaranteed by a private subscription. In due course they would fly this plane across the Pacific.

Anderson was aghast. The flight that had so long existed in his mind was now being spoken of publicly as a reality – and it belonged not to Smithy and Anderson but to Smithy and Ulm. His indignation simmered throughout the dinner, and it gained voice in his after-dinner speech. He and Smithy – he announced – were 'ambitious to be the first aviators to make a trip from Australia across the Pacific Ocean'. It was a project that that they had been designing for several years, and for which they had made many arrangements. 'We hope to make the flight at an early date,' he added firmly. Watching the journalists taking notes, he knew that his words would circulate around Australia.

On 8 July, Anderson and his two passengers flew through a snowstorm before landing safely at Mascot. Their circumnavigation of the continent had taken fourteen days, which, in view of the obstacles they had met, was highly creditable. Moreover, the Sydney welcome was all that they could have wished: a giant marquee, speeches and compliments, and a visit afterwards to Government House. But none of this public praise consoled Anderson. Had Smithy betrayed him?

A COUPLE OF
SPENDTHRIFT DREAMERS

Thanks to their round-Australia flight, Ulm and Smithy were now celebrities. In the week following their arrival home, Ulm negotiated a string of deals that might otherwise have proved impossible. Sun Newspapers bought the exclusive press rights to the flight, Australasian Films bought the film rights, and a well-known accountant, Charles Le Maistre Walker, set up a private subscription. Since the flight now had commercial backing, the Commonwealth government withdrew its offer of assistance, but the New South Wales government more than made up for this disappointment. Premier Jack Lang liked and trusted Smithy – he called him 'a good Australian' – and he offered a government guarantee of £3500 towards the purchase of a plane. Furthermore, the Vacuum Oil Company promised to locate a suitable machine for them in the United States, after which the company would advance them the sum that Lang's government had guaranteed.

Most of these activities were unknown to Anderson as he confronted Smithy at Bon's father's house at Collaroy. Had he known, he would probably have been twice as angry. As it was, his anger was sufficient to fuel the legal suits that he and Hitchcock subsequently pursued, and it lives on in the pages of their transcripts.

Confident that he had valid cause for complaint, Anderson unleashed his accusations. Notwithstanding their five years of close partnership, Smithy had gone behind his back and granted Ulm the sole right to fly with him across the Pacific, and this right had been granted entirely on the grounds that Ulm had a talent for organisation and publicity. Responding to the charge, Smithy declared that since Ulm's skills had transformed their languishing scheme, such a reward was 'only fair'. Anderson replied that it was anything but fair. Moreover, it was suicidal for Smithy to entrust his life to an assistant who was not even a licensed pilot. To this, Smithy could offer only the lame reply that negotiations had gone too far to be undone.

It now seemed to dawn on Smithy that their animosity was growing out of hand. In a feeble attempt to restore goodwill, he clumsily presented a compromise. Would Anderson consider flying across the Tasman Sea instead of the Pacific Ocean? A crossing to New Zealand was waiting to be made, and Smithy himself would be glad to help with it. And so would Ulm: in fact, Ulm would organise it! Anderson responded to this suggestion with angry disdain.

The quarrel continued the next day at the Carlton Hotel in Castlereagh Street, and this time Ulm was present. He told Anderson that he had no intention of standing down. Taking Ulm's part, Smithy argued that they owed the success of the round-Australia flights entirely to Ulm, who consequently had every reason to expect to cross the Pacific. Moreover, sponsors such as the Sydney *Sun* were expecting it, and to substitute Anderson for Ulm would seem like a breach of trust.

Anderson would have none of this. He pointed out that he and Smithy were quite capable of organising and financing their own flights – they had been doing so for years. As for the claim that the Sydney *Sun* expected Ulm to accompany Smithy, that very morning Anderson had heard from the manager's own lips that there was no stipulation by the *Sun* that Ulm should participate in the flight. Anderson then made an offer. If Ulm agreed to drop out, he and

Smithy would give him a third of the proceeds. The offer was refused.

In this fashion they argued for some time, until Ulm started to give way. The most likely cause of his change of heart was Smithy, who was beginning to regret their treatment of Anderson. Sensing that Smithy was ceasing to wholeheartedly side with him, Ulm reportedly said: 'I do not mind including Anderson and making the party three and we will sink or swim together.' This would mean abandoning their choice of the two-seater Ryan, for they would need a plane that seated three people, and this heavier cargo might, in turn, lessen their chances of success. However, all three were so desperate to break the stalemate that they agreed to the deal, and drank to it.

It remained to break the news to Bob Hitchcock – an unenviable task, because Bob had convinced himself that he was part of the team. When Hitchcock's name came up, Ulm shouted, 'Damn Hitchcock!' and muttered that there was no question of taking him, because they were already taking too many people. Smithy replied more temperately that he felt sorry for Bob, but he agreed that there was no way they could take him. Although they dreaded the confrontation, they arranged to see Hitchcock at the office the next day.

It was a wretched interview. According to Hitchcock's recollection, Smithy said: 'We have not sufficient finance for the four of us to go. Andy and I must go, Ulm insists on going, therefore we will have to leave you behind.'

'Is that right?' asked Hitchcock.

'No,' replied Smithy, 'but we have agreed that on our return to Sydney we will pay you a thousand pounds for not taking you, for dropping you out.'

Whether Hitchcock's recollection was accurate one cannot be sure, because it was later challenged by Smithy and Ulm.

The following day, when their ship, the SS *Tahiti*, was due to sail for California, there were more exchanges with Hitchcock. Smithy and Ulm had been drinking at the Carlton Hotel; indeed, in Ulm's phrase, they were 'half sozzled'. Neither was in a condition to speak tactfully. Grinning widely, Smithy is reported to have said: 'We are very, very, sorry having to leave you behind. But it can't be helped, old man, but on arrival back there is a thousand hard for you.'

Ulm is supposed to have added, 'Yes, you will hear a knock on the door and I will say, "Sir Charles, Robert awaits outside for his thousand," and Smith will say, "Show the boy in."'

Hitchcock was affronted by their patronising attitude, and two years later repeated the conversation when he sued them in a court of law.

The *Tahiti* was to leave from Darling Harbour at four pm. Smithy and Ulm had booked first-class cabins the week before, but Anderson's was still to be booked, and all three passages were still to be paid for; rumour had it that Anderson paid for Ulm's. Securing their tickets and passports, to say nothing of farewell drinks and last-minute business, so detained them that Smithy and Ulm almost missed the sailing. They are said to have run all the way to the dock.

Once aboard, they made the most of the luxury. As first-class passengers, they were invited to dine with the captain and ship's officers, and it was at the captain's table that they met Robert Haylett, an American executive of the Union Oil Company. Haylett was drawn to them from the start, admiring Smithy's 'quiet determination and cheerful manner', Anderson's sportsmanship and Ulm's unfailing courtesy. Ulm 'was always a gentleman and at home in any company', said Haylett, and especially so when playing bridge. He considered all three of them 'splendid boys', and when they told him about the Big Feat, he was infected by their enthusiasm. He would prove a valuable friend.

Also sitting at the captain's table was the second officer, William Angus Todd, a mountainous man who was all too fond of the pleasures of dining. An experienced navigator and wireless operator, he agreed to give them navigation and wireless lessons. Haylett recalled that at first Smithy and Ulm required twenty-seven minutes to calculate the ship's position, a task that Todd could perform in seven minutes. By the end of the voyage Smithy and Ulm were as quick as Todd.

On 5 August they reached San Francisco, to be met at the dock by Harold Kingsford Smith, who was always ready to do his best for adventurous young Chilla. Also there was Herbert Dickie of the Vacuum Oil

Company, ready to advance the £3500 that Premier Lang's government had guaranteed. The air race to Honolulu sponsored by James D. Dole was due to start from Oakland's airfield in eleven days' time, and Smithy had half a mind to enter. Accordingly, he motored with Dickie to the airfield to see a plane that had been recommended, but quickly rejected it.

Fortunately, Smithy also rejected the idea of flying in the Dole Air Race, for over the next couple of weeks the contest would turn into a tragedy. Fifteen planes were entered, eight reached the starting point, four managed to set out and only two arrived in Honolulu. Ten lives were lost, a source of anxiety to the Big Feat's Australian supporters. Campbell Jones urged Smithy and Ulm to abandon their flight, and the Australian government congratulated itself on having backed out of so dangerous a scheme.

Smithy refused to be daunted. Since going to war, he had held almost a casual attitude to personal danger. He had known of, or witnessed, so many deaths in the air that he had learned to shrug them off with a fatalistic 'I guess his time was up'. And yet the Dole Air Race affected him deeply. Shocked by the death toll, one suspects he was also shocked by the public's reaction to it. Not really interested in safety for himself – he cheerfully took risks that few others would have taken – he was beginning to see that if he wished for public approval, he must be seen to be safety-minded. The prudent Ulm agreed with him.

Lindbergh insisted that a plane with one engine was the safest choice for long-distance flying. The disproportionate losses in the Dole Race – in which most of the planes had one engine – seemed to disprove this theory, and spurred Smithy into undertaking research of his own. By analysing long-distance flights, he came to the conclusion that power and endurance were the main requirements, and that the safest machine was one with several engines, multiple fuel tanks and room for a co-pilot, wireless operator and navigator. The plane he came to favour had three engines and was manufactured by the Dutch aircraft designer Anthony Fokker, whose planes had been effectively flown against Smithy and his comrades in the Great War.

Unfortunately, a Fokker plane of such power was hard to find, and considerably more expensive than any machine he had hitherto considered. Nevertheless, soon after the completion of the Dole Air Race, Smithy learned that a three-engine Fokker monoplane was for sale in Seattle. It belonged to the Australian polar explorer George Hubert Wilkins, who already had a tenuous relationship with Smithy. A second lieutenant in the Royal Flying Corps at the same time as Smithy, he had participated in the England to Australia air race. After Smithy's exclusion from the race, Wilkins joined the crew of Blackburn's *Kangaroo*, and was aboard when it crashed in Crete.

Wilkins was invited to San Francisco to discuss the price, and he explained that he was selling only the fuselage and the wing, for which he was asking £3000. Smithy himself would have to provide the engines and the instruments, and it soon became clear that more money would be needed than the £3500 pounds of Lang's guarantee. Wilkins offered to take half the purchase price now, and to receive the rest in stages. Even so, Smithy decided to ask Lang to raise his government's guarantee. Urged on by a deputation from the Returned Soldiers League, Lang promised a further £1000, but this still failed to meet the cost of the engines. Smithy had selected Wright Whirlwind engines, each of 220 horsepower. They were expensive but renowned for their reliability: Lindbergh's Ryan was powered by a Wright Whirlwind. It was at this moment that Smithy, Anderson and Ulm chanced on Sidney Myer, Melbourne's most dynamic retailer, who was spending the summer near San Francisco. Myer warmed to these young aviators, and although he shuddered at the risks they were running, he generously gave them £1500.

Fate seemed to favour Smithy and his crew through September 1927. With the help of Harold Kingsford Smith – who was still working for the American-Hawaiian Steamship Company – Keith Anderson sailed to Hawaii to inspect prospective airfields. And in the Boeing Aircraft factory in Seattle, where Wilkins' Fokker was thoroughly overhauled, it now only remained to fit the three new engines. And here they met a setback. The factory supplying the engines was behind in its orders and Smithy's only hope for a quick delivery was to jump the queue.

It came to his notice that Rear Admiral Christian Peoples in San Francisco might be persuaded to allow him to take the Navy's place in the queue. Smithy obtained an interview with the admiral, who bombarded him with warnings until he heard that the proposed plane was a Fokker trimotor with room for a crew of four. Impressed by Smithy's choice of an aircraft and his insistence on safety, the admiral was won over. Smithy received the Navy's place in the queue, and from then on – to quote his words – 'the State Naval Office in San Francisco was at our disposal with every possible source of information'.

Once the engines were fitted, it was time to make the test flight. Smithy would have liked to perform this alone but, recognising that he was a novice with trimotors, he deferred to Lieutenant George Pond, the US Navy test pilot. Seven years older than Smithy and a highly experienced pilot, Pond agreed that they should test the Fokker together, and that Smithy should have time at the controls. Pond marvelled at his young companion's skill and intuition. The plane was unknown to him and yet he flew it with ease.

Years later, Patrick Gordon Taylor – a friend of Smithy's and himself a great aviator – revealed the secret of controlling planes as powerful as the Fokker:

> You don't fly them. You go with them. Giving them a guiding hand, reacting to their demands and wishes and acting in their interests ... You know they need you, vitally, for the exact and accurate reaction to their needs: and so you become sensitively part of the great patient monster you control.

The pilot of such machines required an imagination able to 'project itself forward to situations well ahead of the aircraft', and a 'strength of character which could retain stability of action in any circumstances'. In Taylor's estimation, Smithy had both those qualities, and they formed the basis of his greatness as an aviator.

Pond and Smithy flew the plane to Mills Field, the aerodrome that would later become San Francisco's international airport. There 'the

Old Bus' – as Smithy was already calling her – was formally named the *Southern Cross*, Anderson having proposed the name, which referred to the constellation on Australia's national flag. Smithy and Ulm adopted the name enthusiastically.

At Mills Field, the navigational and wireless equipment was installed, which raised the question of a navigator and a wireless operator. Should the navigator be Second Officer William Todd of the SS *Tahiti*? Ulm objected to Todd because he weighed about 270 pounds – over 120 kilograms – and on a long-distance flight every pound mattered. Smithy nevertheless insisted on inviting him. Todd took leave from his ship and joined Smithy and Ulm at the Roosevelt Hotel in San Francisco.

It was now mid-October, and fate was about to deal the young aviators an unexpected blow. At the New South Wales parliamentary election, Jack Lang's Labor Party lost. The new premier, Thomas Bavin, regarded Smithy and Ulm as a 'couple of spendthrift dreamers'; his open disapproval plunged them into a state of uncertainty, which was worsened when Wilkins requested the rest of his money. They were saved by the kindness of their 'very good pal' Locke Harper, the west coast manager of the Vacuum Oil Company. He had grown fond of these courageous young Australians who often shared his office, and when Wilkins began pressing for payment, Harper arranged a mortgage on the plane. He wished he could have done more, for he sensed how badly they were suffering. Smithy seems to have sought solace in the arms of local girls, but Ulm, with his new wife waiting in Sydney, found comfort in work, though a cold comfort it proved. He tramped from interview to interview, seeking money for a project 'which nobody in the world apparently thought was possible'.

In November a gleam of hope beckoned when the Associated Oil Company of California offered a prize of $15,000 for an unusual aerial exploit. A pair of aviators was required to break the world's non-refuelling endurance record by flying continuously above San Francisco for more than fifty-two hours and twenty-two minutes. While an endurance flight was primarily a publicity stunt, useful money could be made from it and useful knowledge gleaned, and Smithy was keen to profit from both. The flight would test his capacity to endure more than two days of continuous flying. It would demonstrate how the plane behaved in a variety of conditions and situations. Above all, it would show how the plane performed when carrying an excessive fuel load.

To break the record, it was calculated that they would need four fuel tanks in the hollow wooden wing, each holding ninety-six gallons, a main tank in the doorway to the rear cabin, holding 807 gallons, a second tank in the rear cabin, holding 330 gallons, and another tank under the pilot's seat, holding 107 gallons. Such a weight of fuel was beyond the *Southern Cross's* present strength, so its body would need strengthening at the Douglas Aircraft Factory in Santa Monica. Notwithstanding these drawbacks, Smithy was determined to try for the prize, and so was Anderson – now back from Hawaii – who naturally assumed that he would be Smithy's co-pilot. Approaching his wealthy uncle Vincent, Anderson elicited $3000 to cover the fitting of the extra tanks and the strengthening of the body.

Before they could begin the flight, more problems arose. Ulm had arranged a second mortgage on their plane from the *San Francisco Chronicle*, and it refused to allow its asset to be used in a hazardous endurance flight. The team would have been grounded but for the generosity of Locke Harper, who took over the mortgage. Then Anderson discovered that George Pond was to be Smithy's co-pilot. There was an angry scene. The problems with Anderson seemed ever-present.

On 3 December, Smithy and Pond took off from Mills Field. They managed to stay aloft for 139 minutes before a violent shuddering forced them down. Two further attempts met similar problems, but on 9 December they managed to stay up for forty-nine hours and twenty-seven minutes. Their fifth and final attempt, ending on 17 January 1928, lasted for fifty hours and four minutes. They lost the contest by two hours and nineteen minutes.

Looking back, Smithy described the flights as a nightmare: 'We were cramped in the cockpit since the passage way to the rear compartment was filled with petrol tanks; we couldn't smoke; we couldn't sleep; we had to maintain our wits at their sharpest, for at our low flying speed we were always near stalling.' When it was time to disembark, he and Pond were so stiff and weak and cold that they had to be lifted down from the cockpit. But more worrying than the physical discomfort was the antagonism that Pond aroused in Ulm and Anderson. They suspected – probably correctly – that he was trying to capture Smithy for his own Big Feat.

Failure to win the endurance prize threatened to send them deeper into debt, and just when it seemed that they could sink no lower, another unexpected blow struck them. Early in the new year of 1928, they discovered a condition of their government guarantee that had, rather surprisingly, escaped their notice. Lang's Labor government had included an expiry date. The flight must be attempted or abandoned by 14 January 1928, after which the financial guarantee became void. On 17 January, Ulm was obliged to cable Premier Bavin that they were ready to abandon the flight.

To have planned the crossing for so long, and to have kept up hope for so long only to have it snatched away by an arbitrary government decree, was more than they could bear. Years later, Smithy described their feelings in five poignant words: 'We had reached rock bottom.'

ROCK BOTTOM – BUT NOT FOR LONG

'We were absolutely penniless, and moreover heavily in debt,' wrote Smithy. 'We were so poor that we had not even loose cash in our pockets to purchase cigarettes and a meal. We were unable to pay our hotel bill and were driven to all sorts of subterfuges to stave off those to whom we owed money.' Moving to progressively seedier hotels 'whose credit clerks watched us with suspicious eyes', they dreaded what 1928 would bring.

Reason told them they should go home, but emotion, far stronger, urged them to hold fast. Smithy's life had been ruled by the Big Feat for the last eight years; to relinquish it now felt like a form of emotional suicide. Though Ulm's attachment had been of less duration, it had become his potential passport to fame and wealth, and he was prepared to hang on as long as he could see even a glimmer of hope. To make the situation bearable, they indulged in wishful thinking. The current wish was that Bob Haylett's company, Union Oil, would buy the plane and assist them with the flight in order to publicise its products. Ulm felt so sure he could swing this deal that he travelled to Los Angeles, but he made little headway in the negotiations. Even so, his efforts were not entirely wasted, because he earned the admiration of the Union

Oil Company. 'He impressed us all with his determination to succeed,' wrote Haylett, 'and with the painstaking care with which he prepared for every contingency.'

In the face of such disappointment, it was no surprise that 'bitter arguments and petty squabbles' broke out among the team. The first rebel was William Todd. Unstable by nature, and irritated by inactivity, he bewailed ever leaving his ship. Drinking too much, he fell foul of Ulm, who was horrified at the debts he was incurring. The final blow came when he crashed the aviators' car. Early in January he left the team and sailed back to Sydney to join a new ship. On arrival, he pleaded Smithy's and Anderson's cause to reporters but, significantly, did not mention Ulm.

Anderson also showed signs of cracking. Excluded against his wish from the test and endurance flights, he felt more than ever the odd man out, a feeling that was reinforced by Bon and his mother, whose letters begged him to return home. Indeed, in December 1927, Anderson's departure had seemed so imminent that Ulm drew up a legal agreement allotting himself and Smithy 'all the assets and liabilities of the partnership' if Anderson withdrew. At this stage, Smithy and Ulm seemed content to let him go, but over the following two months their attitude altered. When, in February, his abandonment of the Pacific flight grew close to certain, his departure seemed like a break of faith, and something they must at all costs prevent.

February proved a difficult month for all three of them, for it brought disturbing news that nobody could have predicted. On 7 February, the Bundaberg-born Bert Hinkler – a former RAF flyer and Avro test pilot – set out from England in an Avro Avian to fly solo to Australia. On 22 February he arrived in Darwin, having taken just over fifteen days. The world was astounded by his flight, and Australia was in a fever of excitement. He was feted across the country, with newspapers calling him 'the Monarch of the Air'.

There can be no doubt that Hinkler's flight was a blow to Smithy and his companions: they were stuck in America, beset by obstacles, while Hinkler was basking in the acclaim that they had dreamed of for themselves. However, once the shock wore off, Smithy and Ulm came to view Hinkler's flight as a challenge, and it caused them to redouble their efforts. On Anderson it had the opposite effect. He was already depressed, and Hinkler's success seemed to push his mood even lower. He gave notice of leaving on 20 February, the day that Hinkler took off from Singapore. He softened his announcement by saying that he would try to raise funds in Sydney, but this was more an excuse than a reason to leave. The truth was that he was desperate to get away.

Ulm was now in Los Angeles, and Smithy was enjoying respite at his brother's house in Oakland, so Anderson's decision was relayed to them by telegram and telephone from Harper's office in San Francisco. Smithy hastened at once to the office, where Anderson was waiting for him, and did his best to dissuade him. But to no avail. A few hours later Ulm rang Harper's office by arrangement and spoke to Anderson on the telephone. 'We three came over together and we three should go back together,' Ulm announced sternly. In response, Anderson 'stammered and stuttered' into the phone that Bon's father, Arthur Hilliard, had sent him money for his ticket home and he was sailing in two days. But Ulm was persistent: 'For God's sake wait till I get there and let us hear what it is all about ... I do not hold with it at all.'

Ulm had no money for his fare from Los Angeles to San Francisco, but Harper obligingly wired him $40 for the train, and next morning Anderson and Smithy met him at the station. Proceeding to Harper's office, they invited Harold Kingsford Smith to act as their arbitrator, because they knew they were going to find it difficult to agree.

Harold took them to a coffee shop and, listening to their initial arguments, advised Anderson to stay. But then, having listened to them for a further hour, he found himself telling all three of them to go home. It had become clear to him that there was no foreseeable way of salvaging the plane or the flight. The *Southern Cross* must be sold, and Premier Bavin repaid. Any money remaining must be used to clear their two mortgages and pay their other creditors. Ulm hoped that, after paying their debts, they might still have enough to buy a Ryan monoplane, which could then be flown across the Pacific, but this seemed unlikely since they owed more than $16,000. 'You have got to realize,' said Harold firmly, 'that the whole show is up.'

That evening Smithy, Anderson and Ulm returned to the room they were sharing at the Roosevelt Hotel. There Ulm painstakingly drew up a legal document to deal with their current situation. It contained Anderson's power of attorney, which they knew they would need when they sold the plane. More significantly, it included a clause to protect Anderson, if by some miracle the Big Feat seemed certain to go ahead. In such an event, Anderson was to have the right to return to California and claim the second pilot's seat.

In view of Ulm's talent for ousting rivals, this was a curious clause – a response, perhaps, to Harold's insistence that Anderson must receive 'a fair go'. There may, however, have been something more behind the gesture. With Anderson gone, there was a strong chance that Pond might try to become the second pilot across the Pacific. Locke Harper favoured Pond, but Ulm was totally against him. Given a choice between Pond and Anderson, Ulm would choose Anderson every time.

Two days after Anderson sailed away on the *Makura*, Smithy and Ulm flew the *Southern Cross* to Rogers Field, on the outskirts of Los Angeles. To save money they dossed down at the airfield, and made frequent visits to the Union Oil offices. 'I shall never forget,' wrote Bob Haylett, 'the dreary afternoon in March when the two boys sat in my office, discouraged and downcast.' He had just informed them that negotiations were at an end, and Union Oil had no intention of buying the *Southern Cross*. Haylett remembered that Smithy, usually so cheerful, made jokes 'in a bitter sort of way', and cheekily enquired whether those gravely in debt in the United States were sent to a debtor's prison.

Sympathising with their distress, Haylett sought introductions on their behalf to Los Angeles businessmen who might be persuaded to help them. One was the president of the California Bank, and Ulm found to his surprise that the charming and powerful man who sat behind the desk had spent his childhood beside the Murray River in Australia. His name was Andrew Chaffey, and he was the son of George Chaffey, a Canadian water engineer, who, with his brother William, had irrigated dry tracts of the Murray Valley into green oases. Whereas William and his family had stayed on in Victoria, George and his family had resettled in Los Angeles, and had diversified into banking.

Andrew Chaffey was fascinated by the idea of the Pacific flight, and came to Rogers Field to inspect the *Southern Cross*. Being the son of an engineer and inventor, he listened eagerly to the technical information that Smithy poured out. Chaffey's circle of friends was wide, and many were extremely wealthy, so when he spoke of enlisting a rich American patron for the flight, he was not boasting.

The friend he invited a few days later to see the plane was George Allan Hancock, a quiet, thoughtful man who liked the title of captain because he was a licensed master mariner. Aged fifty-two, a founder and director of the California Bank and a millionaire many times over, Captain Hancock held interests in railways, ships, ranches and other real estate. Later he would add aviation. The genesis of his fortune was his childhood home, Rancho La Brea, a sprawling cattle ranch situated between what are today Wiltshire and Sunset boulevards in Los Angeles, some of the most coveted property in the world. Covering 4438 acres, the ranch was dotted with tarpits, which in 1907 were turned into seventy-one seemingly bottomless oil wells. Hancock could afford to take risks, and the challenge of the Big Feat excited him.

As Haylett remembered it, Hancock climbed into the cockpit of the *Southern Cross* beside Smithy, while Chaffey and Ulm and Haylett took seats in the rear cabin. They cruised over the city, with 'the plane running at times with only one or two motors to demonstrate its air stability in the event of trouble'. It was well known that Hancock took pleasure 'in helping others to help themselves', and he was delighted by Smithy's and Ulm's intelligent approach to their work and their emphasis on safety. With a reliable plane like the *Southern Cross*, and astute flyers like Smithy and Ulm, he believed the Big Feat was possible, and in his quiet, courteous way, he told them so.

All in all, he warmed to these young visionaries, who probably reminded him of his own boy. Two years before, in Santa Barbara, Hancock and his 22-year-old son had been trapped by an earthquake.

Hancock had been badly injured, and his boy had died in front of his eyes. The slight impairment that Smithy noticed in the millionaire's speech was a legacy of that 'frightful moment'.

Soon after his visit to Rogers Field, Hancock invited Smithy and Ulm to sail with him to his ranch near the Mexican port of Mazatlán. The young Australians were surprised by the invitation. They had realised that the *Southern Cross* might prove a useful acquisition for someone with a far-flung empire like Hancock's, but now it seemed that it was they – and not their plane – that had caught his fancy. Ulm supposed that his interest in them 'was not unmixed with a degree of compassion'.

After their recent privations, twelve days of luxury aboard Hancock's steam yacht, the *Oaxaca*, was a gift from the gods. But their clothes, threadbare by now, were an embarrassment. 'To board a millionaire's yacht in patched pants,' wrote Ulm later, 'would be bad for prestige and damaging to pride. Better to go in borrowed trousers.' They hired two pairs of an identical size. According to Ulm, 'we had to reef up one pair for Smithy, and let the other down for me'.

In Mazatlán, lying just below the Gulf of California, they swam and sunned themselves and raced about the harbour in the captain's small speedboat. They also discovered that the yacht itself had 'as fine a set of navigation instruments one could find anywhere', and that their host had a passion for navigation. Since this was also Smithy and Ulm's latest passion, their enthusiasms combined wonderfully. But uppermost in the young men's minds was the opportunity to sell the *Southern Cross* to Hancock: or, in other words, to persuade Hancock to finance their Pacific flight. They dared not speak of the sale openly for fear of offending, but at the same time they dared not let an opportunity pass. The cruise was almost over before the sale was mentioned. As Ulm recalled, the captain broached the subject nonchalantly:

'How much money,' he asked, 'is required to put you boys on the right side?' I gulped, my tongue stuck to the roof of my mouth. The query came as a shock. Then I told the frank truth. '16,000 dollars.'

'I'll buy your machine for that sum,' said Captain Hancock. I fought down my mad desire to dance around the chart room.

Hancock's yacht docked in Los Angeles on 2 April 1928. Smithy and Ulm raced to the airfield, only to find that the *Southern Cross* had been seized by bailiffs and moved elsewhere. The following twenty-four hours were passed in feverish anxiety, but they need not have worried. Hancock bought the plane outright. 'He had paid for it a sum sufficient to free us from all local liabilities,' Smithy later remembered, 'and he had set our ambitions flaming as brightly as ever.'

With Hancock backing them, events moved quickly. The captain demanded the best of everything, and sent the plane to the Douglas Aircraft factory in Santa Monica to be refitted. The old fuselage was replaced by a new one, and the massive wooden wing, spanning seventy-two feet, was reconditioned. Smithy and Ulm were now intent on making the flight a model of safety as well as efficiency, so they requested a valve which would empty the four petrol tanks in the wing in fifty seconds, and then reseal them. This was a safety measure: if they were forced to land on the ocean, the hollow wing – detached by a saw from the motors and fuselage – could double as a life raft. Sealed in the wing were a watertight radio and balloon aerial, and a small distillery for converting seawater into fresh water.

Hancock insisted that the best technicians should perform the over-hauling. Cecil Maidment – late of the Royal Flying Corps, and known as 'Doc' – was the Wright Aeronautical Company's most experienced engineer. He had serviced the Wright Whirlwind engine on Lindbergh's plane, and was now in charge of the *Southern Cross*'s three engines. And it was Captain McMillan of the Hydrographic Office in San Francisco who oversaw the installation and testing of the *Southern Cross*'s compasses and navigational equipment.

After several changes of plans – one of which had routed them via New Guinea – their plan of attack was confirmed. They would make the flight in four hops: from California to Honolulu, Honolulu to Suva, Suva to Brisbane, and Brisbane to Sydney. The longest stretch, from

Hawaii to Suva, would be a frighteningly long 3128 miles, but they had estimated that the sum of the fuel in their tanks would enable them to fly for 3645 miles. It seemed like a reassuring margin of safety – but not if the winds were contrary or the navigator made an error.

Hancock, impatient of delay, demanded that the flight begin within a month, but Anderson was the obstacle not yet faced. Smithy and Ulm had tried to keep him abreast of developments. On 16 March, while he was still on the SS *Makura*, they had cabled Arthur Hilliard's office in Sydney to say that the deal with Union Oil was off. A few days later they had cabled again, to tell him that they were 'spending a fortnight with a millionaire friend of Haylett's who was likely take over whole proposition'. Now that Hancock had bought the *Southern Cross*, they were legally bound, by the terms of their latest agreement, to summon Anderson to California to act as the second pilot.

Were they sure they really wanted him? One suspects they were not, although their first cable was direct and to the point. Written in the usual cablegram style, it simply said: 'Finances completely arranged everyone's satisfaction hence essential Keith returns first steamer advise leaving date – Chillachas.'

This cable raised many questions in Anderson's mind, and even more in the mind of Hilliard, who, as well as being Bon's father, was a capable solicitor. Who was providing the finance? Was the plane to be the *Southern Cross*? If not, would the partnership agreement still apply?

A couple of weeks previously, Australian newspapers had carried reports that George Pond and Smithy were about to fly around the world together and would soon reach Hawaii. The reports were quickly denied but they shook Anderson's confidence. He was reluctant to return to California without a fuller explanation. On 12 April he cabled back: 'Require details what machine who owners date departure route crew my status explain Pond world flight you furnish fare SS Tahiti sailing 19th – Anderson.'

Impatient for an answer and irritated by Anderson's 'shilly-shallying', Ulm had no intention of making a lengthy explanation. On 18 April, he cabled: 'Status co-pilot personally urge you come Tahiti failure so to do naturally loses your interest which I will consider tremendously disloyal – Charles.' When Anderson protested that he did not have the fare for the SS *Tahiti*, Ulm retorted that he was in no position to supply a boat fare, and that he was out of patience with this wavering. 'Are you coming per Tahiti or not,' he demanded. 'No reply by nineteenth April considered as your total withdrawal.'

Being a lawyer, Arthur Hilliard was able to shrug off these exchanges with professional detachment, but Anderson's reaction was emotional. Feeling hurt and angry, he tried to stand on his dignity. On 20 April he cabled:

> Presume your failure to answer following my justifiable enquiries in my cable 12th means that my participation in flight undesired I therefore require compensation my third interest and claim as creditor stop will take 13,000 dollars in full satisfaction otherwise I refuse to withdraw and hold you both personally responsible for any action prejudicial to my interests.

On 23 April, Smithy and Ulm cabled an answer which contained some of the information that Anderson and Hilliard had been seeking:

> Mortgagees grabbed plane later sold for 16,000 dollars with your sanction stop we since repurchased from private backing personally secured and offer you third our interest provided you come by first steamer paying own fare or working passage as we cannot raise fare stop honour demands reply immediately, stating coming or not this our last word. Chillachas

Anderson decided to return but, still standing on his dignity, demanded a first-class ticket, for which he expected the others to pay. Meanwhile, in California, Hancock was complaining about the delay, and Smithy and Ulm were running out of time. By 11 May they felt compelled to act decisively, even though one suspects that this must have been painful for Smithy. Their cable was short and to the point: 'Failure to reply our

last cablegram and failure to catch Niagara conveys your non accep-
tance of our offer of a third interest useless catching later steamer as
flight commences shortly hence we now formally withdraw previous
offer Smithulm.'

Having sent off their cable, they composed a calmer letter to
Anderson. They told him that they had reluctantly concluded from his
cables that he did not wish to return, and if this was his wish, they were
prepared to accept it. However, they could not accept his entitlement to
a one-third interest in the newly planned flight, since he had not contributed to its organisation and financing. Moreover, his indecision had
cost them 'great trouble and considerable expense, for we both believed
you would return and planned accordingly'. Even so, they wished him
to know that they 'were both sincerely sorry' he was not to cross the
Pacific with them, and they offered to talk the matter over with him
back in Australia. The letter was dispatched on 16 May.

Anderson blamed Smithy and Ulm for his supposed rejection,
but it was Hancock who dealt the final blow. Wisely concluding that
it would be suicidal for them to cross the Pacific without a first-class
navigator and a first-class radio operator, Hancock began to search for
suitable candidates. At this, Ulm – until then the navigator – hastily
promoted himself to the co-pilot's seat. No longer was there any place
for Anderson in the Fokker, even if he had been present in California.

On the advice of Captain McMillan, Smithy and Ulm selected a
naval navigator called Harry Lyon, a 'burly man of moderate height,
square build with a strong face and the direct manner of a seaman'.
When McMillan introduced Lyon to Smithy and Ulm, he said, 'This is
Harry. He has a lot of faults – two or three teeth missing – but he's a
good navigator.' Smithy liked him instantly.

As well being a highly competent navigator, Lyon was a sharp and
amusing companion with a larger-than-life personality. The 43-year-old
son of US Rear Admiral Henry Lyon, he had, on his own admission,
been expelled from Annapolis Naval Academy 'for all round hell raising', and that had set the tone for the rest of career. After a further
year at the upper-class Dartmouth College – which also failed to tame

him – he ran away to join the merchant navy as an ordinary seaman, or, as he humorously expressed it, as a 'gentleman rope hauler'. For ten more years he served in sail and steam, working mainly in the Pacific and eventually gaining his master's certificate. After America's entry into the First World War in 1917, Lyon transferred to the US Navy, where he rose to the rank of Lieutenant Commander before returning to his old adventurous ways.

When asked if he knew a capable radio officer, Lyon immediately recommended his former shipmate James Warner, a modest Midwesterner in his thirties who had a shy smile and a 'quick nervous manner'. Warner had been raised partly in an orphanage, and had done menial work until joining the US Navy a few years before the war. There he had trained in the brand-new science of radio, and had served as a radio operator on Allied convoys in the North Atlantic, meeting Harry Lyon aboard the US cruiser *St Louis*. By chance, he and Lyon had recently become reacquainted in San Francisco, where, having retired from the Navy, Warner was selling gentlemen's suits. He explained that he had sold twenty-three suits in three weeks, even though 'peddling pants' was not really to his taste.

Only after considerable indecision did Lyon agree to join the Pacific flight. He had flown once and disliked it, but he was impressed by the young aviators' 'extreme earnestness and confidence', and by the 'uncanny thoroughness of their preparations'. Warner, who had never flown, and was well acquainted with the fatalities of the Dole Air Race, wanted no part of the undertaking. He did his best to talk Lyon out of it, but in the end he was talked into it.

As neither Lyon nor Warner knew anything about planes, they found their equipment bewildering. On a chart table just behind the large oval petrol tank in the doorway of the rear cabin, Lyon had an electromagnetic earth-inductor compass, powered by a windmill that revolved outside, on the roof of the cabin. He also had a curious instrument with three parallel wires – used for measuring drift – dangling outside the cabin door: he was obliged to poke his head into the slipstream each time he wished to take a reading. Furthermore, he was obliged to poke his head through a hatch in the cabin roof to take a sextant sighting. Designed especially for navigating by air, his sextant had a fragile 'bubble' – similar to that in a spirit level – to represent the ocean's horizon.

Lyon declared he had never before 'seen a bubble sextant, let alone used one', and wondered how he could gain experience. It seemed to him that riding in a bumpy, fast-moving car conveyed something of the sense of being on a plane, so on a stretch of road near San Francisco he would be driven at breakneck speed towards a friend seated in a parked car about a mile away, with a star in his sextant sights. As Lyon's speeding car drew level with his friend's car, Lyon would take a split-second sighting, and would later compare it with his friend's sighting for accuracy. 'We got in quite a bit of practice at this,' wrote Lyon, 'but after the third arrest it became a bit expensive.'

Jim Warner's two radios – placed on metal shelves on the port side of the rear cabin – were suspended by rubber cords to protect them from vibration. They were powered by two wind generators, like tiny windmills, attached to the outside surface of the cabin, accompanied by a retractable 120-metre radio aerial that trailed behind the plane like a festive streamer. One radio, for short-wave transmission, spoke to shore stations. The other, a long-wave radio, would contact ships at sea in order to ascertain the plane's exact longitude and latitude. Both the receiver and the transmitter could operate over a distance twice as far as earlier radios.

This was comforting knowledge for Lyon, for despite the modern instruments, he was planning to navigate by the old-fashioned method of 'dead reckoning', or calculating one's position by taking into account compass direction, speed, wind drift and the time required to fly from a previously calculated position to one's present position. It was important to Lyon to be able to speak to the ships that sailed below them, in order to confirm his calculations.

Smithy also needed to acquire new skills. He had persuaded George Pond to teach him instrument flying – often called 'blind flying' – which was then a little-known skill, but was evolving rapidly. Previously, like other pilots of the time, Smithy had based his judgement on what he saw and felt and heard. Now he wisely realised that when a pilot was flying in darkness or fog, or over a vast ocean devoid of visual clues, the human senses could become confused: the pilot needed the bank-and-turn indicator, the rate-of-climb meter and the earth-inductor compass, instruments not prey to human confusion. Theoretically it sounded easy, but in practice it was difficult. However, Smithy believed that 'until a man can fly a plane in a black void for hours, seeing these instruments and nothing else, he is not a safe pilot to fly a plane over long stretches of water'.

Smithy, Ulm and Hancock had scheduled the Pacific flight for the end of May, which was fast approaching, so it was time to put the *Southern Cross* through its paces. They flew from San Francisco to Los Angeles, with the captain and Smithy in the cockpit, and Ulm in the rear cabin, observing Lyon and Warner. When Ulm saw how well the Americans worked together, he was reassured. Less reassuring were the communications between the cabin and the cockpit, for ordinary speech was impossible, the doorway being blocked almost entirely by the huge petrol tank. At first they tried to speak through an electric telephone, but the noise of the engines obscured the words: indeed, after an hour in the air, the noise became so painful that they were forced to plug their ears with cotton wool. Ingeniously, Ulm devised a solution. He took a long wooden stick, to which he attached a written message by means of a paper clip. The stick and its message were forced through a small triangular space near the top of the tank and into the cockpit, where it poked Smithy in the back of the neck. Messages were sent to the rear cabin by reversing the process.

On 23 May they flew from Los Angeles to Oakland Airport, near San Francisco. The airport had been built for the Dole Air Race, and in 1928 it had the longest runway in the world, ideally suited to the heavily laden *Southern Cross*, which needed a long run to gain speed for take-off. At Oakland, the airport authorities and local Chamber of Commerce had everything in readiness. All the aviators needed now was a full moon and a favourable weather forecast.

There were some discordant notes in those last hours. Ulm produced a legal document for Lyon and Warner to sign. It described himself and Smithy as the owners of the flight, and stated that, in preparing for the enterprise, they had incurred a 'personal liability of up to $50,000'. Lyon and Warner were to be awarded $40 a week until the flight commenced, to which was to be added $1000 in bonuses and a first-class boat passage home once the *Southern Cross* landed successfully in Sydney. It also indemnified Smithy and Ulm in the event of the Americans' injury or death. An overriding provision debarred Warner and Lyon from sharing in any money generated by the flight, or from writing or publicly speaking about it until they regained America. An even more restrictive provision decreed that they were to leave the flight in Fiji.

It would seem that Ulm's competitive spirit, along with his desire for an all-Australian crew, had made him as antipathetic to Lyon and Warner as he had been to George Pond. He hated the prospect of arriving triumphantly in Australia with two Americans as part of the crew. Nor did he feel that their navigational skills were necessary after Fiji, since a landmass the size of Australia was too prominent to be missed.

Neither Lyon nor Warner took kindly to the contract. Both were senior to the young Australians in age and experience, and had held positions of responsibility in the US Navy. They believed that they should have been regarded as members of a team rather than as hired hands – particularly in an enterprise as risky as this. 'It was quite a work of Art,' said Warner sarcastically. He and Lyon concluded it was the work of the domineering Ulm, and had nothing to do with the fair-minded Smithy. However, not wanting to be seen as cowards, or to miss out on the adventure, they both signed – with misgivings.

What Smithy thought of the document no one knows. Having witnessed Ulm's treatment of Bert Pike and Keith Anderson and George Pond, he should not have been surprised. It was wonderful to be allied to a twin soul whose needs dovetailed with one's abilities, but there was a price to be paid and he must pay it. All he could do was be as generous to the two Americans as the circumstances allowed, and tactfully hold his tongue.

The other discordant note was a sad cable from Keith Anderson. 'You have the best possible machine,' wrote Anderson, 'if your companions skills equal your determination you will succeed good luck regret failure furnish my fare ensuring my participation was not first consideration expect a little justice and compensation upon your arrival here.' In those last, crucial hours, Smithy did not dare to dwell on Anderson's wounded feelings.

By the evening of 30 May the moon was full and the weather report was favourable; 'there was nothing,' wrote Smithy, 'further to delay us. I retired to bed that night ready for the departure in the morning, filled with confidence.'

His confidence may not have been as wholehearted as he pretended. Over the past eight years, many had accused him of being foolishly obsessed with the Big Feat. He had refused to hear them, believing that an aerial crossing of the Pacific was a feasible proposition. Now he was about to discover if his belief was justified – and if it were not, it would probably cost him his life. However, he was in no doubt that this was the greatest adventure he would ever know, and that part of his nature that craved danger and excitement was well satisfied.

WE ABSOLUTELY
WON'T FAIL

Grey curtains of mist competed with sunshine on the morning of 31 May 1928, as a throng of several hundred supporters and reporters gathered at the Oakland airfield to wish the *Southern Cross* and its crew godspeed. Captain Hancock, so unassertive as to be almost anonymous, stood among them, as did Harold Kingsford Smith. Also present were two couples who had befriended the young Australians when their fortunes were at their lowest: a Glendale surgeon, Dr F.T. Read, and his wife, Ann, and a furniture retailer named Robert Wian and his wife, Cora. Ann Read presented Smithy with a bunch of Californian poppies and a good-luck horseshoe, which he added to his lucky photograph of Nellie Stewart and a Felix the Cat brooch that he wore on the front of his cap. The brooch had been given to him by a 'romantic' young lady, he blushingly told a reporter.

As might have been expected, romantic young ladies were in the crowd; one reporter noticed three of them pressing kisses on Smithy as he prepared to enter the cockpit. Almost at that same minute, the mother of Alvin Eichwaldt, a flyer killed while searching for the missing Dole Air Race pilots, tearfully presented her son's silver ring to Smithy to wear. 'The incident touched me deeply,' he wrote later, 'and I gladly assented to her wish.' Turning to the eager reporters, he confidently made a final statement: 'I've waited many weary months for this day. We are fully prepared and if we fail, I haven't a single regret. We absolutely won't fail.'

At about eight forty-five, Smithy and Ulm climbed the steep ladder and clambered into the cockpit, while Warner and Lyon entered the rear cabin through the door in the fuselage. Cecil Maidment cranked up the engines and the Fokker began to move. A minute or so later, one of the engines faltered. Smithy halted the plane briefly, then started it up again, and this time all three engines 'roared in harmony': 'no Wagnerian chorus,' wrote Smithy, 'could have given us greater pleasure.' The overladen *Southern Cross* rose ponderously into the air, pursued by a fleet of small planes carrying movie cameramen. 'I experienced a sensation of relaxation and relief,' Smithy remembered, 'and at the same time a tremendous elation at the prospect before me.'

The sun now broke through a height of mist, giving 'rosy promise' of fair weather. As they flew at 1100 feet past the city skyscrapers and on towards the Golden Gate Bridge, they observed twirls of mist drifting seawards, looking like 'light smoke'. During take-offs on the endurance flights, Pond and Smithy had joked that angels sat atop the bridge, ready to scoop up their souls if the *Southern Cross* crashed. Today, Smithy saw no angels but 'felt that the gods had smiled on us'. He turned to Ulm seated beside him, and they shook hands 'in mutual congratulations'.

Despite the euphoria, both were aware that twenty-seven hours of flying lay ahead of them, during which neither could expect much in the way of comfort. To lessen the weight, they sat on wicker chairs, bought from Wian's furniture store in Glendale. In the custom of the time, the chairs were not bolted to the floor, and the occupants did not wear seatbelts. In turbulent air, men and chairs bounced around the cockpit and the cabin.

Up in the cockpit, Smithy and Ulm sat side by side, with the dual controls in front of them. Theoretically, Ulm was the relief pilot, and he had actually given himself the title of co-commander, but with only basic training and no pilot's licence he was not allowed to take off or to land the plane. He had sufficient skill, however, to maintain its course while Smithy slept – which was fortunate, since Smithy, as the true commander, needed regular rest to stay alert. Far shorter than Ulm, he did better than his co-commander at sleeping in his wicker chair, although he later admitted that his sleep was fitful.

Often it was too windy to sleep. The small, silken Australian flag, which hung between the petrol gauges as a good-luck charm, was ripped to tatters by the air that rushed through the cockpit. The cold could also bite deep. Although he and Ulm were dressed semi-formally for take-off, wearing leather jackets, ties, riding breeches and boots, they gratefully pulled on fur-lined flying suits and caps and goggles once the sun went down.

The rear cabin was more spacious than the cockpit, being eight and a quarter feet long and six feet high. Its occupants could huddle on the floor and sleep, or stand upright and walk, and both actions were accounted great blessings. The air, however, was as cold as in the cockpit, for the cabin was no more than a tubular steel frame covered by thin fabric held in place by criss-crossing bracing wires. How bitterly Lyon and Warner regretted their naive assumption that they should dress for the tropics. The blue serge suits and smart straw hats they had chosen to wear were no match for the freezing winds. Nor was cold their only enemy. In cockpit and cabin alike, 'the roar of the motors assailed their ears with the pound, it seemed, of sledgehammers'. Not even plasticine earplugs could make the din tolerable. On the other hand, those thundering engines were keeping them aloft, so it was also a comforting sound.

The absence of a toilet created a problem. The aviators made do with bottles, the contents of which they tipped out the window. Possibly on this long trip, and certainly on later ones, Smithy took binding medicine as a preliminary to the flight. But perhaps their keenest privation was being forbidden to smoke. Deprived of cigarettes for fear of fire, they suffered the gnawing pangs of nicotine withdrawal. In a joking Morse code message sent to the editor of the *Los Angeles Examiner*, Ulm wrote: 'Smoke two cigarettes at once for me – Smithy and I both crave the odd smoke.'

Through that first morning, calm weather and the excitement of being airborne kept their discomfort at bay. However, with each passing hour flying became more 'monotonous', which did not suit Smithy. The 'blue sea below us', he wrote, and 'the blue vault above us, and the overpowering roar of the engines oppressed one's spirit'. Fortunately, he and Ulm were enlivened by 'cheery little messages' sent by Lyon, 'who maintained a regular delivery of notes which told us where we were, and at other times asked us most unexpected question'. These notes, mainly 'personal chaff of a forcible kind', came pinned to the message stick. Ulm found it 'a strange experience for four men to sit in such a confined space and so near to one another, and yet be unable to exchange a spoken word'. Even so, many written words were exchanged inside the cabin, and many more were sent in Morse code, the era of radio voice-messages lying in the future. Every hour, Lyon calculated their navigational position and sent it by message stick to the cockpit, and at the same time Warner broadcast it to the world in Morse code, on the *Southern Cross*'s special radio call sign of KHAB.

Now that they were familiar with their exciting new transmitter and receiver, Warner and Lyon became radio addicts. Between them, they knew naval radio operators across America and right around the Pacific, and scores of friends were called up, even though Ulm had expressly forbidden private transmissions. Pressing away at the transmitter keys, they cracked jokes, exchanged gossip, flirted with former girlfriends and described scenes aboard the *Southern Cross* in thrilling detail. Smithy and Ulm also sent messages to their families; Smithy also sent a special message to an Emma Myers of San Francisco, who may have been the romantic donor of the Felix the Cat badge. Charmed by the radio chatter and the banter that Lyon kept sending on the message stick, Ulm wrote in the plane's log: 'We're as happy as hell cracking "wise cracks" ad lib.'

News in Morse code went regularly to the *Los Angeles Examiner* and, of course, to the *Sun* newspapers, which received it by way of the Amalgamated Wireless Association's receiving station at La Perouse, in Sydney's south. There, operators relayed messages to the Sydney *Sun*, and to radio station 2BL, which, as the flight progressed, began to share its news with the Melbourne station 3LO and the Queensland station 4QG. The notion of exclusive newspaper rights was nonsense, for anyone with a sufficiently powerful receiving set could listen in. Associated Press – among others – was poaching and spreading the news freely. Ulm found himself apologising to the *Sun* and the *Examiner* for circumstances beyond his control.

At the radio stations, the Morse dots and dashes were not only passed on to the listeners as they came across the airwaves, but were also translated into words so that the station announcer could read them aloud. As one reporter put it: 'Thousands of persons, comfortable before their own firesides, were thus brought into direct touch with the four gallant airmen who were battling their way across the uncharted air.' Listeners stayed glued to their receiving sets for hours, and as the flight progressed it was not uncommon for groups of friends to spend the night together 'listening in'. One woman claimed to have taken her headphones to bed with her, only to be awakened, a few hours later, when 'cheers given at the broadcasting station came through the pillows'. In America, Australasia, the South Pacific and even South Africa, dots and dashes from the *Southern Cross* became a gripping focus of attention. Smithy and his crew were making social history as well as aeronautical history.

Since Smithy's parents could not afford a radio, William spent his nights listening to his neighbour's receiving set. At important moments, he would run next door to fetch Catherine, who – though equally anxious – preferred the quiet of her room to the tension of a public vigil. She was a woman of strong religious faith, and she put her trust in God. Ever since Smithy's near-drowning at Bondi, she had wondered if the Almighty had saved her boy's life for a purpose. Was it his destiny to cross the Pacific? She spent much of the night praying for his safety and success.

As the hours passed, the aviators' early confidence gave way to unease. Smithy had hoped to maintain an even height of about 600 feet for most of the crossing, which would help him conserve fuel, but it was proving harder than he had foreseen. Great buttes of cloud lay in their path, and rather than fly blindly through them, he chose to climb above them, even though this meant that 'the thirst of the three engines became as keen – as insatiable – as that of a man lost in the desert'. He began to fear that those thirsty engines would exhaust their fuel before they reached Hawaii, and terminate their adventure fatally.

Smithy and Ulm made hasty calculations about their fuel reserves, but these proved too inconclusive to calm their nerves. Fear began to unsettle their emotions. 'We ran a gamut,' Ulm remembered, 'between black pessimism and wild exhilaration.' Meanwhile, the light was fading and it was time to climb again, for, in that era, low night-flying was deemed dangerous, especially above an ocean. 'It was safer up there in the darkness,' Ulm remembered, 'particularly if we swept into a patch where blind flying was necessary.'

At the height of 4000 feet, they saw scenes of rare beauty. An excited Warner described some of them to his radio listeners. 'The moon shining down is casting our shadow into the clouds,' he reported in Morse code: 'I'm sending this as I see it.' The burning contents of the Fokker's exhaust pipe inspired another of his Morse descriptions. It resembled 'a livid trail of flame', spitting from the big blue monoplane as it tore its way over thousands of miles of ocean.

Just before midnight, they flew into a squall. For a quarter of an hour the plane bumped and jolted, scaring the wits out of Lyon and Warner, who, though they were used to rough seas, were not used to rough air. Lyon was playing cards when the squall struck and he was hurled across the cabin. Hurled also was the bubble sextant, which broke its bubble in one fierce jolt. To escape the storm, Smithy climbed to 5000 feet – into a moonlit realm where the eerie shadow of the *Southern Cross* moved like a ghost plane at their side. Lulled by the unearthly beauty of the scene, they fell into a kind of reverie. 'There was no world,' Ulm remembered: 'We were sailing lazily on the Milky Way.'

Further worries led them back to reality. Hours previously, they had lost contact with the US Army radio beacons, in California and in Hawaii, which projected beams towards each other, thus assisting ships and planes to remain on course. Since the sextant was broken and the beams were lost, it was crucial to check their position by contacting the ships at sea beneath them. Although the shipping lanes were said to be busy, it was well past midnight before they sighted a vessel. Fortunately it was the *Maliko*, whose captain knew Lyon. Through their Morse banter, the two old shipmates calculated that the *Southern Cross* was on course. A little later Warner contacted the steamship *Manoa*, which also happily confirmed their position. From the same ship Lyon also learned the latest American baseball scores.

In this fashion the night passed, and by six am they were exhausted, deaf and chilled. They would have given almost anything for a smoke – and more than anything for a sight of the Hawaiian islands. Their fear of running out of fuel returned. As the sun came up, Smithy descended to 2000 feet to look for land, and seeing none, became increasingly anxious. One of the wind generators had failed, and the radio batteries were fading.

By nine o'clock everybody was on edge. Smithy saw what he hoped was an island, only to receive Lyon's emphatic rebuke that he knew the Hawaiian islands 'like the back of his hand'. As it certainly was not an island, an argument ensued. In the close confines of the cabin, a nervous Warner, seeing Lyon's angry face, snatched a piece of paper and wrote on it: 'Are we lost?' Grabbing the pencil Lyon scrawled in reply: 'YES'. Ignoring the low batteries, Warner began to transmit radio messages to the world, announcing that they were lost and almost out of gas.

The island proved to be a cloud. Confusion reigned until Lyon made an excellent sun shot with his damaged sextant, and Warner managed to receive a faint but reassuring radio bearing from Hilo, on the island of Hawaii. They were on course and almost there. Just before eleven am, on Warner's insistence, Smithy climbed to 4500 feet to extend their observation, and Lyon saw a brown and white bulge in the distance. Taking a quick bearing, he realised it was the snow-capped volcano Mauna Kea, on the island of Hawaii. Overjoyed, he sent word to Smithy, who quickly replied: 'Damned good work Harry, old Lion. Keep on doing your stuff – Smithy and Chas.' Lyon replied with a joking version of Smithy's RAF slang: 'If we get a cigarette and a cup of coffee we'll feel like flying back – ha, what, Old Top.'

Soon after, they sighted the island of Oahu, green and fresh in the morning sun, and high above Diamond Head they spied army planes waiting to escort them in. A loudspeaker at the aerodrome, Wheeler Field, which an hour or so before had kept the 15,000 welcomers on tenterhooks by broadcasting Warner's anguished messages, was now hailing their imminent arrival.

Just after midday – ten am, Hawaiian time – the *Southern Cross* swooped low over Honolulu. Twelve minutes later it touched down. The crossing had taken twenty-seven hours and twenty-five minutes, leaving them – although they did not then know it – sufficient fuel for another three hours of flying.

As Smithy and Ulm climbed down from the cockpit, they were deluged by reporters. Amid the crush, an attractive young woman attached herself to Smithy. Had he been frightened that they might go down in the sea, she asked him sweetly. He was exhausted, dazed and so deaf from the engines he could scarcely hear her, but he could not resist her innocent charm. Perhaps remembering his near-drowning in childhood, he gave a brilliant answer. 'Hell no, madam,' he retorted. 'I was born to be hanged, not drowned.'

A SHOT AT A DOT

Adorned with leis and 'thoroughly dead-beaten', the *Southern Cross* crew endured a kind but lengthy welcome from the governor of Hawaii. Most were thinking of cigarettes and 'real food and a decent sleep', but not Ulm: his mind was on the brown exercise book that contained the log of the flight. He intended to send a description of their journey to the Sydney *Sun* and the *Los Angeles Examiner*.

The account he transmitted to the papers that night was entirely his own, even though, when it appeared in print, it also bore Smithy's name. Joint attribution was a penalty he was obliged to pay for having so newsworthy a partner; it happened often, and sometimes annoyed him, but mostly he accepted it. One small item transmitted to the papers that night, however, came genuinely from Smithy's pen, and it pleased Ulm inordinately. Ulm's 'expert knowledge', wrote Smithy, 'of every department connected with the flight, and his tireless efforts, are largely responsible for any success we achieve. And I could not hope for a better mate as a relief pilot on the long hop.'

The following day, Smithy flew the *Southern Cross* to a beach named Barking Sands, on the island of Kaui. On his trip to Hawaii the previous year, Keith Anderson had recommended this long stretch of beach as a suitable site for their departure, since the heavily fuelled *Southern Cross* required an especially long runway for its take-off. During the past weeks, labourers had been clearing away the bushes and compacting the sand into a safe airstrip.

Awaiting them at Barking Sands was a stack of petrol drums that had been ferried across from Wheeler Field. The *Southern Cross*'s tanks were to be filled to capacity with 1300 gallons, every drop of which might be needed, for it was impossible to know what winds and storms awaited them. Moreover, since they were aiming for 'a small dot on the map' – a 'shot at a dot', they called it – they knew their destination might take some time to find. In public, Smithy and Ulm spoke of the 'thrill and charm' of aerial exploration, and cheerfully compared themselves with Magellan and Columbus. In private, they asked about beaches on the Canton and Enderbury Islands, which might serve as an emergency landing place. In their hearts, they knew that this section of the flight was indeed a 'long shot' – maybe even a suicide mission.

The next morning, Smithy and his crew assembled before dawn for what everyone assumed would be a stressful take-off. At twenty past five the *Southern Cross* lumbered down the beach and failed to rise. Only on the third attempt did it become airborne, and perilously so, for it soon dropped towards the sea, skimming the tops of the waves. All Smithy's skill was needed to coax it upwards. He aimed to take it up to 850 feet so that Warner could unfurl his radio aerial, for radio was an essential aid in this uncharted airspace. The US Army beacon from Wheeler Field, which had helped to guide them from Oakland to Hawaii, had now been redirected to guide them through the initial 700 miles of their journey towards Suva. Warner soon picked up its comforting buzz.

Alas, the comfort did not last. The first scare came just before breakfast, when Ulm noticed a trickle of liquid flowing from the lower part of the wing into the cockpit. He instantly imagined that petrol was leaking from the wing tank, which, as well as wasting fuel, put them at risk of fire. As it happened, fire was prominent in his mind that morning because, in an effort to satisfy their nicotine cravings, Lyon and Warner had started to chew the two halves of a cigar. Not content with chewing, Lyon rashly lit his half, and would have smoked it if Warner had not stopped him. If Lyon repeated that trick in the presence of a petrol leak, they would go up in flames. Meanwhile, Ulm was gesturing urgently to Smithy, who reached up to touch the trickle. Placing his finger in his mouth, he grinned and wrote on his notepad: 'Glory be. It's only water. Cheer up Old Tops.' The relief was sweet, and Ulm would always remember it.

After breakfast, a scribbled note, sent forward on the message stick, disclosed more trouble. The portside wind generator was failing, as were the transmitting and receiving radios. During the next couple of hours, Warner sat on the bucking floor and stripped and reassembled both radios. Thanks to his skill, the transmitter was restored, but not the receiver, which thereafter functioned only faintly and intermittently. By ten am they had lost the signal from the Army beacon, and were fast losing responses from the ships at sea and the stations ashore. This was a serious blow to their navigation, because the beacon and the passing ships were their only means of checking the accuracy of their course. Psychologically, too, the loss of contact was a shock. 'We missed our chats with the world,' wrote Ulm in the cockpit. Shut away in the rear cabin, the Americans missed them even more.

By midmorning, dark thunderclouds were towering above them. Then came a savage wind, which buffeted the plane. They had entered the first of a succession of storms that would assail them through the rest of the journey, and would force Smithy into making unwelcome decisions. Intent on saving fuel, he was loathe to climb above the storms, and he dared not duck beneath them as it was dangerous to fly an over-loaded plane too close to the sea. Sometimes he saw no alternative but to risk his fuel and climb, but mostly he tried to forge straight through, flying blindly into 'a thin opaque world of nothingness'. He hated blind flying. 'To relax the mind in such a way as to doubt your own senses is not a joke,' he wrote ruefully. 'Your whole world is reduced to an instrument board.' He and Ulm tried to remain optimistic and told each other that the storms would soon pass, but they were fooling themselves. Many hours of wild weather lay ahead. They did not know that they were entering what is now called the Intertropical Convergence Zone, where northern and southern trade winds meet, often erupting into fierce storms.

By now it was early afternoon, and Lyon and Warner, exhausted by the storm, crouched in the cabin in their underwear, their outer clothes having been discarded in the steamy tropical heat. In the cockpit, the co-commanders were also partly undressed, and sat with their feet and lower legs in water, the seals on the windshield having been breached by the force of the rain. Smithy's neuritis-prone left foot was giving him pain, amplified by the effort of pressing the heavy rudder. He and Ulm were aching all over, the result of the physical effort of manhandling the heavy plane through the violent turbulence.

At three-thirty pm, to deepen their troubles, the starboard engine gave 'a tremulous cough' followed by 'a splutter and a kick'. Since leaving Oakland, the engines had never missed a beat, and the cough truly alarmed them. For eight minutes – it seemed like an eternity – the engine coughed and spluttered intermittently. Then it resumed its regular sound. Since all the gauges now registered correctly, Smithy tried to dismiss the spluttering as an aberration. To his mother he transmitted a soothing radio message: 'Will be in Suva tomorrow afternoon all well love Chilla.' One doubts whether he really believed it.

Soon after six pm the wind and rain intensified. So strong was the buffeting that one of the plane's exhaust pipes was ripped away, and Smithy began to fear that the fragile Fokker would be torn to pieces. In the plunging rear cabin, Lyon became so desperate that he grabbed his notebook and managed with difficulty to scribble a message for the cockpit: 'Where the hell are you going – are you turning back?' When an inkpot spilled across his precious charts, he sent a second message: 'What the hell are you doing now?'

Since it was evening, Smithy decided to climb above the worst of the storm. 'I felt,' he later wrote, 'that we were burning up our precious fuel and getting no nearer to Suva.' At 2000 feet he cleared the lower clouds, but, looking up, saw a mass of black vapour swirling above him. Climbing in spirals through the seething vapour, he forced the machine upwards, with courageous Warner all the while transmitting a dramatic commentary to the world. At least, Warner hoped that others were hearing him, for he could not hear them, owing to the failure of his receiver.

At 7500 feet, Smithy and the listeners gained their reward. Suddenly they were flying in a clear, moonlit sky. On their port bow shone the stars of the *Southern Cross*. They twinkled 'like a shower of diamonds in a vault of deepest blue', and 'seemed to hang there beckoning their namesake'. A magical sight, it was long remembered.

The respite from the storm provided a few merciful hours. At eleven thirty-three pm they crossed the equator, and Smithy, and then Ulm, managed to doze. But at five o'clock in the morning the rain and bumps resumed. An hour later they peered nervously at a black cauldron of cloud, lit up by bluish flashes of lightning, rising to about 12,000 feet. Torrents of rain lashed the plane, the lights failed, and in the pitch dark Lyon and Warner were flung about the cabin. For twenty-four hours now, Smithy had been flying through, or close to, hell. He wondered how much more they could endure.

After trying without success to duck under the storm, Smithy climbed upward, and the 'rotten night' gave way to a sullen dawn. Anxiety again racked the crew, for they feared they might soon run out of fuel. They had wasted at least forty gallons in their attempt to avoid the storms; 'fuel is alarmingly low', logged Ulm. They worried about their navigational position, their manoeuvring having fairly certainly led them off course. If only they could sight the Phoenix Islands. If only Warner could receive a radio bearing, or Lyon take a sextant shot. 'Harry', wrote Smithy anxiously, 'can't you get a star position at all? Or can't Jim get a radio bearing. Getting pretty serious.' But the anticipated islands remained hidden, the receiving radio was down and the stars seldom came out of the clouds. To cap it all, the favourable tailwind of the previous day had turned into a strong headwind, and they were still encountering squalls. In Smithy's understatement, they were 'out of luck'.

There were dark moments but there were also rays of hope. Lyon sent a cheeky note via the message stick that warmed Smithy's heart. The two men in the bumpy rear cabin had resolved that Smithy be elected President of the United States as a reward 'for the way he had ploughed through the cloud-banks and squalls'. In view of the terrifying night the Americans had spent, it was an extraordinarily generous sentiment. It was followed by a brief sight of the sun, and Lyon was able to use his sextant. He calculated that they were only slightly off course, although, according to Smithy's recollection, the calculation was 'plotted more in hope than in faith that it was correct'. At about ten am, while pumping petrol from the wing tank to the main tank, Ulm realised that they possessed more fuel than they had imagined, and could fly for another seven hours. They had been nearly thirty-two hours in the air and were 'ragged and weary', but their exhaustion was suddenly forgotten. On hearing the news from Ulm, they all whooped with joy.

When would they sight land? That was the persisting question. As time passed, they began to strain their bloodshot eyes, looking for islands. It was a grim game for which Warner drew up sporting rules. 'If no member of the team sights land before the gas gives out,' he wrote, 'your side loses and you don't need an umpire!'

Just after one pm a small brown dot sprang out of the sea. When it first appeared, Ulm was flying while Smithy was dozing. At the sight of it, Ulm swung the plane so violently that Smithy awoke 'swearing like a trooper'. But peering through the windshield, he too rejoiced. 'Land!' he cried triumphantly. Quickly he took over the controls and dropped to twenty feet so that Lyon could use the sextant. This was dangerously low, but Warner had to be close to the sea in order to calculate their position accurately.

The land was actually the north coast of Vanua Levu, close to the International Date Line. As they crossed it they passed from Sunday to Tuesday, and the dot was joined by other brown dots. The 'excitement and relief were immense', and Smithy and Ulm hugged each other. As they gazed earthwards, they marvelled at Fiji's beauty: 'the light green of the cane-fields, the bronze green of the rubber plantations, and other fields splashed with splotches of red and yellow and purple'.

Their landing field was in Suva, the capital of the British colony of Fiji. Old Royal Flying Corps friends had advised them that the Albert Park Sports Oval was the safest place, however it was barely 4000 yards long, and from the air appeared to be 'about as large as a pocket hand-kerchief'. Ulm had earlier arranged for the telegraph lines to be removed and some of the trees and bushes cut, but it was still bounded by a hotel and road at one end and a sharp, bushy rise at the other. Smithy tried to shrug off his weariness, knowing that he would need all his wits to avert a crash.

He swung in from the bay, flew low over the road and dropped suddenly as he entered the park. The plane touched down, bounced and touched down again, all the time speeding towards the dangerous rise. Smithy turned the plane sharply in a dramatic loop and halted abruptly, just short of two trees. It was a landing that only a master could have made. The next day, echoing comments made by astonished witnesses, the London *Times* praised his 'magnificent display of nerve and judgement'.

Smithy and Ulm had dressed for the descent, but Lyon had not bothered, while Warner, having accidentally soiled his clothes while attending to a call of nature, had nothing clean enough to wear. Consequently, as they came in to land, Lyon was in his singlet and underpants and Warner was naked. To distribute the weight for land-ing, Smithy had ordered them both into the tail of the plane, and as the *Southern Cross* jolted to a stop Warner was thrown through the fuselage. A British nurse is said to have found him lying on the ground, and cov-ered his nakedness with her coat. It would be many years before this incident was reported to the general public – they were heroes, and not to be dethroned. Not until Lyon and Warner returned to Sydney on the thirtieth anniversary of their flight was their embarrassment revealed.

'Dazed, deaf, unshaven and oil-stained', the co-commanders stum-bled down from the cockpit. Exhaustion had rendered them almost incapable of thought, but could not weaken their sense of triumph. They had completed the longest nonstop flight so far made across an ocean, and one of the greatest flights in the history of aviation. Ulm wrote in his log: 'It's hard to realize it's over and that at the moment we are exceedingly famous.'

KINGSFORD SMITH, AUSSIE IS PROUD OF YOU

The aviators' arrival in Fiji led to an avalanche of festivities: an official lunch, a civic reception, a welcome by the Fijian chiefs, a lavish tea hosted by the 'ex-soldiers of Fiji', even a grand ball. What to wear was a dilemma, for the aviators had brought no changes of clothes. Warner made do with a hastily cleaned suit, Ulm and Smithy wore their riding breeches and boots and borrowed shirts and jackets, but the more flamboyant Lyon bought a dashing white tropical suit. Both Americans wore new bow ties.

Smithy had more insistent problems than clothes. As soon as he saw the cramped size of Albert Park, he knew that his heavily laden plane could not leave from it. He must find a long stretch of beach and ensure that the sand was compacted into a runway. Next morning he sailed to inspect the most promising prospect at Naselai, twenty-five miles east of Suva. Ulm also had problems, chief of which concerned publicity. Their landing in Suva had made front-page news around the world, and their arrivals in Brisbane and Sydney were promising to be huge events. Ulm was determined to keep a tight rein on the publicity arrangements.

Ulm's mind was also turning to Warner and Lyon, who, by the terms of their contracts, were due to return to America from Suva.

Much as he hated sharing his and Smithy's glory, Ulm was reluctantly coming around to Smithy's view that that they needed the Americans. Certainly they needed Warner and his radio; whether they needed Lyon he was not so sure. Lyon's perpetual wisecracking irritated Ulm, and he suspected the American was careless. Back in Hawaii, instead of having his sextant repaired, Lyon had sneaked away to drink with his naval pals. But undoubtedly there were strong reasons to keep Warner until Brisbane, not least because he had acquired an enthusiastic following over the airwaves. Thousands of fans tuned into his Morse code messages, and some – including Smithy's father and the British consul in Hawaii – wanted to know why Warner and Lyon were being offloaded when they, too, had a right to a triumphant welcome in Australia.

To answer such questions, Ulm usually cited the requirements of weight: with four people aboard, the *Southern Cross* would have difficulty taking off from a short runway, like the one at Albert Park. Most questioners seemed to accept this explanation, but Smithy had now vetoed Albert Park, so the explanation was no longer valid. Ulm saw that he must reassess the situation. Knowing that tact was essential, he tried to conceal his true feelings. In speeches in Suva, he spoke appreciatively of his navigator and wireless operator. 'I am telling you,' he informed his audience, 'on behalf of myself and Smithy, that we would have been very cold meat but for our two American friends.'

Behind the scenes, it was a different matter. That night, after the ball at the Grand Pacific Hotel, Ulm privately informed the Americans that he was agreeable for Warner to remain, but not Lyon. Since both Lyon and Ulm were jetlagged, exhausted and somewhat drunk, it was not surprising that a fight ensued. Lyon pushed Ulm against the wall, and blood might have spilled had Smithy not intervened. The next morning, feeling somewhat more conciliatory, Ulm paid a solicitor to include new provisions in the Americans' contract. Smithy and Ulm were now prepared to offer Lyon and Warner an extra £100 as 'an act of grace', and to guarantee them a place aboard the *Southern Cross* as far as Brisbane. The document concluded with the cheeky statement that Lyon and Warner wished to record their 'appreciation' of Smithy and Ulm for paying them the extra money and inviting them to Brisbane.

On reading the new provisions, the Americans had considerable misgivings, but their zest for adventure was strong, so they took up their pens and signed. They were truly fond of Smithy, describing him as the only pilot who could have guided them through the perils of the flight, but for Ulm, whose hand they saw all over the document, they felt something close to contempt. Two years later Warner published the contract, and its history, in America's popular *Liberty Magazine*. 'Yes we signed it,' he concluded bitterly. 'You know the saying. "When in Rome act like a Roman candle!"'

Later that day, Ulm cabled the Sydney *Sun*, explaining that, as a mark of 'sincere appreciation for their services', and as 'a sportsman-like act', he and Smithy were inviting Warner and Lyon to stay until Brisbane. 'We cannot speak too highly of their skill, initiative and resource,' he wrote. The following morning – 7 June – all four members of the crew flew in the *Southern Cross* to Naselai beach, where Ulm supervised the plane's refuelling, watching the petrol drums' journey from the boats through the surf to the plane. 'He was strung at high tension,' remarked a reporter, who also noted that Smithy did not join him, instead spending a couple of hours resting in the cabin and reading 'press accounts more wonderful than the flight itself'.

The plane's departure was scheduled for four pm, but it was soon clear that the high tide would prevent a take-off and they must wait until the next day. Warner agreed to watch the plane overnight, but he slipped off to drink kava at a native village. When he returned, he found that the Fijian guards had built a bonfire so close to the plane that sparks were flying onto the fuselage. While sand quickly quenched the fire, it was an incident Warner would long remember.

Early the next afternoon, the *Southern Cross*, carrying 900 gallons of fuel, rose effortlessly into the air. 'We were all,' wrote Smithy, 'in splendid spirits.'

As it transpired, their spirits were rather too splendid. The crew believed their place in history was already assured, and that the flight to Brisbane was a mere formality. Such overconfidence was foolish but understandable. They had flown 3180 miles to Suva, and it was only half that distance to Brisbane. In '19 or 20 hours', Ulm informed the Sydney *Sun* by radio, 'we will be in dear old Aussie again'. They were scarcely airborne when trouble appeared. Their earth inductor compass ceased to function – a serious blow, since it was the most valuable instrument they carried – and once again it seemed Lyon was at fault, for it had not been oiled in Suva. Navigation now depended on the steering compasses, which, being housed in the cabin and close to metal objects, were liable to error. Nevertheless their Morse code message to the world at 5.50 pm gave no hint of trouble: 'We're happy as Larry up here, cooee!'

At about six pm, Smithy and Ulm began their night-time climb to 4000 feet. The air was cold and they donned their fur-lined flying suits, and Lyon and Warner put on their sweaters. By seven pm it was colder, and visibility shrank to a mile, and then to a few yards. A strong wind arose, and rain and wind so lashed the plane that it bucked and plunged and slid. In cockpit and cabin, the occupants were hurled off their wicker chairs. Attempting to escape the storm, Smithy climbed to 9000 feet but found no respite. The turbulence intensified, with drops at times of 400 feet. Icy rain cascaded through the weakened windshield, saturating their suits and boots. So numb were their feet and hands that they could scarcely work the rudder or grasp the controls.

'For four solid hours,' Smithy recalled, 'we endured these terrible conditions.' For three of those hours, he was forced to perform the 'the blackest of blind flying': it required all his skill and courage. At midnight came a slight abatement, and he was able to descend to 4000 feet, where he dodged in and out of storms until about four o'clock. In the rear cabin, tossed about in pitch darkness, Lyon and Warner time and again literally 'hit the roof'. Their terror can be imagined, and their feeling of helplessness too. Cut off from decision-making, all they could do was hang on and suffer. To his credit, Lyon had sufficient presence of mind and generosity of spirit to send a vote of confidence to the cockpit. 'Keep up the old courage,' he wrote to Smithy, 'and you will get us through.'

On Smithy's orders, Warner had stopped transmitting, since anything he described would distress their listeners in far-off lands. Instead, Warner screwed down the key on his transmitter, which thereafter emitted a doleful whine, indicating that they remained airborne. For those at home who had gathered around their radios, it was a tense time, not least for Smithy's parents. Responding to the crisis, the popular Sydney announcer Basil Kirke stayed on air for eighteen hours to bring his listeners the latest intelligence.

Captain Hancock, on his yacht near Mexico, was also deeply troubled. One of Smithy and Ulm's last messages before they encountered the storms had been to him, publicly acknowledging their debt to their benefactor. On hearing their message, the unassuming captain had flushed with embarrassment. Now, realising their peril, he resolved that if they got through, he would make a free gift of the *Southern Cross* to these courageous young Australians.

As the storms abated, the crew realised they were lost. Lyon had been expecting to check their position at the small city of Noumea, which had agreed to keep its lights on, but, lost in the storm, they had failed to find it. When the sun rose, Lyon took a sextant shot and discovered, to his intense relief, that they were barely off course. Warner also received a radio bearing from Brisbane. At 9.50 am they joyfully watched a long grey shadow unroll along the horizon: it was Australia!

They hoped the shadow was the coast near Brisbane, but it was Ballina, 110 miles to the south – the home town of Ulm's wife. Surely this was a happy omen. As they prepared to land in Brisbane, they noticed it was 10.13 am; although it had seemed like an eternity, they had been airborne for less than twenty-one hours. All told, since leaving Oakland, they had spent eighty-three hours and eleven minutes in the air.

On the ground at the Eagle Farm aerodrome, 15,000 people were said to be waiting. As one reporter put it: 'There was a dramatic tension, you could feel the bigness of the occasion.' When the plane touched down, Smithy's helmeted figure could be seen in the cockpit, and at the sight of his waving hand, the fans exploded. Ignoring the still-whirling propellers, they launched themselves towards the plane. Men wanted to shake his hand, women wanted to kiss him. There were cries of 'Smithy, Smithy!' and 'Kingy, you darling!'

It was not Smithy but the two Americans who were the first to disembark. Dressed in their blue serge suits – Warner carrying his straw hat – they tried to run inconspicuously for the fence, determined not to steal their commander's glory. But it was impossible for the Americans to remain anonymous. Sighting the two figures, the fans pounced and raised them shoulder-high, chanting, 'Look, it's the Yanks.' Then, above the seething mass, Smithy, with a 'wide, happy smile', began to climb down from the cockpit. Exhaustion and triumph had raised his mood sky-high. 'Hello, Aussies,' he shouted above the cacophony. 'My kingdom for a smoke!'

The aviators were now propelled towards an official enclosure, where the Governor and Premier of Queensland were waiting to greet them, but nobody had reckoned on the strength of the crowd. The press of bodies pushing against the barricades splintered the wood, sending movie cameras, microphones and reporters flying – whereupon excited fans snatched up Smithy and Ulm and the Americans and carried them to the open cars. About 30,000 spectators lined the route as they made their triumphal way to the Brisbane Town Hall.

The aviators' excitement was now giving way to exhaustion, which made the Town Hall reception something of an ordeal. They had difficulty keeping their eyes open during the long, formal speeches – but then, suddenly, an announcement jolted them awake. The federal government was awarding them a prize of £5000. Smithy raised a laugh with his quick response that the prize was very welcome, since flyers were always short of money. Gratitude, however, could not banish their weariness, and what they really wanted was a bath and a bed. The popular songwriter Jack O'Hagan had engaged luxury suites for them at Lennon's Hotel, and they could not wait to reach them.

Even at the hotel, there was little sleep for Smithy and Ulm. Armed with paper knives and packets of mints, they made straight for the hotel lounge to read the cascade of messages that had come from near and far. A reporter discovered them there, sprawled in easy chairs and opening envelopes, and he has left an arresting portrait of their contrasting personalities. Captain Smith, he wrote, had 'an easy frankness of manner and dry humour', along with a refreshing modesty that seemed 'thoroughly Australian'. With his powerful, lean frame and sunburned, jutting jaw, he could have stepped straight from the pages of an Australian novel. Ulm, on the other hand, was a man of business: brusque and direct, articulating with the 'curt rapidity of a machine gun'.

By six pm all four aviators were bathed and dressed and granting interviews. One reporter caught Ulm in his bath, and was amazed that such a short sleep had refreshed him. Like a 'playful boy', Ulm soaped his curly head and ducked it under the water, then gurgled and surfaced, blowing water off his face. The reporter was enchanted, but Ulm was less so. To be rid of his annoying inquisitor, Ulm jumped out of his bath and, putting on his breeches and flying boots, explained that he and 'Kingy' had brought only the clothes they stood up in, and were obliged to wear them to the banquet, along with borrowed jackets and clean shirts.

Down in the hotel lounge, Lyon, in his blue serge suit, encountered a reporter named Norman Ellison, the Motor and Radio Editor on Sydney's *Daily Guardian*. Having seen Lyon and Warner disembark surreptitiously at Eagle Farm, and suspecting there was more to this than met the eye, he pressed for answers. Lyon admitted that in Suva he had been drunk and 'blotted his copybook', and that Ulm had insisted that he and Warner should leave the plane in Brisbane and not share in the welcome in Sydney. In any case, added Lyon, it was well understood that an Australian welcome was for 'Australians only'.

Ellison was appalled. Surely, if one shared the perils of the flight, one should also share the reward. His indignation on the Americans' behalf was so deep that he contacted Smithy's secretary and sought an interview. When this was refused, he insisted on knowing whether Lyon and Warner were flying in the *Southern Cross* to Sydney – and if not, why not? He also delivered an ultimatum. If they were not allowed to fly, the *Daily Guardian* would charter a plane and fly them to Sydney, to arrive at the same time as the *Southern Cross*. Ellison added that he would wait half an hour for an answer. If he had not heard by then, he would phone his interview with Lyon to the *Guardian* and set about chartering the plane.

Ellison's tactic worked. A press conference was called almost immediately, and Ulm assumed the role of spokesman. He said that there seemed to be a story circulating 'that the Americans were being left behind in Brisbane'. It was not true. Of course they were coming to Sydney. And, sure enough, when the *Southern Cross* took off the next morning, the Americans were aboard.

By now the accolades were pouring in from overseas. The London *Times* proclaimed they had secured a 'great triumph'; the *New York Times* declared that no praise was too high for them; and the *Washington Post* called their flight 'the last word in the art of aviation'. Captain Hancock announced to the press that he was giving Smithy and Ulm the *Southern Cross* as a gift, 'to commemorate the magnificent achievement of yourselves and your brave American companions in bringing our two great countries closer together'.

Meanwhile, at Mascot Aerodrome, a vast crowd was gathering, headed by the governor-general, Lord Stonehaven, and Prime Minister Stanley Bruce. To crown all this, the crew's songwriter host of the previous night, Jack O'Hagan, had composed a catchy song that countless Australians would soon be singing. 'Kingsford Smith, Aussie is proud of you,' ran the chorus. Not great poetry, perhaps, but it captured the feeling of the nation. Smithy and his companions were national heroes.

EXCEEDINGLY
FAMOUS

Among the 20,000 people waiting at Mascot were Smithy's and Ulm's parents and Ulm's wife. They had been riding a wave of emotion for the past week, and had difficulty controlling their feelings. Observers noted that seventy-year-old Catherine Kingsford Smith's hands shook and her eyes grew wet as she spoke about 'her boy'. She waved her white handkerchief at the descending plane, but once her son's face appeared, the same handkerchief was used to dry her tears. Ulm's wife, Josephine, in a jade green hat, was similarly affected. Reporters noted that she clenched and unclenched her hands as Ulm came into view.

In front of a battery of cameras and microphones – the biggest, it was said, ever to confront an Australian – Smithy hugged his parents and held his mother's hand throughout the speeches. When his time came to speak he seemed almost overcome, and said, with deep sincerity, that it was the happiest moment of his life. But the restless fans could not contain themselves. Smithy had scarcely concluded his speech when they began to storm the official enclosure; soon the cheeks of the aviators bore the lipsticked imprints of numerous kisses. Rescue came in the shape of a large truck, which carried them on its back for a circuit of the aerodrome. Eventually, a procession of open cars conveyed them to the heart of the city.

At Longueville, Smithy found the streets ablaze with flags and bunting, and the family house crowded with friends and relatives, who chaired him through the front door and stayed till well past midnight. Smithy's brother Leofric would recall that, after the last guest left, 'Chilla sat by the fire quietly reading a book for an hour before turning in'. It was, Leofric said, 'his way of doing things'.

During the following weeks Smithy and his crew were guests of honour at more than fifty receptions, lunches and dinners. They were feted by the Lord Mayor of Sydney, the Sydney *Sun*, the Returned Soldiers League, the Aero Club, the Journalists' Institute, the Queensland Residents in Sydney, and many more. At the Sydney Technical School Old Boys' Dinner in the Australian Hall, Smithy's arrival triggered a 'stupendous scream of greeting from 300 young voices'. On 12 June 1928 they flew in the *Southern Cross* to Melbourne, where 50,000 are said to have turned out to greet them. On the return journey, they touched down in Canberra for an official welcome at Parliament House, where Prime Minister Bruce declared they had secured a 'niche in the temple of the immortals'. His government's prior lack of faith in them was now forgotten.

Smithy's mother accompanied them on that journey, as did Ulm's wife – and so, of course, did Lyon and Warner, since a chorus of highly placed voices insisted on it. The public display of American–Australian friendship was an attractive novelty, and encouraged by the prime minister. Many were eager to further Australia's relationship with the 'great sister democracy beyond the Pacific', and Lyon and Warner became favourites with the public.

Bowing to popular demand, Ulm praised the Americans generously in public. In private, he was anything but generous, reminding them that, by the terms of their contract, they were forbidden to touch any of the money now being donated to the aviators. Rumours of Ulm's meanness were beginning to circulate, and it was presumably to capitalise on these that, towards the end of June, the Americans called a curious press conference.

Lyon was the witty spokesman. 'Boys,' he told the reporters, 'Warner and I have been having a bit of a talk, and we think by the look of things that we are going to get even more money on our return than you have got here. Now, we do not think that this is fair. We want to tell you that when our fund in America reaches a total of yours here in Australia, we propose to split everything over that amount with you fifty-fifty.' One assumes that Lyon was mocking Ulm, or maybe trying to shame him; perhaps he thought he could make Ulm relent. Refusing to be drawn, Smithy and Ulm declined the offer. However, they declared themselves 'deeply touched by this fine example of palship'.

Certainly, money was waiting for Warner and Lyon in America. The Hearst newspapers had started a fund for them, but it amounted to about $12,000 – a small sum compared to what, had it not been for Ulm, they might have anticipated as their share in Australia. As well as the Australian government's £5000, Smithy's old friend Lebbeus Hordern gave £5000, a Victorian pastoralist named Marcus Oldham gave £4000, the *Daily Guardian* newspaper raised nearly £5000, the Sydney *Sun* raised £450, the New South Wales government gave £2500, a celebration at the Longueville Wharf Reserve netted £262 – and so the list went on. The final count, if one includes the gift of the *Southern Cross*, was close to £50,000, an immense sum for the time. Years later, when Lyon and Warner wrote memoirs of their time in Sydney, they bitterly decried Ulm's lack of generosity. In retrospect, however, the public and private attacks on Ulm seem rather excessive, for his contribution to the success of the flight was far greater than that of the American latecomers.

Smithy's newfound wealth made surprisingly little difference to him. He bought an expensive Studebaker car, and later a speedboat, but otherwise he returned to his simple life in Arabella Street. His greatest personal extravagance was to buy the family house for his parents, and to equip it with the conveniences and comforts they had been unable to afford. He also bought them an annuity that assured them of an income of £4 a week. One of his greatest satisfactions was to be able to say to his mother and father: 'You are now free of debt for the rest of your lives.'

Smithy's father, who was old now, and frail, was almost overwhelmed by his son's generosity. 'He's a wonderful boy,' he told reporters. 'Look at all the improvements in this house – he's done all this. Not a penny will he let me spend. That's the sort of man he is all over.'

Catherine Kingsford Smith had no difficulty putting the money to good use. Still the backbone of the family, she coped with a constant flow of sons, daughters, friends and relatives, who now included eleven grandchildren. She even managed to cope with Smithy, who was finding it hard to deal with the stress. 'He cannot keep still,' she confided to a journalist. 'If there is nothing to do, he starts roving up and down the house.' She was relieved when he discovered the calming effect of reading. 'He reads novels as fast as he can get them,' she told the journalist.

While it was true that Smithy loved a simple life, he was certainly not averse to lionisation. Since childhood he had enjoyed being centre stage, and he possessed the charm and wit and easy friendliness to carry it off. As the press noted, Smithy had many different styles. There was Smithy the 'cobber', who liked nothing better than a 'smoke night' with his mates, knocking back beers and telling bawdy stories; there was Smithy the public man, smart in his evening clothes and fluent on a public platform; there was Smithy the squadron leader, spruce in his uniform and born to command. There was also Smithy 'on the job', his mind centred on his plane: curt and incisive in speech, terse with pressmen, but always pleased to see a pal and spare him time. And there was Smithy the brother, the son, the friend: lovable, generous, full of fun and boyish charm, sometimes irresponsible but easily forgiven. Whether he was chatting to the governor-general or joking with the lowliest mechanic at Mascot Aerodrome, Smithy scarcely seemed to put a foot wrong.

Maybe what Smithy liked best were the theatrical performances at which all four aviators were guests of honour. In mid-June, the entrepreneurs Benjamin and John Fuller staged a benefit performance at the St James Theatre of *Rio Rita*, a popular musical comedy starring the Queensland soprano Gladys Moncrieff. On the night of the performance, as Gladys stood on stage with the 'gallant four', she kissed

Smithy. This seemed to license the other women on stage to shower the aviators with kisses. As the curtain fell, the audience's last sight of them was 'four heads above a struggling crowd of chorus girls'.

The kissing was repeated a few days later at Melbourne's Theatre Royal, when Smithy and his companions attended a packed performance of another musical, *The Girl Friend*. Moving pictures of the aviators' arrivals in Brisbane, Sydney and Melbourne were shown to the audience before the performance, and at the conclusion, a team of chorus girls hoisted Smithy on to their shoulders and chaired him boisterously around the stage. Smithy's blushes were noted by many onlookers. By the standards of the time, his sexual morals were rather too free and easy, but his capacity to blush counted heavily in his favour.

Nellie Stewart was invited to these galas, but she was too ill to attend. Instead she sent a public letter to the Sydney *Sun*, declaring that she was the proudest woman in Australia. Her beloved 'Kingsford' had carried her photo with him as his mascot – even though 'it is true that he sat upon me in the cockpit', she added cheekily. 'And who knows,' she concluded. 'Perhaps my staunch faith, which never once wavered, that he would succeed with anything he attempted, has helped just a wee, wee bit ... God bless you Kingsford.'

Nor were these the only women to succumb to Smithy's charm. He and Ulm were so inundated by fan mail that the Atlantic Union Oil Company gave them a city office, at 66 Pitt Street, and an enthusiastic secretary named Ellen Rogers to help them deal with their correspondence. Most of the letters to Smithy were from women, some offering marriage, others making less respectable proposals. 'How wonderful it would be,' wrote one female admirer, 'to have such a wonderful man as you for a husband. It would be heavenly!'

The inspiration for the offers of marriage possibly came from the news that Smithy was seeking a divorce. It was now three years since Thelma's departure, and he could apply to the New South Wales court for a divorce on the grounds of desertion. He had always intended to free himself as soon as possible, and the possibility of a knighthood added an urgency. Ross and Keith Smith had been knighted after their flight from England to Australia in 1919, and many believed that Smith and Ulm deserved knighthoods. But in the straitlaced climate of that time, Smithy's uncertain marital state was a mark against him, and made a knighthood unlikely. It was a pity, because in other ways he and Ulm received gratifying honours. Both had been awarded the Air Force Cross, and Smithy had been made an honorary squadron leader in the RAAF, and Ulm an honorary flight lieutenant.

With the divorce petition underway, Smithy turned his mind to his next project. Even before leaving for California, he and Ulm had planned to cross the Tasman Sea. They expected it would be dangerous, for the Tasman's weather was wild and passing ships were few; a rescue at sea would be almost impossible. Two New Zealand aviators, Captain George Hood and Lieutenant John Moncrieff, had been lost over the Tasman only the previous January. An experienced crew was essential, and Smithy urged Ulm to retain Lyon and Warner, who, rather surprisingly, declared themselves ready to make the flight. But the death of Lyon's mother put an end to Smithy's plan. Lyon sailed home, and Warner went with him.

It was tricky to replace the Americans, because Ulm was still insisting that the radio operator and navigator should be classed as employees. A quick replacement, however, was needed, and Harold Litchfield, formerly the third officer on the SS *Tahiti*, seemed suitable and was agreeable to Ulm's terms. For their radio officer, they engaged Tom McWilliams, the head instructor at the Union Steamship Company's radio school in Wellington. He had been recommended by the New Zealand government, which added to his desirability. Since neither of the new recruits had worked in the air before, Smithy proposed a training flight within Australia to acquaint them with their duties. He was keen to make a nonstop flight to Perth, an adventure that no aviator had previously performed.

Meanwhile, the redoubtable Cecil 'Doc' Maidment – especially imported by Smithy from the United States – made a careful inspection of the *Southern Cross*'s engines in preparation for the flight to Perth. On 8 August they flew from Richmond Aerodrome, near Sydney, to Point

Cook Aerodrome, near Melbourne, from where they were to make their nonstop attempt. The flight was not without its drama. Halfway to Perth the plane encountered such turbulence that 'the new boys' wondered if they had made the right decision; and when they reached Perth, not a soul was there to greet them. It was their own mistake, as they had miscalculated their flying time. Even so, they felt short-changed, especially since their flight had set a record. They had covered the distance from Melbourne to Perth in a mere twenty-three and a quarter hours.

For the following two weeks, the *Southern Cross* was marooned by rain at Perth's boggy Maylands Aerodrome. This was no hardship for Smithy, who hired a light plane and flew to visit his friends at Carnarvon, but for the others it was a dismal time. When he got back to Perth, during a lull in the rain, he and Ulm were finally able to coax the *Southern Cross* into the air. They managed this by drastically reducing the load, and by abandoning McWilliams and Litchfield, who rejoined the plane at Tammin, on the Perth–Kalgoorlie railway line. From there, after a fleeting visit to Adelaide, they flew on to Sydney.

The fast flight to Perth reassured Smithy, and plans for the Tasman crossing went ahead. The *Southern Cross* was again overhauled, with new American propellers fitted and the seating improved. The pilots' seats were now bolted to the floor and carried lap belts, but, believing that the cabin crew must move about freely, Litchfield and McWilliams still sat on unbolted and unbelted wicker chairs.

As they flew off on 10 September, Smithy and Ulm made light of the hazards ahead of them. Ulm said to his wife, 'Cheerio, Old Bean. Keep your head up till we meet again.' And Smithy told his mother, 'We'll be over in New Zealand for breakfast. Everything's going our way. We'll make it OK.' That night, his parents and other relatives sat beside the two new radio sets at Arabella Street, ready for the broadcast of the flight. His mother said, 'I feel so confident of success, I would like to have gone myself.'

Five hours out from Richmond Aerodrome, the aviators met an electrical storm, quite as severe as anything encountered over the Pacific. 'All around us was a black chaotic void,' Smithy remembered, 'punctuated every few seconds by great jagged rents of lightning, like vivid green snakes.' In the rear cabin, Litchfield and McWilliams were 'tossed about like peas', and almost killed when lightning struck the radio aerial, burning out the transmitter. Desperate to stay in contact with the outside world, McWilliams sat on the pitching floor, trying to repair the transmitter by torchlight. Perhaps fortunately, radio listeners could hear nothing across the airwaves – and so the aviators' friends and relatives were spared much anxiety.

After the lightning came freezing, flooding rain. 'I do not think,' wrote Ulm, 'that either of us has ever been so cold before.' Ice began to form on the plane's surfaces, so increasing its weight that the combined strength of the co-commanders was required to handle the controls. Ice also played tricks with the instruments, and Smithy was flying entirely by the instruments. Seeing the airspeed needle fall to zero, he assumed that the engine had stalled. Thinking to restart the engine, he began a steep dive, only to discover that the three motors – still running – were propelling the plane downward at a suicidal rate. He pulled out of the dive just in time.

The near miss shook him severely. Later, he would write in his autobiography:

> Our instruments on which we relied had played us false. Below us was the roaring main. Above and around us was the roaring storm. The *Southern Cross* was encrusted with ice; our radio had failed, and we were alone in the midst of the deserted Tasman Sea. It was pitch black; we could see nothing, hear nothing. We did not know where we were. I think that night I touched the extreme of human fear. Panic was very near and I almost lost my head. I felt a desire to pull her round, dive-climb – do anything to escape. We were like rats in a trap – terrified, dazed with fear.

During past perils, he had usually been able to summon a surge of adrenaline that made him feel that he could beat the odds. That night, over the Tasman Sea, he discovered what it was like to feel helpless in the face of the elements. All he could do was will himself to keep flying, and hope that somehow they would survive.

Gradually the storm abated, the sun appeared and Litchfield managed to take a bearing. Then the clouds parted and, looking down, the crew saw the sheen of water. The shining water was Cook Strait. At about six am they circled the capital city, Wellington, and had the fun of seeing the residents run out in their nightclothes to look at the *Southern Cross*. One man stripped off his pyjama top and waved it like a flag.

They were flying now almost entirely over land. At 9.22 am, after fourteen and a half hours in the air, they landed at Christchurch's Wigram Aerodrome, where about 30,000 admirers had assembled to welcome them. The *Southern Cross* was the first international plane to reach New Zealand, and the excitement was so intense that the exhausted aviators had difficulty coping with the melee.

Over the following month, New Zealanders took the aviators to their hearts. Prime Minister Gordon Coates called them 'our own kith and kin', and having McWilliams, a true New Zealander, among them greatly added to the crew's popularity. They made a grand tour of both islands in Bristol Fighters loaned to them by the New Zealand Air Force, and they were feted at dinners, receptions and balls. Wherever they went, as the newspapers noted, Smithy was a magnet for female attention.

By the end of September it was time to return to Australia, a journey that filled them with foreboding. In preparation, they ordered new propellers to replace those damaged by ice. They also searched for a suitable departure site, for although Christchurch had the largest aerodrome in the country, its runway was too short for the heavily laden *Southern Cross*. Luckily, they learned of a flat paddock sometimes used by pilots at Fairhall's farm, four miles from the town of Blenheim, north-east of Christchurch. On this long, bare stretch – which today is covered in grapevines – the New Zealand Air Force speedily constructed a hangar and an airstrip of more than a mile in length. Each day crowds came to inspect the project's progress, and to gaze in wonder at the famous plane.

The date of their departure now depended on the weather, for Smithy dared not make the journey unless the outlook was calm and fair.

Housed in Blenheim's 'homely' Criterion Hotel, waiting for the strong winds to fade, he and his crew passed their time in various ways. Ulm took flying lessons and a flying test from the New Zealand Air Force. It was an inspired decision, because he had long been embarrassed by his lack of proper certification. He – and no doubt Smithy too – had been chastened when voices were publicly raised against his appearing in his Australian flight lieutenant's uniform with wings sewn on the tunic. As a pilot licensed by the New Zealand Air Force, he could display his wings without discredit, and his self-esteem seemed to grow.

Meanwhile, the others soothed their nerves by forming a sort of jazz band. Smithy played the ukulele, McWilliams played the mouth organ, Litchfield blew the Swanee whistle, and Ulm, when present, banged a tin tray in place of a drum. When they performed selections from the popular musical *Rose Marie*, a crowd often gathered to listen, although some shunned them. Despite the public adulation, there were those who disapproved of Smithy and his crew. They took offence when the *Southern Cross* arrived in New Zealand on the Sabbath. They took more offence when they heard tales of the aviators' boozy parties, and Smithy's indiscreet sex life – for, like the rock stars of today, Smithy found it difficult to resist his female adorers. Of course, most young New Zealanders probably accepted their 'goings-on' as simply the high spirits of four courageous young men.

It was late on the evening of 12 October when Smithy received a favourable weather forecast. He told his crew: 'We'll give it a go. Go to bed and sleep till 3.15 – then tuck into ham and sausages and eggs.' They climbed aboard at five am, rather later than hoped, because 4000 followers and 600 cars had blocked the runway. Intrepid fans climbed to the top of a haystack and were the most vocal. When Smithy climbed into the cockpit and shouted, 'Cheerio, New Zealand!' a thunderclap of cheering rang around Fairhall's paddock.

Smithy's high spirits quickly sank when he found that the weather forecast was mistaken. The headwind produced such turbulence that they almost turned back to Blenheim. And perhaps they should have, for they flew on at a reduced speed for another twenty-three and a half hours, fearing towards the end that their fuel would run out. Finally, at two-fifteen am, they approached Richmond Aerodrome only to find the runway shrouded in fog. Smithy called for a searchlight but it was slow to shine. Luckily, the owners of parked cars heard of his request, and their headlights guided him in.

Smithy's mother was waiting for him, as were thousands of others, wrapped in coats and blankets and eiderdowns. Catherine, embracing him, was shocked by his appearance. At the end of his tether, he had once again narrowly avoided death. At the moment of landing, the tanks of the *Southern Cross* held just three gallons of petrol.

THE ORDEAL OF
COFFEE ROYAL

To found a successful airline was still Ulm's ambition, especially after the triumphant flight to Christchurch. 'I could perceive,' he wrote in his autobiography, 'that it was my job to lay plans for translating transient glory and temporary fame into terms of permanent solidarity.' That the public was not responding to air travel as enthusiastically as many had predicted did not dismay him. A cramped and noisy aircraft bumping through squalls with a cabin reeking of vomit was as repellent to Ulm as to anybody. He meant to ensure that his airline – which he proposed to call Australian National Airways – would be safe, comfortable, clean and reliable: as good as a train service but much faster. He hoped that ultimately his planes would fly around and across Australia, to New Zealand, maybe to New Guinea and possibly even to England.

Ulm spent many hours creating Australian National Airways in the last months of 1928. His secretary, Ellen Rogers, remembered him holding 'countless interviews and dictating countless letters'. There was much to be done because he was not seeking a government subsidy. Instead, he secured Frederick Stewart, the largest owner of private buses in Sydney, as his chairman, and under Stewart's chairmanship the new company offered £1 shares to the public, hoping in this way to raise capital of £85,000. Smithy, the co-founder and co-managing director of the company, gave it less attention. Ellen remembered him spending 'a lot of time chatting to his cousins and friends' at the office in Martin Place. He had no head for this type of business and was thankful to leave decisions to Ulm, who preferred to be in control anyway.

Smithy had other concerns to occupy his mind, one of them rather surprising. No sooner had he decided to seek a divorce in Sydney from Thelma on the grounds of desertion than a Mrs Ethel May Ives sought a divorce in Perth against her estranged husband, Aubrey Leonard Ives, on the grounds of adultery. The photograph of an unnamed woman, alleged by Mrs Ives to be the co-respondent, was produced in court and shown to witnesses. The 'dark, buxom woman' in the photograph was identified as Mrs Charles Kingsford Smith.

The identification caused a stir in the newspapers, and Ives' connection with the Corboy family was soon revealed. Ives had been a partner of Thelma's mother in a goldmine in Nullagine, and when he moved to an orchard at Maida Vale, near Perth, Thelma and her brother, Terry, had gone to live with him. Thelma's mother insisted that her daughter had nothing to do with Ives, and told the court that Thelma lived in Perth and supported herself by 'doing needlework'. Witnesses, however, testified that Thelma was living at the orchard as Ives' wife. Appeals lodged by the petitioner and the respondent caused the case to drag on for over a year, ending with a judgement in the High Court in September 1929. The divorce was granted, and costs of £2500 pounds were awarded against Ives, making it the most expensive divorce action in Western Australia to that time.

While in Perth on his inaugural flight across the Nullarbor, Smithy had visited Thelma and gained her agreement not to contest their divorce, 'so long as it is for desertion only'. He, in turn, agreed to make her a 'donation'. Realising that, in view of his newfound wealth, Thelma could have demanded a large sum but had not done so, he felt he should be generous. He suggested £250 and, since he did not want to be accused of collusion, sought the court's permission to make the gift.

The youngest of seven children, Charles Kingsford Smith (Smithy) was indulged by his older siblings. This photograph of Elsie, aged eighteen, and her baby brother, aged five, was taken shortly before the family emigrated to Vancouver.

A year after enlisting in the Australian Imperial Force, Smithy (right) transferred to the Royal Flying Corps. In August 1917, he lost part of his left foot to a German bullet. He is shown here with a fellow injured serviceman.

In 1920, Smithy was a stunt flyer in America. Elsie and Smithy, sailing back to Sydney, took part in the ship's fancy-dress ball: Elsie (right) is in Quaker attire and Smithy (left) is dressed as a woman.

In his mid-twenties, Smithy was a pilot at West Australian Airways (WAA) and married to Thelma Corby, whose parents ran cattle near Marble Bar. He longed to 'aviate across the Pacific'. Both his work and wife would be sacrificed to that longing.

Keith Anderson (right) worked at WAA with Smithy (centre) and shared his dreams of flying across ocean. In 1926, the two were joined in this ambition by Charles Ulm (left). Later, Anderson would abandon the crossing.

'Ulm had similar ideas to mine,' wrote Smithy. 'He wanted to do something that would make the world sit up.'

Above: In 1927, Smithy and Ulm flew around Australia in an old Bristol Tourer. Here the New South Wales premier, Jack Lang, greets them at Mascot Aerodrome. He was so impressed by the feat that he promised money for their Pacific adventure.

Left: Thanks to the generosity of American Allan Hancock, Smithy and Ulm gained an aircraft, the *Southern Cross,* and crew for the Pacific crossing. Harry Lyon, their navigator, stands on the left; James Warner, their wireless operator, is on the right.

Taking off from Oakland, California, on 31 May 1928, the aviators' first stop was Hawaii; their second, Fiji. Smithy deftly landed the *Southern Cross* on a sports field in the capital, Suva, and a host of Fijians turned out to watch the 'bird-ship' set down.

Thousands of Australians kept in touch with the perilous flight through the wireless. Smithy's parents, William and Catherine Kingsford Smith, sit on their Sydney verandah and read a message sent from the plane in Morse code.

On 9 June 1928, the trans-Pacific flight ended at the Eagle Farm aerodrome in Brisbane, before a crowd of 15,000. The *Southern Cross* had been in the air for almost eighty-four hours.

After their Pacific crossing, Smithy was made honorary squadron leader and Ulm honorary flight lieutenant in the RAAF. Ulm provoked criticism when he insisted on wearing wings on his tunic, though he was not a licenced pilot. In September 1928, he gained his licence.

Top: In September 1928, Smithy and Ulm (left) became the first Australians to fly to New Zealand. Accompanying them in the *Southern Cross* were Harold Litchfield, their navigator (next to Ulm) and Tom McWilliams, their wireless operator (far right).

Bottom: A Fokker monoplane with three engines, the *Southern Cross* drew excited crowds wherever it went. When Smithy used it to give joyrides, thousands bought tickets to ride with him. Here it flies high above Sydney.

In 1929, the team of four took off for England but were forced down in north-west Australia on a desolate mudflat they nicknamed 'Coffee Royal'. When pilot Bertie Heath rescued them after thirteen days, they carried him jubilantly on their shoulders.

In June 1930, Smithy completed his circumnavigation of the world by flying from Ireland to America. His co-pilot, Evert van Dijk (far left), was Dutch; his navigator, Patrick Saul (centre right), was Irish; and his wireless operator, John Stannage (far right), was a New Zealander.

Top left: After crossing the Atlantic, the aviators were given a ticker-tape parade in New York, down Broadway to City Hall. Some Smithy biographers claim this parade never happened, but newsreel footage disproves them.

Bottom left: Following the circumnavigation, Smithy and John Stannage travelled to England and Europe. Close friends, their relationship would deepen when Stannage married Smithy's niece, Beris.

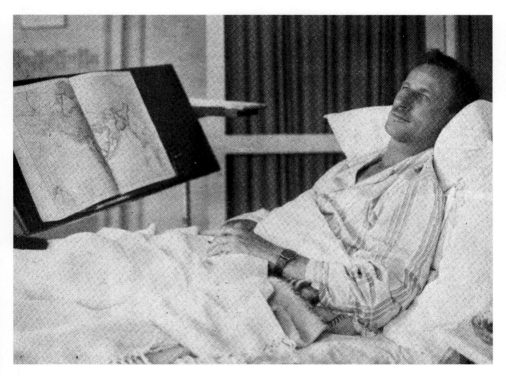

In Holland, Smithy fell ill and his appendix and tonsils were removed. He was an impatient patient, obsessed with achieving the fastest flying time between England and Australia. Here he plans his flight home on his new Avro Avian, the *Southern Cross Junior*.

One reason for Smithy's impatience was his approaching wedding. He had met Mary Powell on a passenger ship in the Pacific in 1929. Here she greets him as he alights from the *Southern Cross Junior*. Less than two months later, on 10 December 1930, they were married.

In December 1932, Mary gave birth to a boy, and photographs show Smithy's devotion to his son. Here he is seen with close associates (left to right) Jack Percival, his publicist; Tommy Pethybridge, his engineer; John Stannage, his wireless operator; and Gordon (Bill) Taylor, his co-pilot.

On 15 December 1933, Smithy, Mary (seated, centre), the Stannages and other friends celebrated the laying down of a case of port in readiness for Smithy's son's twenty-first birthday. The party was at Leo Buring's cellars in Sydney.

By 1934, Smithy's health was failing. The stress under which he had lived for two decades was overwhelming him, and on his record-breaking flights he suffered panic attacks. To lessen their force, he tried a strict health regime and worked out regularly in the gym.

Mary feared for Smithy but felt powerless to help. Here she is with him at Mascot Aerodrome, outside the hangar of his and Ulm's financially troubled airline. Today this airport still honours the name of Kingsford Smith.

Right: Mary and his mother, Catherine, were the strongest influences in Smithy's life. His mother, he said, 'has always been my inspiration and my best friend'. Here in 1934 she tenderly pins a boutonnière of wattle to his coat before he begins a trans-Tasman flight.

Below: On 6 November 1935, Smithy attempted to set another record time between England and Australia. Here he is in the forward cockpit of the *Lady Southern Cross*, with Tommy Pethybridge in the other cockpit. The plane never reached its destination; it was last seen above the treacherous Bay of Bengal.

One of the last photographs taken of Smithy,
at the Lympne airport on 6 November 1935, just
hours before he took off on his final, tragic flight.
He looks remarkably fit and relaxed.

The judge granted permission and the case progressed swiftly. On 26 October 1928 Smithy received the preliminary 'decree nisi', which six months later was replaced by a 'decree absolute'. By May 1929 the Kingsford Smith marriage was over, and Thelma was able to marry Ives when his divorce was granted.

Smithy found his appearance in the divorce court upsetting, but Thelma constituted the least of his legal troubles. In the last months of 1928 he was harassed by other painful and costly conflicts in court. And they, too, were battles with former friends. Keith Anderson had watched Smithy's and Ulm's growing fame with envy and pain, tormented by the knowledge that it might also have been his. At first he seems to have been hoping for some sort of reconciliation, sweetened perhaps by a sharing of the publicity and the prize money. His thoughts centred on their last letter to him, in which they had expressed a willingness to discuss their disagreement once the flight was over and they were back in Australia.

In that spirit he had watched the landing at Mascot and participated in some of the celebratory dinners. At the Sydney *Sun* dinner, his hopes must have risen when Ulm, in the course of a speech, mentioned him favourably. 'I am very proud,' said Ulm, 'that Anderson was associated with us in the early stages.' Perhaps Ulm was hoping that a few kind words would solve a tricky situation, which every moment was becoming trickier, since Anderson's sympathisers were beginning to join forces with those of Lyon and Warner. But no real peace-making or financial solutions were offered.

Encouraged by Bon's lawyer father, Arthur Hilliard, to believe that he had a legal claim to one-third of the profits, Anderson called on Smithy and Ulm. He told them he came as a 'creditor' to claim what he was owed. He also reminded them that his mother and uncle were creditors too, since they had contributed at least $3000 towards the flight.

Smithy was not ready for this confrontation and left the task of responding to Ulm. Equally ill at ease, Ulm dealt with the claim brusquely. He insisted that by refusing to return to California to participate in the flight, Anderson had forfeited his place in the partnership.

He admitted, however, that Anderson's family had contributed $3000, this sum representing, according to Ulm, $1000 for each of the three partners. Ulm wrote out two cheques to settle what he concluded was his and Smithy's share of this debt: each was for the Australian equivalent of a thousand US dollars. Anderson refused to accept the cheques.

Anderson now resolved to bring a lawsuit against the co-commanders. The case had not yet come to court when, in mid-July 1928, Smithy and Ulm announced that they were planning to fly across the Tasman Sea. The news spurred Anderson and Hilliard into action. Afraid that Smithy and Ulm might abscond with the *Southern Cross* on the forthcoming flight to New Zealand, the two hurriedly visited Mr Justice Owens at his private house and won an injunction to restrain the defendants from 'alienating, disposing or otherwise dealing with the plane'. When the time came for Smithy and Ulm to depart for New Zealand, they were obliged to obtain the court's permission to leave.

Did Anderson, in his fantasies, see himself sitting in the cockpit beside Smithy on the *Southern Cross*'s next flight, with Ulm – the unlicensed pilot – relegated to the rear cabin? It seems he did, because on 3 August 1928, Anderson's counsel submitted that 'in any active flight by the defendant, Anderson should be allowed to take part'. The Chief Judge in Equity, Mr Justice Harvey, declined to accept the submission, and made a witty reply that provoked much laughter. 'I will not rule anything of the kind,' he said. 'I would not ask the defendant to fly with Anderson. There might be a dispute in the cockpit.'

Five days later, Bob Hitchcock also entered the fray, suing Smithy, Ulm *and* Anderson for £1000, which he claimed had been verbally promised him as compensation for being denied a part in the Pacific crossing. To Smithy's and Ulm's embarrassment, the writ was served at a Returned Soldiers League dinner, being thrust under their noses while they were autographing menu cards. Because much of Anderson's evidence concerned events and conversations in California, an adjournment was ordered until statements could be furnished from overseas.

Although now technically opponents, Anderson and Hitchcock continued their close friendship, and just before Smithy and Ulm crossed to New Zealand, they decided to make their own adventurous flight. They chose to fly in one of the old Bristol Tourers which had previously been used for the flight around Australia, and which still belonged to Interstate Flying Services. Their destination was to be London, but they did not manage to leave Australia. Taking off from Mascot on 6 September 1928, they were forced back by storms. On their second attempt they abandoned the flight at Pine Creek because the Bristol's engine overheated and seized up. Although they longed to try again, they had no money to procure a fresh aircraft.

The court cases dragged on well into 1929, absorbing more time, energy and money than either side could afford. As his case progressed, Anderson grew increasingly pessimistic. Having abandoned his place in the crew, and refused the co-commanders' efforts to reinstate him, he realised he had very little chance of establishing a claim to the prize money or the plane. Worn out, and impoverished by the struggle, he capitulated. On his lawyers' advice, he sought a settlement.

He was prepared now to concede that by voluntarily abandoning his post as co-pilot and returning to Sydney, he had dissolved his partnership with Smithy and Ulm. He was prepared to accept that ample chances had been given him to join their new flight, and that Smithy and Ulm were under no legal obligation to pay his fare to California. And he was ready to give up his claims to a share in the prize money and profits of the Pacific flight, and to ownership of the *Southern Cross*. What he needed in return was money, for his legal costs were about £400 and he had no money to pay them.

Smithy was more than ready to settle. Indeed, he was deeply troubled, because the evidence given in court had forced him to review his own conduct, and much of it seemed less than honourable. At Smithy's urging, Anderson was offered £1000 in settlement. This would pay his costs but it would not restore his public reputation, so Smithy and Ulm were obliged to offer more. They agreed to state publicly that Anderson's failure to cross the Pacific 'was not the result of any lack of personal courage'. This satisfied all parties, and on 22 February, in Sydney's Supreme Court in Equity, Mr Justice Harvey dismissed the case.

It was a happier outcome than Keith Anderson had dared to expect. Feeling betrayed, he had gone to court demanding justice, but what he really wanted was to turn the clock back to that time when he and Smithy were best mates. Smithy's generosity restored his faith in their friendship, and that gave him peace. Not only could he now pay his legal fees, he could also fulfil his heart's desire and buy a new plane. He chose a sporty little Westland Widgeon III, which he named the *Kookaburra*. When at the last minute he realised he was £300 short of the purchase price, Smithy gave him a promissory note to bridge the gap.

Hitchcock was now preparing his case, which was to be tried by Mr Justice Davidson and a jury of four. He was ready to testify that he had left Western Australian Airlines to join Interstate Flying Services in response to a promise that he could participate in the Pacific flight. When, a year later, that promise was rescinded, Hitchcock claimed he was offered a compensatory gift of £1000, to be delivered when the flight was completed and the aviators returned to Australia. Since these promises were not in writing, and Smithy and Ulm denied making them, the claim was easily undermined by Mr Richard Windeyer, the fearsome barrister engaged by the defendants. On 21 March the verdict went against Hitchcock.

Neither of the co-commanders had expected the court cases to last so long. For months they had been planning a flight to England in the *Southern Cross*, and were dismayed at the continuing postponements. Ostensibly, the trip was to purchase a fleet of Fokker trimotors that was being built under licence at the A.V. Roe factory near Southampton for Australian National Airways. For Smithy, however, the flight had strong personal significance. He longed to be the first pilot to circumnavigate the world by flying across the equator, and this flight to England rep-resented the second sector of his route. Thereafter he had only to cross the Atlantic Ocean to America and he would have completed the circle. He also had hopes of beating the fifteen and a half days taken by Bert Hinkler to fly between England and Australia the previous year.

Having re-engaged Litchfield and McWilliams, the co-commanders announced in February that they would take off early in March. They organised a farewell party at the Hotel Australia, writing the invitation in their own inimitable style:

A Spot of Food and Drink on Wednesday 20th February 1929 at 7pm. Dress: Dinner Jackets, Pyjamas, Mess Kilts, Kilts and/or BVDs. [BVDs were a fashionable brand of underpants.]

   We would particularly like you to join us, as this may be our last 'spot' before we reach London pubs. Replies to 'Smithy and Charles' at the Australian National Airways Office at Challis House in Martin Place.

The party was premature. February ended, the second lawsuit showed no sign of finishing and their flight to England remained in limbo. The pressure of the preparations, the uncertainty of the departure date and the emotional strain of his court appearances exhausted Smithy. He contracted flu. By the time Hitchcock's case ended, Smithy was running a temperature of 104 degrees Fahrenheit, and Ulm was also showing the effects of stress. Ulm was obsessed by the need for a spectacular aerial triumph to wipe away the negative publicity of the court cases. He refused to countenance procrastination.

On Monday, 25 March, with Smithy confined to his sickbed, Ulm was at Richmond Aerodrome testing the *Southern Cross*'s propellers and speaking to reporters. He assured them that Smithy would be well enough to leave by the Thursday. When that day arrived and Smithy was still ill, Ulm moved their departure date to Saturday, 30 March. On 29 March – Good Friday – Catherine Kingsford Smith gave an inter-view to the press. 'I am much more worried about my son's health than I am of the flight,' she announced. 'Our main anxiety is Charlie's health. That is worrying us considerably.'

The flight's first stop was to be Wyndham, in the north of Western Australia. A week earlier, Captain Clive Chateau, a senior pilot employed by the Atlantic Union Oil Company, had flown to Wyndham on his company's behalf to inspect the terrain and the weather. The wet season had lasted longer than usual, and the only suitable landing places were waterlogged. On 29 March, Chateau reported that the grounds were at last drying out, but that care must be taken. He warned the aviators by telegram that they should not set out until they received his 'final OK'.

Ulm read the telegram with irritation. He thought that Chateau was overcautious. Desperate to leave, he was ready to take the risk if Smithy agreed – and Smithy, despite his illness, gave his consent, even taking the plane up for a quick test flight. They both longed to be off, and were buoyed by a day-old forecast in *The Sydney Morning Herald* which predicted fair weather in the far north-west. That night they sent Chateau a telegram: 'Southern Cross leaving ten o'clock Saturday morning unless you advise ground definitely unsafe for landing.' Chateau's telegram, advising them to remain at Richmond, did not arrive until after midday the next day. By then they were in the air.

The *Southern Cross* took off from Richmond, as promised, on Easter Saturday. It was farewelled by a vast crowd of well-wishers, Smithy's parents among them. Now that their leave-taking was at hand, his mother exhibited a stoicism for which she would soon become famous. An affectionate smile as she kissed Smithy goodbye, a fluttering of her handkerchief as he walked to the plane and a smile of confidence as he rose into the air were the only signs of emotion that she allowed herself to show. In the words of her second son, Wilfrid, 'She fought down the pleas that any mother would long to make in the hope that her calm confidence would help him achieve further successes'. She placed her boy in God's hands, and prayed that his God-given skills would protect him in times of trouble.

Trouble soon appeared. Less than an hour into the flight, Litchfield accidentally knocked the long-wave radio aerial, causing it to unfurl and break off, thus depriving the crew of the ability to send messages, although they could still receive them. Smithy and Ulm now faced a dilemma: should they dump an almost full tank of petrol and return to Richmond, or continue the journey without a properly functioning radio? They chose to forgo the radio and press on.

Before their departure, Smithy had confessed to his cousin Ray Kingsford that he still felt decidedly ill. After twenty hours in the air he felt considerably worse, and was not helped by having to fly blind in low cloud and heavy rain. By dawn, Wyndham should have been close by, but what they glimpsed through gaps in the cloud looked nothing like the port they remembered. As visibility was too poor for a sextant shot, Litchfield gave an opinion, based on drift readings, that they were too far north. For several hours they flew south, before, having failed to find a landmark, they started to retrace their course.

Suddenly, miraculously, they saw habitation. Far below lay a collection of huts that was later identified as the Drysdale Mission, a remote Catholic outpost run by Spanish Benedictine monks in the far north of the Kimberley. Litchfield, however, failed to recognise the settlement, because he carried only British Admiralty charts and the *Times Atlas of the World*, neither of which showed mission stations.

Ulm, ever practical, dropped a weighted message containing the words 'Please point direction Wyndham', and after repeated circling at low altitude, the crew saw a man appear below them and point to the south-west. This caused astonishment, because they supposed that Wyndham lay to the east. They had no means of knowing that the message they had sent groundward had not yet been found. The man was pointing towards a possible landing ground, a short distance from the mission.

Hopefully, they turned south-west. Three hours later – it seemed an eternity at the time – they came upon another cluster of buildings. They had no idea what these were, but later learned that they were the Port George Presbyterian Mission, inhabited by two missionaries, their families and a group of Aboriginal people. Smithy circled so low that the excited spectators were able to identify the *Southern Cross*, and they shouted and waved. Another message was dropped from the plane, asking for three white sheets to be laid on the ground with an arrow pointing in the direction of Wyndham, and the mileage to Wyndham written on the centre sheet. After some delay, three white sheets were placed on the earth with the figure 150 on the middle sheet. The sheets pointed east.

They assumed that this meant that they were 150 miles from Wyndham, and that it lay to the east – a direction which left them 'dumbfounded', to use Smithy's term. An urgent discussion was held in the cockpit. Should they try landing in one of the mission's rocky paddocks, or did they have enough fuel to reach Wyndham? Smithy doubted they had enough fuel but, ill and exhausted, was in no mood to insist. Instead, he shut himself off from the arguments. When, weeks later, he was asked why he had decided to press on, he replied that he simply could not remember. Obeying his colleagues, he turned east and flew for almost an hour, but saw no sign of Wyndham. By now they had been in the air for twenty-seven hours and their fuel supply was dangerously low. Alarmed, they turned back towards the haven of Port George. By now the cloud was lower and thicker, and they could not find the mission.

If Smithy had been in possession of his health and his wits, he would probably not have made such muddled decisions. Weakened by illness, he lacked the will to override Ulm's determination to push on. Nor was he fully aware of McWilliams' and Litchfield's inexperience. If Lyon and Warner had been the radio operator and navigator, one imagines, Smithy would never have flown for so long in the wrong direction, and almost certainly would have landed at Port George. By now their petrol was almost exhausted and there was no alternative but to set down. Smithy spied a boggy stretch of grass beside a mudflat in the mouth of a river, and down they went.

Thanks to his skill, the *Southern Cross* landed safely. 'The wheels ploughed into the mud,' Smithy remembered, 'and as we forged through it, she tilted dangerously forward. I thought she would turn over on her nose. But she sank back on a level keel – and came to rest. Again the dear Old Bus had saved us.'

Their landing place was 'a great swamp, bordered by tidal inlets', where the tide rose and fell by forty feet: fortunately they were above the high watermark. Beyond the swamp lay tropical bushland, and beyond that a row of low hills. On the other side, a tidal river snaked its way to the Indian Ocean, its muddy banks the home of mangroves, mosquitoes and crocodiles. As Litchfield discovered when he eventually used his sextant, they were in latitude 15 degrees, 35 minutes south, and longitude 124 degrees, 45 minutes east. The river was a tributary of the sprawling Glenelg River, and the Port George Mission was not far away – but how far, or in what direction, they had no idea. That they should try to reach the mission quickly occurred to them, but was just as quickly rejected. The *Southern Cross*, with its silver sheen and large wing, was conspicuous from the air – or so they thought – and they were confident that it would soon be sighted. For Smithy, who had a deep-seated dread of going down in a remote place where he might not be found, this was a comforting thought.

As soon as they had checked everyone for injuries, they examined their tools and food supplies. Since they had considered that they had no need of a life raft on this trip, they had dispensed with the axe and saw, and removed the freshwater still and the emergency radio in the wing. This was disastrous, because they needed an axe and a saw to cut wood for their signal fires, and fresh water because the river was brackish. Most of all, they needed the emergency transmitter. As soon as they landed, McWilliams set up the aerial and had the receiver working. But the ability to transmit depended on a supply of power, which a wind generator provided during the flight. On the ground they had no means of generating power, so their radio transmitter was useless.

A worse discovery confronted them when they opened the food locker. The chocolate, dried fruit and malt that constituted their emergency rations were missing. Ulm vowed he had checked the locker a few days before, so presumably the food had been stolen. Their only food was seven stale sandwiches, a flask of coffee, a bottle of whisky, and several tins of Allenburys Food, a type of infants' gruel which they were carrying to the postmaster's baby at Wyndham. That night they ate some of the gruel, and consoled themselves by sipping a small drink of coffee laced with whisky – a drink which, in high-class restaurants, was sometimes called 'Coffee Royal'. Soon they had nicknamed their wretched landing place Coffee Royal. It was an amusing notion, and the name took hold.

Worn out by illness and exertion, Smithy slept six hours that night; the others, thanks to the heat and the mosquitoes, scarcely closed their eyes. Nevertheless, they all arose on Easter Monday ready to make the best of their prison. Finding a spring of fresh water on a nearby knoll, which they named Darlinghurst Hill, they felt more hopeful; they might even have been mildly content had it not been for the insects. 'The flies would start at 6 am in the morning,' Smithy remembered, 'and would pester us all day and at 6 in the evening, just as they were quitting, along would come battalions of mosquitoes and not ordinary ones either, but great cannibals that sounded like three engined Fokker planes.' As a defence, they put on their heavy flying suits and gloves and sweltered in the steamy heat.

Once they had collected firewood and assembled their signal fires, listening to the radio became their first priority. Much of the Morse chatter coming through the receiving set concerned their where-abouts, for no message transmitted during the flight had conveyed more than a vague notion of where they might be. Their final message, sent by Ulm, had omitted a particularly vital piece of information. 'Have been hopelessly lost in dense rain storm for ten hours,' Ulm had told their listeners. 'Now going to make forced landing at a place we believe to be 150 miles from Wyndham in rotten country.' He said nothing of being *west* of Wyndham – an omission that would gravely impede the searchers.

Back in civilisation, many people were speculating on their where-abouts. Colonel Brinsmead believed they might be at the Port George Mission, while others believed they were at the Drysdale Mission, but since neither mission possessed a radio, there was no quick way of knowing. Many called for an aerial search, with the Sydney *Sun* the first off the mark. Having contacted Norman Brearley – who was ready to supply as many aircraft as were needed – the editor of the *Sun* char-tered a plane and pilot and dispatched them to the Drysdale Mission. Another plane, hired by the Australian National Airways Board, was sent towards the Port George Mission. Over the course of the next few days, two more planes entered the search. But it was guesswork. What the searchers needed was a reliable indication of the stranded plane's latitude and longitude.

Had Smithy and his crew had a viable radio transmitter, they would have been quickly found. And there was actually a method of converting a radio receiver into a transmitter: instructions went out again and again from the Sydney station 2FC, and from the Amalgamated Wireless stations in the north-west, explaining in detail how to perform the conversion. A nervous McWilliams was afraid to attempt the change, for fear that he might fail and leave them with neither receiver nor transmitter. How Smithy and Ulm must have regretted the loss of resourceful Warner.

Since McWilliams was not prepared to risk the conversion, a new plan was tried. This required raising and cranking the wheel of the plane in order to create sufficient power to drive the generator. Having no tools other than two screwdrivers, a pocket knife and a leather belt, it was uphill work. 'While Litchfield and I turned the big wheel,' Smithy recalled, 'Ulm held the generator to it, making a friction drive, and we were actually able to generate half the normal current.' Alas, the signal it produced was so feeble that nobody in the outside world could hear it.

Easter Tuesday dawned and the men were hungry and depressed. The 'four of us', Ulm wrote in his diary:

grow visibly weaker, and are growing incapable of serious exertion.
Smithy and Litchfield pushed their tired way to the hilltop and started a
fire. Then more listening in on the radio, varied with fruitless attempts
to shoot birds with an automatic revolver. McWilliams endeavoured to
cheer us with his mouth organ. Rain – and then more rain.

By now the missing aviators were worldwide news, and the Australian public was angry that the federal government was taking no action. On the Wednesday, Prime Minister Bruce was accosted at his holiday home near Melbourne by a *Sun* reporter. Bruce – who was still in his dressing gown – explained that the *Southern Cross* was making a commercial flight, and that an air search by the government was therefore not appropriate. The reporter was outraged, and a mischievous ditty found its way into print:

Bruce was in the bath-tub,
Blowing soap and bubbles,
Didn't care to comment on,
Other people's troubles.
Somewhere in the desert,
Somewhere on their own,
Four Australians missing,
Fighting all alone.

Since the federal government was useless, the citizens of Sydney took action. A Southern Cross Rescue Committee was formed by an old friend of Smithy's, the city's Civic Commissioner, John Garlick. He attracted donations of £4000 in a matter of hours, a sum which almost doubled over the next couple of days. A six-seater de Havilland DH.61, named the *Canberra*, was chartered with a pilot, Les Holden, and a radio operator, mechanic and doctor, ready to find and assist the lost plane and its crew. Unfortunately, the *Canberra*'s departure was so delayed that it did not reach the Kimberley until the following week. This further infuriated the public. Eventually, the prime minister felt forced to act. Reluctantly, he ordered the seaplane carrier HMAS *Albatross* to make the long voyage to the north-west coast with six planes on board to conduct the search. Again there were delays, and it was 11 April before the *Albatross* finally sailed from Sydney Harbour. In the meantime, hunger and disappointment were dragging the aviators down.

'I think it was on 4 April – our fourth day in that awful spot,' wrote Smithy, 'that we really began to starve.' They had exhausted most of the gruel and were now relying on a species of sea snail they found in the swamp, and a type of bean that grew among the mangroves. There was little nourishment in either, and the snails, which they stewed in large quantities, made them nauseous. Wrote Ulm miserably in his diary:

Heat, flies, mosquitoes, light fires, pull down trees, pull up grass
for smoke, walk for water, eat a few snails, drink water and a very
thin cup of gruel, listen to radio, turn generator until every ounce of

energy is gone; then lie down to be eaten alive by mosquitoes – that is our day – when will it end?

Later it would be said that Ulm had exaggerated their sufferings, but his companions testified that what he wrote was, for the most part, true.

On 7 April two unexpected people entered the search, and when Ulm and Smithy heard the news over the radio they could scarcely believe their ears. These new searchers were Keith Anderson and Bob Hitchcock. Down at their favourite drinking place, the Customs House Hotel at Circular Quay, there had been much speculation about the *Southern Cross*'s location. Like Brinsmead, Anderson had studied the radio messages, and deduced that the aviators were at, or near, the Port George Mission. His arguments convinced the publican, Jack Cantor, who took up a collection and offered him £500 to search the Port George area. When he was advised to take a mechanic with him, Anderson chose Hitchcock. And such was Hitchcock's devotion to Anderson that he agreed to go, even though he had no reason to care about Smithy and Ulm, and had recently had surgery on his leg, leaving him with a wound that had not yet fully healed.

Why did Anderson respond so generously? The simple answer was affection for Smithy. 'I am dead keen on going on account of our old friendship,' Anderson told Cantor. He hated their estrangement, and now that justice was done and the slate was wiped clean, they were mates again. In that same spirit of mateship, some years previously, Smithy had saved Anderson's life in West Australia. Now he could return the favour.

Thanks to Smithy's generosity, Anderson owned a plane, and since he had hoped to use it for an endurance flight over Sydney, its tanks were capable of carrying sufficient fuel for twenty-four hours in the air. In a flurry of excitement, he and Hitchcock took off from Richmond on 4 April, with scarcely any preparation. They had no radio, very little food and water, insufficient tools and a compass known to be inaccurate. This did not seem to perturb Anderson, who believed he knew the north-west coast so intimately that he could not fail. Anyhow, he was much too excited to worry about tools or maps or compasses. Bon, still his fiancée, was indignant at his departure, mostly because she considered the journey unsafe, but also because he was looking for Smithy. She appears to have held a persisting grudge against Smithy, and was so angry that she is said to have threatened to end their engagement if Anderson flew off on his rescue mission.

Anderson's first stop was Broken Hill, which he reached with difficulty, the faulty compass having taken him off course. His next stop was the South Australian inland town of Marree, although he had been aiming for Oodnadatta; the compass was again the culprit. The following day, despite engine troubles, he reached Alice Springs, where, to his joy, a telegram from Bon assured him that she still loved him. On 10 April he left Alice Springs, expecting to spend that night in Wyndham. He was already off course when more engine trouble forced him down south-east of the Wave Hill Cattle Station, in the Northern Territory.

Hitchcock soon repaired the engine, but the plane could not take off. Heavy with fuel, it needed a long runway to become airborne; without an axe and spade with which to clear the way, they were obliged to remain on the ground until a rescuer came. Like Smithy and his crew, they had no radio to transmit their position. But their predicament was far worse than Smithy's. Anderson and Hitchcock had no source of water or food. If help did not come quickly, they would die.

A MOST PERPLEXING
PARADOX

'I suddenly heard the noise of a plane,' Smithy would remember. 'We stoked the fire up, hoping and praying that they would see us; but imagine how we felt when the plane came no nearer.' That was on 6 April 1929, almost a week after they were stranded. On 9, 10 and 11 April more planes passed without spotting them. 'I was beginning to feel pretty down-hearted,' Smithy confessed, especially as 'the last of the baby's gruel was petering out, and there was absolutely nothing else to eat'. Each night on the radio they heard reports of the day's fruitless searching. 'I know from my own personal thoughts,' wrote Smithy, 'we all were afraid that the only thing found would be our dilapidated old bus and some bones.' Unrescued, they would die the miserable, isolated death he had so long feared.

The twelfth day of April was now dawning, the thirteenth day of their ordeal. The Southern Cross Rescue Committee's plane, the *Canberra*, piloted by Les Holden, flew out from Wyndham, and just before ten am the watchers at Coffee Royal saw it approaching from the south-east. Immediately Smithy and Litchfield hurried up Darlinghurst Hill to light their signal fire, and this time the aircraft did not turn away. Looking down, the radio operator, John Stannage, saw two white patches on the edge of the swamp, which, as they flew lower, turned out to be rocks. But just beyond them he spied the wing of a plane. As he watched, two figures emerged from beneath the wing, and two more came running down the hill. Had he been closer, he would have seen that the runners were Smithy and Litchfield. Both were crying with relief and whooping with joy. Excitedly, Stannage transmitted the words 'Found, Found'. The message was received at the Australian Amalgamated Wireless's radio station at La Perouse, in Sydney. Minutes later the news was dispatched across the waiting nation, and then to the waiting world.

In Sydney, Catherine and William Kingsford Smith heard the news almost as soon as it went to air. For thirteen days they had tried to maintain hope, a severe ordeal for Smithy's ailing father, whose only recourse was to keep active. Just before lunch on 12 April he turned his radio to the shipping news. 'Oh, don't waste time on that,' Catherine called from the kitchen, but he replied, 'You never know.' Almost immediately the announcement came bursting across the airwaves. 'I can't describe my feelings,' William told reporters. 'I know we felt like collapsing but we didn't. I went straight outside and hung out the flags' – an Australian flag from the flagpole and a Union Jack from the window.

At least fifty telegrams arrived that afternoon, and countless phone calls. Hordes of relatives and wellwishers knocked on the door. 'We belong to the public today, Father,' cried Catherine as she hurried her husband outside to be photographed by reporters. She was trying to stay calm for William's sake, but she had to confess that she was beside herself with joy.

All that afternoon emotion washed across the city of Sydney like a spreading wave. Church bells rang, taxi drivers tooted, ferries gave 'cock-crows' on their sirens; in the streets, 'complete strangers told each other how good it was'. The stock exchange halted to cheer, a court case stopped to applaud, schools called special assemblies, churches held special services. At the Darlington Home for the Deaf, Dumb and Blind, where Smithy regularly took the children for rides in his Studebaker car, there was joyful 'pandemonium'. The news, wrote the Sydney *Sun*, 'filled everyone with a wondrous sense of comradeship'. Nor was Sydney alone in its reaction. Not far from Melbourne, Prime Minister Bruce at his holiday house expressed the feelings of the entire nation: 'I am sure the whole people of Australia, as well as those overseas, will be relieved that these gallant men have been found.'

At Coffee Royal, the starving men were eating. The *Canberra* had circled above them for about an hour, dropping bags of food and a note saying 'back tomorrow'. For the next five days a series of planes flew over them, dropping tea, sugar, salt, tinned meat, mosquito nets, citronella, tobacco, cigarettes, shirts, trousers, fishing lines, a revolver and bullets, green vegetables, fruit and 'medical necessities' – enough, it was said, to open a general store. Smithy remembered that they 'tore open the packages and gulped down the food like hungry wolves'.

But public attention was soon diverted elsewhere. On 11 April, Jack Cantor – Keith Anderson's financial backer – had announced that Anderson and Hitchcock and their rescue plane had not arrived at Wyndham. At first nobody was much concerned, for it was presumed that they had stopped at a nearby cattle station, but when, two days later, they were still missing, the alarm was raised. Assistance was at once requested from the Southern Cross Rescue Committee but it did not trust Cantor, and refused to search for the second missing plane. Days were wasted in verbal battle, until the Commonwealth government intervened. By 16 April the *Canberra* and five RAAF planes were following Anderson's supposed route. These were soon joined by a seventh plane owned by the Queensland and Northern Territory Air Service, known as Qantas.

Back at Coffee Royal, Smithy and his crew were now eager to leave, for drying winds had hardened the ground sufficiently for a plane to land and take off. On 13 April, pilot Bertie Heath of Western Australian Airways managed to land his plane beside the *Southern Cross*, and a few days later he and pilot Tommy O'Dea were flying in sufficient fuel to enable the Old Bus to fly to the nearby port of Derby.

Raw from sunburn, swollen from infected insect bites, painfully thin and bearded like 'wild men', the aviators were a disturbing sight, and their pent-up emotions were almost as wild as their appearances.

Heath had brought with him a reporter from the *West Australian* news-paper named J.A.K. Tonkin, who would soon dispatch a copy of Ulm's diary and a vivid account of the aviators to the Sydney *Sun*. He wrote that Smithy and his crew laughed, talked and ate with 'overflowing aban-don', stopping at intervals to pat their stomachs and exclaim: 'Gee boys, isn't it great to have a full stomach.' Tonkin described how, from time to time, they would burst into song, giving excited renditions of 'The More We Are Together' and the *Hallelujah* chorus. He also described their boisterous horseplay, and snapped a photograph of them hoisting Heath onto their shoulders. Having little understanding of the effects of stress, Tonkin was surprised by their physical and mental vigour. It was so different from what he was expecting.

The next day another reporter arrived in Tommy O'Dea's plane. He was John Marshall, the local correspondent for Smith's Newspapers, the parent company of the aggressive Sydney papers the *Daily Guardian* and *Smith's Weekly*. Marshall and Smithy had been on good terms since 1927, when Marshall and his wife were passengers on the Perth–Sydney flight, but on this occasion Marshall made trouble.

The Presbyterian missionary from Port George, George Baird, and three Aboriginal guides also arrived at Coffee Royal that day, having traversed about eighteen miles of dangerous swamp to reach the avi-ators' camp. Learning that the mission was geographically close – and not taking into account the forbidding nature of the terrain – Marshall started to feel there was something suspicious about the aviators' failure to reach it. He was also puzzled that they had been unable to construct a fire conspicuous enough to be seen from the air. And why had nobody sighted a plane as large and easily visible as the *Southern Cross*? He began to wonder if the aviators had deliberately hidden themselves in order to generate publicity.

The more Marshall analysed the situation, the more dubious he became. He found it hard to believe that resourceful men like Smithy and Ulm had almost starved to death. Surely they could have caught fish, or killed birds, or searched for roots and yams. As he looked around the site, now so well equipped, it reminded him more of a holiday camp than a lonely place of privation. Indeed, with three planes present and thirteen people milling around, Coffee Royal seemed 'quite a township', and a 'lively one' at that.

Marshall wove his suspicions into an intriguing article for the *Daily Guardian*, hinting at but not openly saying what he suspected. He described Coffee Royal as a 'paradox' of a place, where expert navigators became lost, where expert operators could not adapt a radio, where signal fires and large aeroplanes could not be seen, and where resourceful men could find nothing to eat. He described the aviators' grim chuckles as they listened on the radio to the nation's anxiety: 'Smithy would grin and say: "It's beginning to get everybody guessing."'

The *Daily Guardian* and *Smith's Weekly* had long been supporters of Smithy and Ulm, but on receiving Marshall's mischievous article, their editors changed their tune. Being aware of the strength of Australian egalitarianism, they suspected that many of their readers probably distrusted 'tall poppies' and would take pleasure at seeing them cut down. Accordingly, they ran more articles in a similar vein, some coming from the pen of Norman Ellison, the reporter who had championed Lyon and Warner and distrusted Ulm.

The public does not seem to have been much surprised to read that Coffee Royal was a 'paradox'. This was an era of advertising stunts, and for some days rumours had been circulating that the *Southern Cross*'s disappearance was a hoax devised by Ulm and promoted by the Sydney *Sun*, which, having exclusive rights to the flight, was enjoying a huge boost in circulation. The rumours increased when Anderson so impetuously flew off and vanished. It was supposed that Anderson would fly to the missing aviators' hideout, proclaim his discovery, be hailed a hero and receive a reward. The rumours gained momentum when the public learned that Smithy had helped Anderson buy the plane.

Meanwhile, at Coffee Royal, the *Southern Cross* was at last preparing to depart. Although damaged in the wing, the plane was deemed airworthy, and on 18 April, Smithy flew it to Derby. There he had planned to rest, but the arrival of an urgent telegram changed his mind. It was from Jack Cantor, who begged him to search for Anderson and Hitchcock.

Over the next two days, Smithy and his crew flew across the Kimberley to Wyndham, searching the country north-east of Port George but finding no trace of the missing airmen. The following day they intended to search again, but suddenly, far to the east, the search was ended. At midday on 21 April, the Qantas plane, the *Atalanta*, sighted a stricken aircraft in the Tanami Desert. As the pilot flew low, he saw a man lying beneath the wing. He dropped a can of water, although he assumed, rightly, that the man was dead. He also assumed, wrongly, that the man was Keith Anderson. It was Hitchcock. In the following weeks, a ground party would discover Anderson's body in the scrub close by.

This graphic evidence of their deaths hit Smithy hard. While he was thankful that he and Anderson had parted amicably, this did not pardon his own previous behaviour. The evidence in court had revealed so much that he was ashamed of, not least his readiness to abandon Anderson when he furthered his plans to cross the Pacific with Ulm. Was Coffee Royal a punishment for his and Ulm's heartlessness? Should he have died instead of Anderson? One cannot know what questions he asked himself, but certainly he was deeply troubled.

The discovery of Anderson's and Hitchock's bodies gave fresh life to the newspapers' insinuations, which quickly turned into outright attacks. As more facts came to light, Smithy and his crew were accused of criminal negligence. They had embarked without adequate tools, rations, maps, weather reports or radio equipment. A huge sum of public and private money had been expended searching for them, and as yet there was no definite proof that they had even been lost. Most serious of all, two gallant airmen had lost their lives trying to find them. 'All Australia with a grieving heart asks why; and all the world looks for an answer,' cried the leading article in *Smith's Weekly*. The editors of *Smith's Weekly* and the *Daily Guardian* wanted blood. They shrugged off writs for libel from Smithy's and Ulm's solicitors, and led a chorus of voices demanding a public inquiry into the tragic episode.

On the long flight back to Sydney on 23 April, Smithy circled low over Anderson's plane. 'I flew around several times,' he recalled, 'and we caught sight of a body under the wing of the plane, clad only in singlet and underpants.' The sight of the body, he later wrote, 'affected us deeply'; and he said 'us' because Ulm also showed signs of distress, although Ulm's distress was mild compared to Smithy's. The body on the ground gave reality to Smithy's longstanding fear of being lost 'miles from nowhere'. In trying to save Smithy from such a fate, Anderson had died the death Smithy most dreaded for himself.

A doubtful welcome awaited the *Southern Cross* when it eventually landed at Richmond Aerodrome. Smithy refused to speak to reporters except those from the Sydney *Sun*. To the rest, he simply commented that 'all the papers seem to misinterpret all we say. I will say nothing in future.' Silence was a wise decision, because other papers were joining the fray, notably the magazine *Aircraft*. Its editor, Edward Hart, bore a grudge against Ulm. Pulling no punches, Hart called Coffee Royal 'a put-up job'.

Just as a wave of joy had washed over Australia when Smithy and his crew were found, now a wave of grief engulfed the nation. King George sent condolences to the dead men's families, while the prime minister promised a state funeral. Out in the desert, Anderson had left a poignant account of his last hours, written on the rudder of his plane. The message heightened the sense of tragedy. 'No water to drink', Anderson had written, 'no take off able to be attempted … thirst, heat, flies and dust'. 'To me personally,' wrote Smithy, 'this last diary of Anderson's approaches in tragedy and pathos the last words of Captain Scott as he lay dying in his tent in the Antarctic.'

The outcry for an investigation into Coffee Royal became so loud that Prime Minister Bruce could no longer ignore it. He agreed to an inquiry. In reply, the loyal *Sun* pointed out that Smithy and Ulm were not the only pilots to rush off recklessly into the outback. Anderson had also lacked tools, a radio, food and water; why, then, should he not be held accountable? It was a fair question, even though *Smith's Weekly* accused the *Sun* of impugning a dead hero. Bruce saw the justice of the question and decided on a compromise. While the main purpose of the inquiry would be to investigate the *Southern Cross's* disappearance and rescue, its secondary aim would be to inquire into the disappearance and deaths of Anderson and Hitchcock.

On 14 May the Southern Cross Air Inquiry began at the Darlinghurst Courthouse, in Sydney; later the inquiry would hold sessions in Melbourne and Adelaide. Three men, carrying between them an impressive sum of legal and aeronautical knowledge, formed the presiding committee: Brigadier General Lachlan Wilson, a distinguished military lawyer; Cecil McKay, the president of the Aero Club of Victoria; and Geoffrey Hughes, a Sydney solicitor and president of the Aero Club of New South Wales. Among the lawyers were the counsel to the committee, Mr J.H. Hammond, who acted as chief examiner, Frederick Myers, who represented the Anderson and Hitchcock families, and Jack Cassidy, who represented Smithy and the crew of the *Southern Cross*. Seventy-four witnesses were marshalled, mostly connected to aviation, some of them highly placed, such as Lieutenant Colonel Brinsmead. Evidence was also heard from reporters such as Norman Ellison, and observers such as Dr Sinclair, Smithy's physician in Longueville. So avid was the interest throughout the nation that major newspapers carried long stories of its proceedings, and each day the public gallery was filled to capacity – mainly with women. One reporter observed that eleven ladies crowded into the jury box, and their 'mauve, blue and brown hats broke the monotony of the brown walnut panelling'.

It must have cheered Smithy to see his loyal female audience, because the courtroom was like a stage and, for a natural actor like himself, the scene before him was stimulating as well as frightening. He was well prepared, having rehearsed his evidence with his lawyers: a formidable ordeal it would surely be, but he had the skills to meet it. If only his reputation and his future life were not riding on the outcome!

Smithy was the first to give evidence, and he acquitted himself well. Speaking confidently and to the point, he withstood a long and rigorous examination from Hammond. Keeping his wits and his temper, he even managed a joke. While testifying that he had been ill just before the flight commenced, he was asked, 'Did you have a doctor's certificate?' His reply produced laughter: 'All I have is a bill.'

Even though Ulm had been much at fault in the lead-up to the flight, Smithy took full responsibility for all decisions, including their disputed departure on 30 March. At all times he expressed faith in his crew, defending Litchfield against an accusation of faulty navigation, and explaining away, as best he could, McWilliams' failure to convert the radio into a transmitter. Questioned whether he would attempt the flight again, he replied, 'Probably yes, with even more adequate provision.' And when asked, 'Why did you take risks on this flight?' he replied with dignity: 'We are in the position of pioneers, and pioneers always take risks.'

Only once did Smithy's composure falter. When asked if the rumours of a stunt were true, and if he and Ulm and Anderson and Hitchcock were fellow perpetrators, he told the court angrily: 'They are absolute and malicious lies that affect dead men's reputation.' The lies disgusted him, and he was glad to have the chance to denounce them publicly.

By contrast, Ulm was not calm. The ordeal of Coffee Royal and its hostile aftermath, and his own troubled conscience over Anderson's and Hitchcock's deaths, had left him sadly unstrung. In the witness box he was sometimes defensive, sometimes pompous, often indignant and occasionally pugnacious. He once said of himself, 'I am quick tempered and sometimes as blunt as the bows of a cargo tramp,' and that quick temper and bluntness alienated his questioners. At times he became so 'heated' when answering that he was obliged to apologise to the presiding committee.

Ulm faced awkward questions about Chateau's telegram advising against the *Southern Cross*'s departure. After some argument, he made the high-handed admission that he had disregarded Chateau's advice because he feared that Chateau might not give permission for some time, and he was impatient to be off. Then came questions about their days at Coffee Royal. Was Ulm's diary an 'absolutely accurate statement' written while at Coffee Royal, or had it been reworked since then to meet the taste of newspaper readers? Ulm replied coldly that the diary was true, but he knew he was on shaky ground. Differing versions of the diary were circulating, which seemed to support the rumours that it had been embellished for publication. In one of the versions, put together in Derby and published in the Sydney *Sun* on 16 April, Ulm had included a dramatic description of their plight on the day they were found. 'Without any exaggeration,' he wrote, 'we were just on the final point of complete starvation. A painful death by starvation and thirst was not more than three or four days away.' This exaggeration would soon be exposed when their rescuers came to testify.

Ulm was aiming to quash the notion that Coffee Royal had been a type of holiday camp and they had been the perpetrators of a hoax. Understanding his motives, the *Southern Cross* crew loyally backed him up. Excerpts from the diary were read out at intervals in the courtroom, and when questioned about their accuracy, Smithy testified that Ulm's story was 'substantially accurate'. But by exaggerating their suffering, Ulm had weakened his own credibility.

On 3 June there came a brief boost to their flagging spirits. This was the day when Smithy and Ulm were summoned to Admiralty House, each to be invested by the governor-general, Lord Stonehaven, with the Australian Air Force Cross. The impressive ceremony was duly photographed by newspaper cameramen, and for the first time in months the aviators were able to resume the role of heroes. But two days later Ulm was tossed back into the turmoil of the inquiry when an unexpected witness appeared in the courtroom. He was James Porteus, the advertising manager of the Shell Oil Company.

Porteus testified that in May 1927 – shortly before the Around Australia flight – Ulm had boasted in the Customs House Hotel in Sydney that an effective way of gaining money and publicity would be to become lost in central Australia. Porteus said that he had assumed Ulm was joking. However, when the *Southern Cross* became lost, he recalled Ulm's words and passed the story on to Jack Cantor, who in turn passed it to the *Daily Guardian*'s Norman Ellison.

There seems little doubt that Porteus was speaking the truth. Ulm enjoyed presenting himself as an astute strategist, and took pleasure in explaining his clever planning, although whether he would have actually put such a plan into operation is highly doubtful. In ordinary circumstances, his words would have been classed as a type of showing off, but coming to light in such a context, they did him serious harm. The only defence he could summon was denial. He entered the witness box abrim with righteous indignation, and denounced the evidence as 'a deliberate fabrication'. 'Mr Porteus and I have never got on well together,' he added. 'I have two or three times pretty forcibly told him I did not like him.'

The following day, just as the committee was retiring to write its report, two unexpected witnesses came forward, inspired by Porteus's revelations. One was Dudley Lush, the local manager of the Neptune Oil Company, who testified that, in May 1927, Ulm had told him that it would be 'a good stunt' to become 'lost in the wilds'. The other was William Todd, whom Ulm and Smithy had briefly engaged as their navigator on the Pacific flight, but who had fallen out with Ulm and resigned. During the preceding weeks, Todd had been in hospital in New Zealand, but he had kept abreast of the inquiry by reading the newspapers. Incensed by Ulm's denial, he had sailed to Sydney and, finding himself too late to give evidence, had approached Arthur Hilliard. With Hilliard's help, the case was reopened, and on 12 June, Todd gave his evidence.

Todd remembered that, at a meeting at the Roosevelt Hotel in San Francisco in October 1927, Ulm had proposed 'getting lost as a device for financing the Pacific flight'. 'We were discussing some method of getting enough money to get the flight started,' Todd recalled: 'Ulm said that if his plan for getting lost in central Australia for four or five days had been followed, sufficient money would have been available.'

The combined testimonies of these last-minute witnesses were so convincing that Ulm was left almost without a defence. Observing his floundering, the chairman tried to throw him a lifeline. Might Ulm have made the proposition as a joke, he asked. 'Never,' replied Ulm defiantly. Instead, he started attacking Todd. 'I say that Todd is a deliberate liar of the lowest order,' he declared, 'and his statement is malice.' Indeed, Ulm became so heated that the chairman was obliged to caution him.

On 24 June the committee issued its findings. Smithy and Ulm were cleared of the suspicion of having perpetrated a deception. They were nevertheless censured for their carelessness over tools, rations, weather reports, maps, signal fires, the radio aerial and the radio. McWilliams was reproved for failing to turn the radio receiver into a transmitter. As for Ulm, his testimony troubled the committee. It was prepared to accept that his proposal to 'get lost' was not serious. It was prepared to accept his diary was exaggerated rather than untrue. But it did feel obliged to pronounce his testimony 'unreliable'. When Ulm saw the newspaper headline 'Ulm's evidence viewed with suspicion', he realised his darkest hour had arrived.

Looking back, he wrote: 'All my castles in the air, based on a foundation of exhausting endeavour, were placed in danger of complete collapse by the misfortunes of Coffee Royal.' Kingsford Smith's castles were also in jeopardy, and he was far less at fault than Ulm.

HE PURSUED ME
QUITE RELENTLESSLY

The inquiry left the crew of the *Southern Cross* angry, exhausted and despondent. Ulm believed he had been pilloried. To have his honesty questioned was devastating enough, but to be told that the flight was ill-organised, when organisation was his most prized virtue, was an even deeper wound. The previous year, with a professional writer's help, he and Smithy had turned the log he had kept on the Pacific flight into a book, and ample space had been devoted to the careful preparations that preceded the crossing. Readers and reviewers had complimented him on the 'thoroughness' of the flight's planning. To have doubts cast on his capacity for planning was the final straw.

To make the verdict more bearable, Ulm began to ascribe it to envy. 'Some of us will always have enemies because we dare to be individual,' he wrote feelingly, 'and individuality is an offence against mediocrity.' He detected 'a peculiar streak in human nature which prompts mankind to seek the flaws in prominent men'.

Smithy was shocked by the findings, but fortunately could turn a calmer face to the public. 'I think our errors were magnified,' he wrote. 'They were human faults and had we been successful it is improbable that any undue comment would have been made.' Running a temperature, he had been too ill to remain in the courtroom to hear the last part of Ulm's evidence, and Catherine escorted him back to Arabella Street.

His physical collapse seriously worried his mother. During his dangerous flights, she had managed to control her fears by telling herself that his skill and determination would carry him through, but whether he possessed the emotional fortitude to endure this current ordeal she could not be sure. She knew his highly strung nature, and sensed his pain. The best she could do was to rally the family around him and hope that their loyalty and love would provide a shield. But his parents and brothers and sisters were themselves now targets of public opprobrium. A few years later he wrote:

> I cannot forget how certain of my countrymen turned from adulants to defaments almost overnight. Scurrilous letters, unsigned of course, were written to me, my mother and to other members of my family. The public which had been 100 percent 'Smithy' when we took off for the flight, now began tossing mud and rocking the pedestal on which they had placed me.

Unlucky Anderson was now the hero. That the inquiry found him as negligent in his preparations as Smithy and Ulm seemed to count for nothing: heroes could do no wrong. A huge state funeral with full military honours was planned for him, with the Mosman council providing a burial place in Rawson Park, overlooking Sydney Harbour. On 7 July a crowd of mourners – some said there were 6000 of them – marched behind the cortege to the burial site, while three squadrons of planes swooped low to drop flowers on the grave. It was one of the most moving funerals Sydney had seen.

By this time, the *Southern Cross* and its crew had left Australia. Desperate to get away, Smithy and Ulm had scheduled their departure for England on 21 June 1929, but since they required fresh permission to land in, or fly across, foreign countries, another four days passed before they could take off. Litchfield and McWilliams were on board, partly because there was no time for new recruitment, but partly also because the ordeal of Coffee Royal had bound the crew together.

They were desperate to break Bert Hinkler's record of fifteen and a half days to England, for Bert was now, officially, the greatest aviator of 1928. On 20 June, at the International Air Congress in Copenhagen, he received the coveted gold medal of the International Air Federation for his flight to Australia; he was also about to receive the Royal Aero Club's Britannia Trophy, for the best British flight of 1928. Remarkable as Hinkler's achievement was, Smithy and Ulm did not believe that it surpassed their own flight across the Pacific. Their disappointment can be imagined. The best way to raise their spirits was to regard Hinkler's success as a challenge. Accordingly, they announced that the *Southern Cross* would reach England in nine days. Brave words, they lacked reality. So shocked were they by the inquiry's proceedings – and by the committee's report, which appeared on 24 June – that they were unable to focus adequately on the task ahead. Some of their preparations were half-made, or overlooked. It was ominously reminiscent of the previous March.

The *Southern Cross*'s departure from Richmond was quiet. Gone were the vast crowds and speeches: only a bevy of aeronautically minded schoolgirls from the nearby Osborne Ladies' College sped them on their way. To his annoyance, Smithy discovered on entering the cockpit that his helmet, gloves and goggles had been stolen, and he was obliged to find replacements. 'Better have a look and see if the rations are still there,' shouted an observer. 'Yes,' replied Smithy, adding with irony: 'The Air Inquiry Committee should be here now!' He had already informed reporters that, in compliance with to the inquiry's findings, they were carrying an additional wireless, plenty of emergency rations, a shotgun and adequate tools.

Wyndham had been abandoned as their first stopping place, and instead they landed at Derby. They were to resume at sunset on 27 June and follow the shipping lanes to Singapore, but as Smithy fired the engine, a pinion in one of the magnetos on the port side broke. The question now arose whether they should wait a couple of days for a replacement or – since the *Southern Cross* was still capable of flying – take a risk and fly now.

It would have been wiser to wait, but the urge to be away overruled wisdom. By midnight they were flying blind through torrential rain, while the port engine intermittently short-circuited. Nevertheless, they managed to reach Singapore, where they were scheduled to land on the racecourse. As he prepared to set down, Smithy was shocked to realise that, because of the shape of the course, he would be unable to take off from it. He aborted the landing and, after a hurried radio call to the RAF aerodrome nearby, touched down safely there.

The next day they hoped to land on the racecourse at Rangoon (today Yangon), in Burma (today Myanmar), but once more they ran into trouble. They had forgotten that they would be arriving on a Saturday – a race day. Hastily, they radioed the Thai port of Signora and made their landing there. The following day they battled blinding rain and the short-circuiting engine as they flew across the Malay Peninsula to Rangoon. There the magneto was temporarily repaired, and the next day, after crossing lower Burma and the Bay of Bengal, they touched down at Dum Dum Aerodrome, on the outskirts of Calcutta (today Kolkata).

The departure from Calcutta on 3 July was a nightmare. The previous night, Smithy had noticed a line of bamboo stakes at the end of the runway, and had ordered their removal. They were cut down overhastily, and as the *Southern Cross* rose into the air, its underbody was gashed. Soon after, they flew into such a wind that there was no way they could reach Karachi without refuelling. Fortunately, Bamrauli Aerodrome at Allahabad assured them of landing rights and a fresh supply of petrol.

Thereafter, all was confusion. At Allahabad, Smithy missed the aerodrome and ended up landing on the parade ground of the Royal Warwick Regiment. The officers thought this a tremendous joke and offered to help, for which Smithy was 'very grateful as the thermometer registered 118 degrees in the shade'. But nobody could restart the *Southern Cross*: at each attempt, one propeller remained stubbornly still. Smithy had had enough: 'I'm at the end of my resources,' he told bystanders. Ulm took over, and couple of hours later managed to start the engine. The crew and some of the officers climbed aboard and they made for Bamrauli, where they enjoyed a lively celebration. At dawn they started out for Karachi, and were lucky to complete the take-off. As they rose into the air, they almost collided with a half-hidden wire fence.

At Karachi – today in Pakistan, but then in the west of British India – the RAF mechanics were waiting to overhaul the plane. 'Never since the *Southern Cross* started on its adventurous journey,' Smithy told reporters, 'has it given so much trouble.' There was no way now that they could complete their flight in nine days: the best they could hope for was fourteen days. However, Ulm gained encouragement from their arrival in Karachi. He foresaw a time when Australian National Airways might fly mail and passengers from Sydney to Karachi, to connect with the Imperial Airways service already operating from Karachi to London. He stored the idea away for future development.

Originally, they had planned to travel nonstop from Karachi to Baghdad, and from Baghdad to Rome, but the frail state of their port engine made this impossible. On 6 July they flew along the desolate Persian Gulf and landed at Bandar Abas, where the British vice-consul helped smooth their way with the unfriendly Persian officials. Taking off the next morning in a dust storm, the port engine again showed signs of failing, although the culprit this time was a valve, not the pinion. Fortunately, they were almost at the RAF Depot near Basra, so down they went to drink beer in the mess while they waited for more repairs. At Baghdad, where they spent the night, they were almost devoured by fleas. Baghdad fleas were remarkable for their size and vivacity, Smithy reported solemnly.

The next evening, landing in Athens, their spirits rose, for they still had a chance of beating Hinkler's time. But they had not reckoned on Athenian red tape. The following morning they were not permitted to leave: the commandant was away and they must wait for his return. 'And when will that be?' Smithy enquired. Nobody knew. Smithy was not in a mood to be thwarted by officialdom. 'Well, since we must wait,' he said craftily, 'I should like to taxi the machine over the ground to see if it is firm enough to carry us.' The officials agreed and helped swing the propeller. 'Now,' shouted Smithy as the engine sprang to life, 'we will see if it will carry all four of us and our baggage.' At this, the crew jumped in with their bags and they took off for Rome. It was 'a shabby trick', Smithy admitted later, but he felt no regrets.

In Rome, Doc Maidment, summoned urgently, was waiting to cure their port engine; and such was his skill that on 10 July 1929, they finally landed at Britain's newly opened Croydon Aerodrome. The director of civil aviation, Air Vice Marshall Sir Sefton Brancker, was there to greet them, but the welcoming crowd was sparse, and confusion as to their identity prevailed. The aviators were amused to overhear an argument: 'I tell you, they flew in from Australia,' said one voice. 'No, they are from America,' said another. Then a third voice chimed in: 'I'll bet they are British because if they had been Americans there would have been a hundred thousand here to welcome them.'

Their welcome may have been quiet but there was pride in their achievement. The *Daily Express* described it as 'a victory snatched by the power of the human will from the very jaws of defeat'. The *Southern Cross* had made the journey from the West Australian coast to Croydon Aerodrome in the record time of twelve days, twenty-one hours and eighteen minutes – more than two days ahead of Hinkler's time.

Over the following weeks, in between celebratory dinners and lunches, Smithy and Ulm purchased, on behalf of their airline, four Avro 10 trimotors. Their Fokker bodies were almost identical to that of the *Southern Cross*, but their Lynx engines came from the British Armstrong Siddeley factory. Describing them as luxurious saloon passenger aircraft, Ulm put their cost at £50,000. Meanwhile, the condition of the *Southern Cross* was causing concern. Smithy was still determined to fly across the Atlantic and on to California, and so complete his circumnavigation of the world, but if the Old Bus was to attempt such a feat, she would need a thorough overhaul. He could imagine no better engineers than those who had constructed the plane, so he arranged for the *Southern Cross* to be overhauled at the Fokker factory in Amsterdam. He knew that such an overhaul might take many months, but he was prepared to wait.

Smithy and Ulm flew to Amsterdam and made a triumphal entry into Schiphol Aerodrome, escorted by two Dutch squadrons of Fokker fighter planes. Intending now also to visit Germany, the two aviators travelled by train to Berlin, where they received an 'enormous welcome' from President Hindenburg. Berlin was memorable also for their meeting with an unusual Australian couple: Lindsay Fabre and Aussie, his boxing kangaroo, who were performing on the European vaudeville circuit. Aussie had a liking for aviators, and charming photos exist of Smithy and Ulm shaking Aussie's outstretched paw.

Returning to England, Ulm arranged for the shipping of the Avros to Sydney and began to interview prospective pilots, while Smithy began to plot air routes across the Atlantic to California. Several of the routes he considered were via South America and inspired by the 1927 crossing of the South Atlantic by Dieudonné Costes and Joseph Le Brix. His final choice was a nonstop flight from Amsterdam to New York, with a further flight to Oakland to complete the circle. Late in September he sailed for New York on the liner *Amsterdam* to make preparations.

Smithy was eager to talk to Anthony Fokker, now a United States resident and president of the Fokker Aircraft Corporation of America. They had met briefly in California in 1927, when the Pacific crossing seemed so hopeless, and he remembered the confidence Fokker had shown in him then. When they met now in New York, Fokker showed an even greater confidence. He handed Smithy a cheque for £1000 towards the Atlantic venture, and offered him the *Southern Cross*'s overhaul free of charge. It was a wonderfully kind gift and most gratefully received, for, notwithstanding the huge sums that had come to him after the Pacific flight, Smithy was again short of cash.

Warmed by Fokker's generosity, he set out for California and his brother Harold's house with a lighter heart. The wounds inflicted by Coffee Royal were beginning to heal, and his mood was reflective. Events had overtaken him so swiftly in the last few years that he had had little chance to take stock of his life. Now, travelling by train across America, he had the time and inclination to evaluate his past – and decide upon his future. He realised he was at a crossroads.

After completing his circumnavigation of the world, where was he to go? He was conscious that aviation was changing. Previously, the great aviators had been men who – often in pairs or even fours – had flown vast distances in conditions so hazardous that few had believed such flights were possible. By late 1929, most of the long-range, pioneering flights had been accomplished, and record-breaking flights were starting to take their place. Strong male – and even female – pilots in their twenties, eager to break speed records, were taking off in small, fast planes, flying solo for twelve hours at a stretch and sleeping barely four hours a night. They faced untold physical and mental strain, but they earned glory for themselves, and at the same time convinced the public that one day passengers and mail would be transported speedily, safely and seemingly effortlessly to the other side of the world. Breaking records was a tough game in which one needed the health and strength of youth, and at thirty-two, Smithy wondered if he might be too old for it. Nevertheless while in England, he had felt sufficiently sure of himself to discuss the purchase of a fast single-engine biplane from the factory of A.V. Roe.

He also knew the time was fast approaching when he must settle down. His upbringing had conditioned him to believe that he should have a wife and children; indeed, he loved children. But since Thelma's departure, the problems of accommodating the demands of flying with the demands of a wife had always seemed beyond him. If only he could find a woman who could learn to understand his ways and accept his needs. For he could not readily give up his dangerous profession, and he might not even manage to stay sexually faithful.

Family life had never seemed more important to him than during that week he spent with Harold in California. He had long been fond of Harold's daughter Beris, a strikingly attractive girl of twenty-five, and he decided that it was sad that she had so little knowledge of her extended family in Australia. Now he proposed to take her on a visit to Sydney to see her grandparents and uncles and aunts. But first she must accompany him to Vancouver, where he was about to visit old friends and his father's old Masonic lodge. Beris found her uncle wonderfully entertaining and quickly accepted his invitation. By mid-October they were enjoying themselves in Vancouver.

Smithy had booked passages to Sydney for himself and Beris on the *Aorangi*, leaving from Vancouver on 16 October, but a lunchtime speaking engagement at the Canadian Club almost caused them to miss the boat. The captain was kind enough to delay the ship's departure; as they hurried up the gangway, he and Beris were aware of the stares of disapproving passengers. Among those staring – although with interest rather than disapproval – were a Melbourne family, Florence and Arthur Powell, and their daughter Mary. Blue-eyed, brown-haired and the possessor of a 'winsome smile', Mary was fresh from boarding school, and accompanying her parents on a trip around the world. Many years later, she told the biographer Ian Mackersey: 'Daddy called me to the rail, saying, "Here's Kingsford Smith at last." It was a bit like saying, "Here's the King." I went over to look. I remember my absolute amazement that, in the flesh, he was such a little man.'

That evening, in one of the ship's spacious lounges, Smithy sat beside Beris and evaluated the 'female talent'. Mary was wearing a red dress, and when Smithy saw her, he confided to Beris, 'The one in red will do me.' From then on, in Mary's words, 'he pursued me quite relentlessly'. She was flattered by his interest. He was by far the most famous person on board, and always the centre of excited attention. But, she added, 'he was so informal and such fun to be with it was impossible not to enjoy his company. He had such a magic personality. I wasn't yet nineteen, naïve as they come, and here was this man of the world, this god person, coming up to thirty-three, who was staking me out.'

Smithy wasted no time in joining Mary and her young friends. 'We'd all gather round him,' Mary remembered, 'and he'd regale us with priceless stories, with brilliantly mimicked foreign accents, and we'd sing to his ukulele.' She marvelled that he never showed even a hint of conceit or self-importance. 'As for his great flights, you had to drag the details out of him.' On their fourth day at sea, Smithy asked if he could meet Mary's parents that evening for coffee. He came in his squadron leader's dress uniform, looking splendid. He tried to explain his unusual choice of clothes by pretending that he had accidentally spilled water on his dinner jacket, but it was untrue. He wanted to impress them, and he succeeded. The Powell parents were not only impressed, they whole-heartedly liked him.

Smithy began to court Mary in earnest, but was careful to behave with the utmost delicacy. Determined to do nothing that might frighten her, he confined their love-making to romantic kisses on the boat deck. As Mary later put it: 'I think it was my artlessness, my naturalness, that he found appealing.' One night, as they steamed towards Suva, he asked her to marry him. 'I was stunned,' she remembered. She was also elated: here was a man who could have had almost anyone he wanted. Although at first she refused him, she soon found herself yielding. Pressing his advantage, he confronted her father and formally asked for her hand.

Her father was less elated, for although the newspapers of the day were too discreet to mention Smithy's sexual affairs, they were common knowledge among many Australians. Mary was Arthur's youngest child by five years, and the apple of his eye. Naturally, he wanted only the best for her, hence her expensive schooling and the trip overseas. And of course he was determined to protect her. He is reputed to have said, 'You've got a damned cheek. You're much too old for her.' Smithy was not deterred; he confidently persisted, and his open friendliness was disarming. By the end of the voyage the Powell parents had so far weak-ened as to permit a year's courtship, with the possibility of marriage at the end of that year – if both were still of the same mind.

When the *Aorangi* berthed in Sydney in early November, Smithy's family was on the wharf to meet him. Almost his first words to his mother were an excited announcement that he had finally met the girl he intended to marry. 'Oh dear, not an American,' his mother is said to have replied, plaintively. To reassure her, he brought Mary to Arabella Street. It was a wise decision, because she and the Kingsford Smiths hit it off immediately. They were 'so full of fun', Mary told Ian Mackersey, so 'happy go lucky'. She could not fail to like the family.

Some have whispered that Mary and her family were above Smithy socially. It has been pointed out that her father was wealthy, and that she had been a pupil at The Hermitage in Geelong, a sister school of the prestigious boys' school Geelong Grammar. Smithy later endorsed this idea of social superiority by telling reporters in America that his future father-in-law was a 'rich rancher' and successful horse breeder, although in fact he was speaking of her uncle rather than her father. This incident has caused some to assume that Mary was the socialite daughter of a rich Western District pastoralist, and a member of Melbourne's highest society. It was a mistake. For the first years of her life, Mary and her parents lived in the working-class suburb of Footscray, close to one of Australia's largest horse and cattle saleyards, and near Australia's most famous racecourse. Both venues would figure importantly in the life and fortunes of her family.

Mary's grandfather, William Hamilton Powell, had been a pioneer of Footscray, arriving in gold rush days to make his name as a builder and hotel keeper. He was also skilled with horses, a talent he passed on to some of his sons. Young Gus Powell – Mary's uncle – earned fame throughout Australia as a champion rough rider, as fine a horseman, it was said, as any in Buffalo Bill's circus. From such a beginning, Gus went on to breed and train racehorses, eventually becoming one of Australia's most respected and successful owner-trainers. In fact, his horse Mosstrooper was one of the greatest jumpers in the history of the Australian turf, winning almost £13,000 in prizes by August 1930. The large Powell family had profited from the horse's success, and the trip around the world undertaken by Gus's younger brother probably owed something to Mosstrooper's career.

Gus, and to some extent Arthur, also exported 'walers', the name given to the Australian horses sold to the British Army in India and South Africa, and to Indian Maharajahs as polo ponies. Until cars and trucks displaced the horse in the late 1920s, the waler trade was highly lucrative, and Gus and Arthur often travelled to India to sell their horses. Arthur's business interests, however, were wider than Gus's. On the electoral rolls he gave his profession variously as contractor, exporter and horse dealer. He also became one of the creators of the flourishing Melbourne engineering business Die Casters Limited, of which he ultimately became the chairman of directors.

Although Smithy was preparing for the opening flights of Australian National Airways in January 1930, his mind was in Melbourne with Mary, and he could not wait to travel down there. Mary was to turn nineteen on 28 November, and her birthday was to be celebrated with a grand dinner at the fashionable Australia Hotel. Unable to curb his impatience, Smithy bought a large diamond engagement ring. With the ring in his pocket, he arrived at the Powells' apartment in the Melbourne suburb of Hawthorn, determined to carry all before him. By the end of that evening Mary was officially engaged to marry him, and the wedding was set for September 1930.

Once the news was out, the reporters garnished it. The Powells were not well known in social circles, and not at all known in Sydney, however Arthur Powell's eldest sister, Jessie, was married to Herbert Gepp, an influential industrialist who lived with his family in Sydney's Elizabeth Bay. The Gepps provided that aura of social prominence that the Sydney newspapers were seeking, and columns of newsprint were devoted to Smithy and his bride.

To convey himself and Ulm to Melbourne for the birthday celebrations, Smithy had commandeered one of their airline's planes, and although the service had not yet officially started, he carried six passengers, charging them ten guineas each for the round trip. Mary flew back to Sydney with them, intending to stay for a time with the Gepps. Dressed in a blue silk dress and brown fur coat, she showed surprising confidence in handling the reporters. This was her first flight, she told them, and it was 'a jolly good trip' and she 'loved every minute of it', although she had to admit that 'it was bumpy in parts.' Perhaps she would have been less confident if she had known more about the relentless scrutiny of the press. 'I lost my privacy forever that day,' she later said. 'I had no idea what it was about to do to my life.'

Smithy watched his future bride with affection and pride. Observing how capably she handled herself in the air and on the ground, he felt sure that he had made the right choice.

RIGHT AROUND
THE WORLD

Ulm had returned from England in October 1929, and was continually busy. He had overseen the building of the Australian National Airways hangar and workshop at Mascot Aerodrome. He had supervised the passage of the aircraft bodies and engines and wings from the wharf to the workshop, and had hired forty-three employees. He had even chosen the colour of the pilots' uniforms, as well as that of the planes. It was Smithy's task, having arrived in November, to check the skills of the pilots – mostly former Royal Flying Corps or RAF flyers – and to test-fly the Avro 10 trimotors once they were assembled. As soon as he was satisfied, he piloted a series of short joy rides to publicise the airline. He also helped Ulm name their planes. Since they were sister ships to the *Southern Cross*, they became the *Southern Cloud*, the *Southern Star*, the *Southern Sky* and the *Southern Moon*. Later they would be joined by a fifth plane, the *Southern Sun*.

The start of the airline's passenger service was planned for eight am on New Year's Day of 1930, and, given the occasion's importance, the departure from Mascot was surprisingly quiet. Smithy arrived early, bringing with him his 77-year-old father, who had scarcely flown before and was in a state of high excitement. 'They say I am old enough to know better,' he told reporters, 'but I am satisfied to go anywhere with my boy.' He was helped into the cabin of the *Southern Cloud*, along with the seven young male passengers, while Smithy and his co-pilot, George 'Scotty' Allan, waved goodbye to a small crowd of wellwishers, many of whom were in evening dress, having come directly from New Year parties. Five and half hours later the passengers were disembarking at Eagle Farm. They had struck bumpy weather along the coast near Ballina, but William remained calm; indeed, he was reported to be 'happy as a sandboy'.

The company's motto was 'safety, comfort and dependability'. Its reassurance was timely because many people were afraid of flying. A mishap on another of their flights on the inaugural day was therefore a serious matter. The *Southern Sky*, piloted by Ulm and Paddy Sheppard, left Eagle Farm at eight am, but received such a buffeting from the wind and rain that Sheppard was obliged to set down in a paddock at Bonalbo, breaking two wire fences and ending up amidst fallen trees. Thanks to Sheppard's skilful handling, no one was hurt, and the bags of airmail and eight passengers – Mrs Ulm and Captain Chateau among them – were motored to Casino and sent on by train to Sydney.

The problem remained of how to return the plane to Mascot. Even after on-the-spot repairs, a take-off seemed impossible in so small a space. Smithy ingeniously solved the problem by tying a rope to the tail of the plane and attaching the rope to the stump of a tree. He advanced the plane until the rope was taut, then revved the engine and signalled to an assistant to cut the rope. At once, the plane was catapulted sufficiently high and fast to be airborne. The skill and daring he displayed in this take-off were to enter the annals of aviation history.

A few mishaps were almost inevitable. In January the *Southern Moon* became bogged on the runway at Eagle Farm, and in April a plane was forced down near Taree, again with no injuries. Smithy was now expert at blind flying, but other pilots were less so – although all were personally checked by Smithy for their blind-flying skills – and stories abound of near misses in fogs and storms. The public seem to have accepted the risks, however, for bookings were multiplying and many passengers expressed themselves well satisfied. A letter from a Mr R.A. Cornell to the *Urana Shire Advocate* painted a vivid picture of the flying experience. He was booked to fly from Sydney to Brisbane in the *Southern Moon*, departing in the Australian National Airways' bus from Challis House in Martin Place at seven-thirty in the morning.

At Mascot he and his bags were weighed – he was allowed thirty-five pounds of luggage – and he was seated in his wicker chair, which had arms and a cushion. The seats were ranged in two rows of four, and none had seatbelts, though at least they were bolted to the floor. Beside each seat were earplugs of cotton wool, sick bags, peppermint sweets, daily papers and route maps. Plate-glass windows ran down the cabin sides and could be opened during the flight; indeed, sometimes passengers threw their sick bags out the window, only to see them blow back in. At the rear of the cabin was a tiny lavatory, with paper towels and a mirror, and a sign forbidding its use when the plane was above populated areas.

Cornell was astounded by the noise of the engines. 'I thought we would all be blown up before we got started,' he wrote. Once in the air, the noise ceased to disturb him, for what magical sights lay beneath: they even made him forget the drought and the low wool prices. Below was farming land set out like a huge draughtboard, with every crop a different colour. Houses and cattle looked like something in a child's Noah's Ark. He was 'spell bound', except for the bumps and drops they encountered while entering Brisbane. One man was sick and everyone was scared, but that was a small price to pay for five hours of wonder. The entire trip was a revelation – 'as near to Heaven,' he wrote, 'as I can expect to get.'

A passenger named Mary Anne also expressed her delight. She had felt apprehensive as she took her seat in the cramped cabin, but Smithy's cheeky banter lifted her spirits. 'Don't worry, girls,' he told Mary Anne and her companion. 'And don't jump until you see me getting ready to jump. The chutes are above you and the paper bags are on your right. You'll probably need them first.' The girls, realising that he was teasing them, giggled so hard they forgot to be afraid. 'I loved him for that,' Mary Anne remembered.

Thanks to Ulm's energy and competence, the new airline boomed. By the end of June 1930, it had carried 2919 paying passengers and

291,840 airmail letters. Of the 426 flights that had been completed – two were abandoned because of bad weather – only eleven ran behind schedule, and no passenger was injured on any flight. Furthermore, on the Sydney–Brisbane route, nine of every ten flights carried their full complement of passengers. With this enviable record, Ulm was hoping to extend the daily service to Melbourne and to Hobart. Smithy and Ulm were constantly congratulated, and the dark shadow of Coffee Royal seemed to be fading.

Although Ulm complained of Smithy's absences from the office, overall he was content. The airline was bringing him the emotional and financial rewards he craved, and the worries of the past few years could be forgotten. Smithy, on the other hand, was not content. Like Ulm, he was hoping to leave past worries behind and enter a new life, but his hopes of happiness were largely pinned on Mary, and here he faced a dilemma.

A difficult flight was the breath of life to him, but she could not bear to think of him undertaking anything that might be truly dangerous. The long-range ocean crossings worried her particularly, and she begged him to give up his projected Atlantic crossing. He refused to give it up. He explained that it was his 'supreme ambition' to encircle the globe, and besides, the flight was already organised, and in a few months he must sail to Europe to collect his plane. Reluctantly, she accepted his answer – but what of the flights that might come after it? She could not spend her married life in a state of perpetual anxiety. She made him promise that the Atlantic would be the last of his ocean crossings. He gave his word, but he doubted he would be able to keep it.

Ulm had assumed that he would be the co-pilot on the Atlantic flight, and Smithy seems to have agreed, although one wonders if in the depths of his heart he really wanted Ulm. He was beginning to see that his co-commander was not the rock of dependability he had once thought him. As the Coffee Royal inquiry had revealed, Ulm's urge to dominate, his misguided attempts at publicity and his uncertain temper were becoming somewhat of a liability. An unguarded comment to a reporter named Robert Thomas revealed something of Smithy's uneasiness. Thomas had asked if Smithy was currently tied to a newspaper. 'Not on your life, boy!'

replied Smithy. 'I've finished tying myself up to newspapers. That was Charlie Ulm's stock bloomer. Far worse than twelve days' starvation at Coffee Royal was being told you'd managed a good story for the "Sun"!'

Fortune was on Smithy's side when Ulm was told by the Australian National Airways board that he could not be spared, being deemed essential – as indeed he was – for the daily running of the airline. Smithy's spoken regrets at Ulm's absence were half-hearted. Nothing was said openly, but there seems little doubt that Ulm realised the situation, and was hurt by it. A number of his pilots were aware of his feelings. Some years later, one of them wrote:

> I do know very definitely that during the ANA days, Ulm became extremely resentful that Smithy wasn't playing fair. It was pretty disgusting the way he left everything to Ulm, then cleared off for months to fly the Atlantic in search of more glory. Ulm used to complain that Smithy was just a passenger.

To this, Smithy might possibly have answered that Ulm would have been angry if Smithy had been more than a 'passenger'.

Smithy sailed on the *Sonoma* on 15 March 1930, bound first for America and then for the Fokker factory in Holland. It was an emotional farewell. Mary wept, and even stoical Catherine was close to tears. Catherine, however, had an additional source of anxiety; William was unwell and his symptoms suggested cancer. She dreaded what might lie ahead, but she accepted that Smithy must go. Like Mary, she knew that Smithy 'wouldn't be happy', until he had 'done it'.

After a quick visit to Harold in California, Smithy was in New York by late April, ready to board the *Statendam* for Rotterdam. The voyage was made memorable by the presence of the eminent Dutch-American writer, scientist and historian Dr Hendrik Willem van Loon. Smithy at once fell under 'the charm and grace of his personality' and a warm friendship developed between them. Nevertheless, as soon as he arrived in Holland he focused wholly on the *Southern Cross*. The plane had been dismantled, overhauled and reassembled, under 'the loving care' of the

Fokker staff and 'the incomparable Maidment'. She was waiting for him, 'glistening and glittering in her shiny new paint of dark blue and silver, emblem of the constellation against the dark blue of the night sky'. 'With a feeling of exaltation', Smithy put his 'new beauty' through her paces by completing a sensational loop over the Schiphol Aerodrome.

His rapturous reunion with his Old Bus intensified his search for a co-pilot. As a compliment to the Fokker company, he decided to engage a Dutch pilot, and turned to the Dutch national airline, KLM. His first choice was a pilot named Biekman, but Biekman's wife forbade it. His next choice was Evert van Dijk, an experienced Fokker pilot with skill in blind flying. A stern, imperious man who spoke halting English, Van Dijk's forbidding manner was softened by his hero-worship of Smithy; he was almost overcome at being chosen by such a celebrated aviator. To select his radio officer, Smithy made a quick visit to London, and there had the good fortune to locate a radio technician he already knew. He was John Stannage, whom Smithy had first met in California back in 1928, and who had been manning the radio on the aircraft *Canberra* when it sighted the stranded crew at Coffee Royal.

A New Zealander by adoption, and a radio officer on several ships before joining the *Canberra*, Stannage had been visiting his parents in England when he heard that the post was vacant, and immediately visited Smithy at the RAF Club in Piccadilly. Smithy gratefully accepted him for two reasons: firstly because he liked him and recognised his competence, and secondly because Stannage was even shorter than Smithy and weighed not much more than nine stone, a bonus given that space and weight were serious concerns.

Smithy and Stannage boarded a KLM flight to Amsterdam, where they collected the *Southern Cross* and flew themselves back to England, there to acquire a new all-wave radio receiver and a powerful 600–800 meter radio transmitter, both of which Smithy considered highly desirable for a safe ocean crossing. On hearing that he was expected to wait several weeks for the transmitter's delivery, Smithy contacted Signor Guglielmo Marconi himself, and arranged for its delivery the next day. Intent on avoiding the mistakes made on his two previous flights,

Smithy was striving for maximum efficiency; and Stannage, who like most had believed Ulm was the vital organiser in their partnership, was surprised by Smithy's managerial skills. Regardless of how easygoing he seemed in everyday life, wrote Stannage, in aeronautical matters Smithy was meticulous, decisive and well informed.

Smithy had elected to make his crossing to New York not from Amsterdam – as previously thought – but from Ireland, so they flew the *Southern Cross* to Baldonnel Aerodrome, outside Dublin. It was there, with the help of the Irish Aero Club, that Smithy found his navigator, Jonathan Patrick (Paddy) Saul, a 'genial and ebullient' master mariner and former Royal Flying Corps pilot. Smithy was relieved to recruit Saul so easily, because he dared not waste any more time. He had just discovered that two skilful French aviators, Dieudonné Costes and Maurice Bellonte, were preparing to make an Atlantic crossing from Paris.

This French attempt should not have surprised him. Over the years many flyers had tried to cross the Atlantic nonstop, beginning in 1919 with the successful duo, Alcock and Brown. Most of the flights had been from west to east – from North America to Ireland or England or Europe – thus reaping the advantage of the strong westerly winds. Only one team had managed to reach North America from the east, and that only partly so: a plane named the *Bremen* had made a crashlanding on an island off the Quebec coast in April 1928. The danger of the east–west route was well known. Half a dozen flyers had lost their lives in the attempt and some of them were airmen whom Smithy had known. How do you rate your chances of success, asked a reporter. Smithy answered: 25 to 1.

Poor weather further delayed the *Southern Cross's* departure. For three weeks Smithy and his crew waited for a better forecast, all the time fearful that the French aviators might leave before them. Admittedly, the waiting was not arduous, for they were grandly entertained by the 'charming ladies of the Curragh'. Smithy had planned to leave from the Curragh, the wide plain of County Kildare, before he discovered the superior virtues of the beach at Portmarnock, just north of Dublin.

It was from Portmarnock beach that the heavily laden plane eventually set out in the early hours of 24 June. Smithy intended to follow the Grand Circle Route, passing over Galway Bay, on the west coast of Ireland, and then crossing 1900 miles of open sea to Cape Race in Newfoundland. From there they would head for the coast of Maine, and then follow the coastline to New York City.

Smithy had been warned of headwinds and fog, but since he was increasingly impatient to leave, he accepted the risk. He could not have been surprised when, at about three in the afternoon, they met a belt of fog that became denser by the moment. By 3.03 pm the sea was lost to sight, as were the sun and stars, thus depriving Saul of his sextant readings. Twelve minutes later Smithy was flying blind. It was fortunate that Stannage could gain radio bearings from the ships sailing below them, otherwise they would have been lost in the fog. Smithy later told the readers of the *New York Times*:

> Stannage would pick up a ship south-east of us and ask for our bearing in relation to that ship. Getting that position, he would then fish out another ship, say, to the north-east or south-east, obtain another bearing, and turn the figures over to Saul, who plotted them on a chart. The intersection of these two position lines would then be our position, not only in relation to the two ships, but approximately to the rest of the world.

Smithy also acknowledged his debt to the liner *Transylvania*, which became in a sense their mother station. 'Other ships,' he wrote, 'were asked to take bearings of the plane's signals at one minute intervals and keep the Transylvania informed of the result. Periodically these were flashed to the Southern Cross and Saul was able to plot an almost continuous check of his position on the chart'. By this means, the *Transylvania* 'practically navigated the Southern Cross across the ocean, by means of her radio'. It was an historic moment: the first time a plane had been steered across an ocean by a ship rather than by the sun and stars.

Meanwhile, the fog remained their enemy, and there seemed no method of overcoming it. When they flew above the fog, the headwind proved so strong that they could make only thirty-five miles per hour.

When they flew below it, they were obliged to drop so close to the sea that their radio aerial hit the water, and they were forced to climb. Never in his life had Smithy been required to fly for so long solely relying on his instruments, and he judged himself lucky to have an experienced pilot like van Dijk sitting beside him.

Hour after hour they pushed on, trusting their instruments, until a message from the rear cabin caused near panic. Saul and Stannage had discovered that the radio bearings were not coinciding with their compass readings. 'I shall never forget,' wrote Stannage, 'the nasty numb feeling when I began to realise that all the compasses were being affected in some mysterious way, and that we were lost.' Desperate to find the cause, Saul began taking apart one of the compasses.

Throughout the ordeal Smithy remained calm, coping as best he could with the faulty compasses, but unwittingly flying in circles around the southern tip of Newfoundland. When they finally managed to gain a true reading from the *Transylvania*, they were 200 miles south-east of Cape Race and their petrol was low. All hopes of reaching New York in one hop were dashed. A nearer landing place must be found.

Their first thought was to contact the Cape Race radio station, but 'the big stiff won't answer', Smithy exclaimed angrily over the airwaves. When the big stiff did answer, they learned that the only landing field clear of fog was at the tiny fishing village of Harbour Grace. But how to find Harbour Grace? Despite their urgent requests, no plane came forward to guide them in. Then, miraculously, through a gap in the fog they glimpsed a rocky coast that Saul identified as the Avalon Peninsula. Flying as low as possible, Smithy saw cottages, boats and people waving white sheets. The *Southern Cross* had found its safe landing place.

They touched down at eight am local time. They had been thirty-one and half hours in the air, and were cramped and exhausted and deaf from the engine noise. Smithy was so stiff and sore that he could barely climb from the cockpit. 'It had been,' wrote Smithy later, 'a long, tedious and at times nerve-racking journey.' He judged it the hardest flight he ever made.

A HIT AND
NO QUESTION OF IT

T he exhausted crew, after sleeping for nearly twenty-four hours, awoke to a cascade of enthusiastic reviews and telegrams. The two that Smithy treasured most were from rival Atlantic flyers. Herman Kohl, of the unsuccessful *Bremen* team, cabled: '*Donnerwetter*! It is a tremendous achievement.' And Dieudonné Costes, still waiting in Paris, generously told reporters: 'It was a magnificent performance.' Two months later, Costes and Bellonte, learning from Smithy's mistake, would fly nonstop from Paris to New York.

Throughout the journey, Stannage had transmitted regular bulletins to the *New York Times*, and the radio networks WOR and WABC had broadcast their progress to stations across North America, which in turn relayed it to England, New Zealand and Australia. Thus the citizens of the world had shared their journey, and none more so than Mayor James Walker of New York. He was preparing them a heroes' welcome in the form of a tickertape parade through the city streets, and for this the aviators were unprepared.

Limits on weight had prevented the four flyers from bringing personal luggage, and the only clothes they possessed were the fur-lined flying suits and boots they were wearing. If they were to do justice to a parade, they must find better clothes. Stannage sent a radio message to their New York hosts, requesting that four blue suits and the appropriate shirts, ties, shoes, socks and underwear should be waiting for them at their hotel. He transmitted their sizes for guidance: van Dijk was six feet tall and weighed twelve stone, two pounds; Saul was five feet, ten inches, and weighed eleven stone, four pounds; Stannage was five feet, six inches and weighed nine stone, four pounds; and Smithy was five feet, eight inches and weighed ten stone, two pounds. Of course, Smithy was exaggerating: according to his Army records, his height was five feet, six and a half inches. But he was sensitive about his height and liked to add a couple of inches – even if it meant that his trouser legs were rather too long.

The aviators needed to be properly dressed because their arrival had set America abuzz. The *New York Times* had already devoted three pages to their crossing, and was preparing three more for their arrival. As they flew over Boston, they were joined by aircraft carrying reporters and photographers, and the moment they landed, more newshounds were snapping at their heels. Begrimed from the flight and sweating in his flying suit, Smithy scrubbed the oil from his face and made a speech to the assembled crowd. 'This is a great and glorious moment,' he told his audience, going on to thank the people of America, Newfoundland and Ireland, and the radio operators of the ships who had helped the crew on their way.

It had been hoped that the aviators would arrive early enough to take part in the normal city parade, ending with a formal mayoral welcome. As most celebrities came to New York by ship, the usual automobile route on such occasions started at Pier A at the docks and followed Broadway up to City Hall, with police escorts and showers of tickertape. An amphibian plane had been parked at Roosevelt Field to convey the flyers to the docks so they could commence their ride. However, their late arrival and their unkempt appearance meant that the mayoral welcome had to be postponed. Instead, a phalanx of police escorted them to the Roosevelt Hotel in Roosevelt Square, where photographers snapped them bathing and shaving and pretending to be asleep in their beds.

Soon after noon the following day, the aviators, dressed in their new blue suits, boarded two open cars for their triumphal welcome. Smithy had refused the invitation to start from the docks, preferring instead to proceed down Broadway from 44th Street to City Hall Square, escorted by thirty police motorcyclists with sirens blaring. Large crowds of spectators had already assembled along the route: 5000 were said to be waiting outside their hotel, and 10,000 more were lining the route to City Hall.

Some have claimed that Smithy did not receive a true tickertape parade. The written reports and photographs and newsreel films prove that claim is false. Clouds of tickertape floated down from the adjacent skyscrapers, prompting one reporter to describe the Woolworth Building as being 'lost in a snowstorm'. All along the way, people shouted and waved and cheered and threw paper, 'like a bunch of kids having a grand time'. Stannage and Smithy entered into the spirit of the festivities, and at cries of 'Charlie, Charlie' Smithy stood up in the car and waved excitedly. But Saul and van Dijk seemed uncertain of what was expected of them, and are said to have looked 'a trifle embarrassed over all the fuss that was made'.

City Hall Square had been packed for hours, with 200 patrolmen maintaining order. At the centre of the square stood a flag-draped pagoda, on which were assembled an array of microphones, for the speeches were to be broadcast across the United States, Canada, England, New Zealand and Australia. As he presented the aviators with their Medals of Honor, Mayor Walker began a laudatory speech. 'It is such deeds as yours,' he solemnly pronounced, 'deeds performed not for vainglory but in the interests of science, that forge new links and the bonds of amity and concord.' He had scarcely uttered these words when a grandstand holding the photographers partly collapsed, and some of the cameras fell to the ground. The mayor, with presence of mind, cleverly ad-libbed about precarious platforms in American politics. And Smithy, with equal presence of mind, joked into the microphone about 'the excellent exhibition of a tailspin' they had just witnessed. The mayor's and Smithy's quick thinking turned an embarrassment into witty repartee, and the thousands of listeners across the world were impressed.

Later that day, at the Roosevelt Hotel, Smithy made a roundabout telephone call to Mary. According to the *New York Times*, it was the first time in the history of radio that such a conversation had been made over an ordinary telephone. Smithy's voice was directed to the International Radio Station at New Brunswick, thence to London, and then picked up, amplified and projected along a radio beam to Sydney, where the vibrations were gathered up and sent by wire to Mary in Melbourne.

'Yes, hello, darling, how are you, I'm speaking from New York,' he shouted into the phone, explaining, by way of apology, that he was 'just a shade deaf' from the journey. He described the tickertape parade – 'they threw paper all over us' – and the Atlantic flight – 'I thought I was going to lose my life!' 'I'm not going to take any more chances,' he told her. 'I want to marry you sometime.'

Close to midnight that same day, after a dinner at the Progress Club, Smithy boarded an overnight train for the General Electric Laboratory at Schenectady, so he could be ready at seven-fifteen am to make a second telephone call. This time he was to speak to his family and to Charles Ulm; their conversation was to be broadcast over the NBC network in North America, and relayed to England, New Zealand and Australia. His mother, his brother Leofric and Leofric's wife, Elfreda, his niece Beris and Charles Ulm were crowded into a small, bare room at the Amalgamated Wireless studios in Sydney, and no doubt they were nervous, for it was to be a very public conversation. And yet once the long-distance conversing was underway, they were surprisingly relaxed.

'Hello, Mum, dear,' Smithy began tentatively, and she replied, 'Hello, son. No more ocean flying, I hope?' To which Smithy answered, 'I told you I had to do this last one. There are no more oceans left to fly. Have you any more stowed away?' Smithy then spoke about Mary, and about his wedding, scheduled for 15 September, and said ruefully, 'I'm afraid she will find it a bit strenuous having a flying husband, but probably the "no more oceans" business will pacify her a little bit.' When he heard that his father was too unwell to come to the phone, he said, 'I'm afraid I cause a lot of consternation in the family. I am sorry about that, darling, but, after all, think what fun it is!'

When Ulm's turn came, they talked amicably about the flight. 'How are you, old man?' Ulm asked. 'It couldn't have been too good around Cape Race?'

'It wasn't too good, Charlie,' replied Smithy warmly. 'I think the old machine rather missed your being there.' But when Ulm began to talk about Australian National Airways, Smithy's warmth waned; his mind was not on the airline.

A plane was waiting at Schenectady's airfield to fly him back to New York, for he and the crew were invited at eleven am to board Anthony Fokker's yacht, *Helga*, at the Columbia Yacht Club. Fokker was enormously proud of the *Southern Cross*'s achievement and determined to fete its crew. Other aeronautical guests had been invited too: James Fitzmaurice, who had been part of the *Bremen*'s flight; Clarence Chamberlin, who had flown the Atlantic in the opposite direction; Bernt Balchen, who had been Admiral Byrd's pilot over the South Pole; and two attractive 'aviatrices', Ruth Nichols and Luba Phillips. After lunch on the yacht, they put on swimsuits and played with Fokker's eight-passenger flying boat, which was parked on the Hudson River, just below Fokker's home on the Hudson Pallisades. Hearing that Smithy could 'swim like a fish', Fokker tested those skills on the aquaplane attached to the yacht, and managed to throw him off. Everyone 'shouted with delight' at what Smithy amusingly called his 'terrible thumping'. That night they dined with one of Fokker's neighbours, Manuel Rionda, the Cuban sugar baron, in his fairytale mansion overlooking the river. Indeed, Smithy must have thought his entire visit to New York was rather like a fairytale.

The next morning, 30 June, Smithy and the crew were flown to Washington in one of Fokker's new planes, the gigantic F-32, which was powered by four engines and could seat thirty-two passengers. Allowed to take turns at the controls, the Australians declared that it was more like driving a ship than a plane. This, thought Smithy to himself, was how Australian passenger planes would be one day. In Washington, more wonders awaited them, for they were to lunch with President

Herbert Hoover at the White House, and later be welcomed in the Senate. The president questioned Smithy closely about Australia, especially about the western goldfields, where he had spent over a year, and which Smithy also knew well.

The following day came the meeting that America had been anticipating: Charles Lindbergh called on Smithy at the Roosevelt Hotel. The historic moment when the two conquerors of the Atlantic came face to face was a splendid success, especially when, with obvious sincerity, Lindbergh declared that Smithy's crossing of the Pacific 'was the greatest feat in aeronautical history'. Smithy replied modestly that whereas Lindbergh had been alone, he himself had 'four men and three motors'. Afterwards, Smithy told reporters that 'Slim Lindbergh' was a 'swell guy'. If there had been lingering doubts about Smithy's popularity in America, that statement would have banished them.

By now – to quote an American observer – Smithy and his boys were 'a hit, and no question of it'. Smithy had lent the word 'cheerio' to the American language, and almost everyone had seen photos of 'Smithy's girl' and 'liked the look of her'. Much of the Australians' success, according to the same American observer, stemmed from their modesty and their 'shunning of fuss and feathers'. Nobody could begrudge them their success because they were 'so human and brotherly and democratic'.

In Canberra, Prime Minister James Scullin was receiving the same message from Herbert Brookes, the Australian Trade Commissioner in Washington. Thanks to 'their enterprise and daring', wrote Brookes, 'Australia has received a measure of publicity that is phenomenal'. Such praise placed Scullin in a quandary. While several newspapers were agitating for Smithy to be knighted, he was still considered such a maverick and so outside the conventional system that a knighthood was deemed inappropriate by many in power. In the end Scullin compromised, elevating Smithy instead to the honorary rank of Wing Commander, although even that honour tended to be labelled 'wrong in principle' by several highly placed officers in the RAAF. Fortunately, Smithy knew nothing of this controversy, and was delighted when the next day he received news of his elevation.

On 2 July 1930 the *Southern Cross* began its journey to California. Smithy had been hoping to fly to Oakland in one hop, but Roosevelt Field was too small for a take-off with full tanks, so a stop was planned in Chicago, and another in Salt Lake City. At both cities the aviators were lionised.

Two days later, on 4 July, Independence Day, they landed in Oakland. Thousands witnessed the landing and emotions ran high. For the first time, a plane had encircled the world by crossing the equator – that is, encircled it at its widest circumference. And for the first time on a round-the-world flight, all sections of the journey had been flown in the same plane. Smithy was hailed as the great circumnavigator of modern times, the aerial Magellan. 'I felt a thrill of satisfaction,' he wrote, 'that I had been able to bring the dear old bus safely round the world. This was our Journey's End.'

THE RIGHT STUFF
IN HIM

T he advertisement in the *New York Times* on 27 June 1930 was
intriguing:

For Sale: Southern Cross, three motored monoplane. Flown about
275,000 miles, but still able to laugh at its one time owner, Squadron
Leader Charles Kingsford Smith, who is entirely satisfied with the
performance, but frightfully hard up and needs money so that he
can be married. Apply Kingsford Smith, Oakland Airport in a week
or two.

It was a joke, but it had a serious side. Smithy was now the outright
owner of the plane, having bought out Ulm's half-interest with his
shares from Australian National Airways. The *Southern Cross* was his
to do with as he wished, and since the days of pioneering flights were
almost over, and since Mary was determined to stop him flying over
oceans, why should he not sell it? He was asking for $50,000 – he main-
tained that the original investment had been $70,000 – and it was true
that he needed the money. A rich Nebraskan had seemed interested but

had only offered $10,000, as had a syndicate of businessmen. Nobody would make a serious bid.

Smithy had strong reasons to sell but this did not stop his conscience from troubling him. Captain Hancock had donated the plane, and he knew that it should therefore be returned to him, especially since Smithy and Stannage and Saul were now bidden to the captain's ranch and flying school at Santa Maria, in California. Not content with owning railways, ships and real estate, Captain Hancock was now the founder and owner of the successful Allan Hancock College of Aeronautics. In its first year, thirty-four out of forty-nine students had graduated. He could not wait to show Smithy and his crew around its excellent facilities.

Smithy asked Hancock to accept the *Southern Cross* as a gift, and Hancock, 'with his wonderful tact', refused the offer. He was happy, however, to store the plane at the college for as long as Smithy required. If Smithy had been hoping that Hancock would offer to buy the plane, he was disappointed, but in every other way the captain's generosity was admirable. For two weeks they enjoyed the freedom of his ranch and his yacht, after which he arranged for them to visit the castle built by the newspaper baron William Randolph Hearst at San Simeon, on the Californian coast. Hearst had first seen the medieval castle on Germany's River Rhine, and had it transported stone by stone to California, not only preserving its original charm, but adding some twentieth-century splendour. The marble swimming pool, the private zoo and the private movie theatre sent the quiet and modest Stannage into raptures. Smithy, writing more temperately, called it 'the most luxurious and magnificent home that I have ever entered'.

Nor did the excitement end there. In Los Angeles they met glamorous Hollywood stars, including Myrna Loy and Mae West. And when they left Los Angeles for New York at the end of July, they flew part of the way in a Lockheed plane provided by the Goodyear Tire and Rubber Company. It averaged 171 miles per hour; once it touched 186! To the aviators, such speed was 'wonderland'.

In New York in the middle of July, Smithy and Stannage boarded the German liner *Europa*, bound for Bremerhaven. Smithy was pleased to employ Stannage as his 'unofficial secretary', because in these weeks they had become close friends. During the Atlantic flight, Smithy had been impressed by the young man's skill, intelligence and good sense, while Stannage, for his part, hero-worshipped his boss. 'He had,' wrote Stannage, 'that precious gift of all born leaders of men – an infinite capacity to inspire confidence.' None of those who worked for him 'would have refused to follow him anywhere'.

The two were also drawn together by the memory of their first real conversation. It had taken place at the Newcastle Waters cattle station, on that fateful day when Smithy and Ulm had flown over Anderson's wrecked plane in the Tanami Desert. Stannage had just made a similar flight in the *Canberra*, and, sitting together in the shadow of their planes with mugs of billy tea, he and Smithy had talked at length. 'Our hearts were very heavy,' Stannage remembered. Their shared distress seems to have created an empathy that grew stronger as they travelled through America. Normally, Smithy was incapable of freely describing his wartime experiences, but while waiting with Stannage at the Los Angeles aerodrome, he shared some of his more troubling recollections.

At Bremerhaven at the end of July, Smithy and Stannage were met by a Fokker representative, who flew them to Amsterdam for a victory parade almost as stirring as New York's. They and van Dijk were 'borne in triumph' through streets packed with admirers, vocal with pride that a Dutch plane had flown so far. Smithy also renewed contact with his shipboard friend Hendrik van Loon.

Smithy had expected to return to England in the following weeks to take delivery of the small, fast Avro Avian biplane he had ordered the previous year. He was intending to fly it home in time for his wedding on 15 September. At the same time, he hoped to break Bert Hinkler's record time of fifteen and a half days for a solo flight to Australia. But suddenly he cast his plans aside and went to stay with Hendrik van Loon at his house in the town of Veere, in the south-west Netherlands. He needed van Loon's help, because he suspected he was on the verge of a serious illness. For weeks he had been trying to ignore nagging pains in his abdomen, and now he feared the worst.

At his host's insistence he was examined by a doctor, who diagnosed appendicitis. Smithy was rushed to a hospital in the nearby city of Middelburg, and immediately underwent an emergency appendectomy. Troubled by Smithy's general health, the doctor also insisted, a week or so later, on removing his tonsils, in a procedure so fearsome that it was fortunate that Smithy could laugh about it later. 'They were removed,' Stannage recalled, 'with a sort of snipping instrument, without any anaesthetic': it was amusing to hear Smithy imitate bearded, old Dr Koch saying, 'Dis vill hurt a liddle'.

Smithy was discharged into van Loon's care on 15 August, and was warned that his convalescence must continue and he must postpone his flight to Australia. This caused much heartache. Mary was dismayed that the wedding must wait, while William and Catherine Kingsford Smith had urgent reasons for wanting him home: William was dying of cancer, hanging on to life in the hope of seeing his beloved boy.

At the start of September, Smithy insisted he was sufficiently well and flew to England to prepare for the flight. But he had not taken into account the strain that making such preparations imposed. Twice in the last eighteen months anxiety had so weakened him before a major flight that he became ill, and now it happened again. His temperature rose and 'his nerves shattered'. On 10 September he gave an interview to a reporter. 'Look at me,' Smithy ordered the startled young man, and as he spoke he 'held out his hand trembling and shaking violently'. 'This,' he said, 'is the result of my three years' pioneer long-distance flying coupled with the operation I had to undergo as soon as I made the Atlantic crossing.' He added: 'I am no fool, and even if the doctor allowed it, I would not attempt to make the solo flight in this condition.'

At the end of the month his father's health worsened. He had no choice but to pull himself together and make the flight. On 9 October he took off from Heston Aerodrome, near London, in the Avro Avian, which had been adapted to his needs. It had a powerful DH Gipsy 11 engine from the de Havilland factory, enlarged tanks that could carry 113 gallons of fuel, and an extra oil tank with a release valve into the engine sump when the oil ran low. Remembering Coffee Royal, Smithy also carried two gallons of water and ample dry rations. He had named the plane the *Southern Cross Junior*, believing that it had the stamina of its older namesake and could take on any manner of competitors.

This was just as well, because there were four aviators in addition to himself hoping to better Hinkler's time. Fortunately, two of them – Charles Pickthorne and Charles Chabot – were flying together, which ruled them out of the solo category. Of the two solo flyers, Captain F.J. Matthews had left England on 16 September, while Flight Lieutenant C.W. Hill had left on 5 October. It pleased Smithy to see them as competitors whom he must try to overtake in the course of the flight, but in fact he had embarked on a time trial. The winner would not necessarily be the first to reach Australian soil, but the pilot who made the fastest time.

Smithy was embarking on a flight unlike any he had made before. Previously on long-distance flights, he had been accompanied by three companions, all of whom had provided physical and psychological support – qualities which Smithy, being still convalescent, particularly needed at that moment. Also, his plane had only one engine, and since the Dole Air Race he had come to believe that a single-engine plane placed its pilot at risk. And now he was about to fly to the other side of the world in such a craft! But if he were to break speed records, he accepted that this was the type of plane he must employ.

Was he well enough to attempt to break a record? He decided to fly as far as Rome, and there assess the situation. To quote his own words: 'In Rome, I said to myself: I'll see if there is not a kick left in the old man.' Having decided that there was 'a kick', he pressed on with the time trial and the next day flew to Athens. Indeed, he felt so well in Athens that he spent a convivial evening exploring the city, although he refrained from drinking alcohol. On his doctor's orders he had not touched a drop since before his operation, and he was determined to abstain until he reached Sydney.

The next afternoon he reached Aleppo, in Syria, without mishap, but on the following morning on the way to Bushire (today Bushehr), he met a dust storm and was obliged to fly for hours without sight of the ground. He found the semi-blind flying unnerving, and ten hours in a cramped cockpit exhausting to his body and mind. He had planned to sleep for four or five hours at each stopping place, but in his keyed-up condition he slept poorly. Nevertheless, he set off for Karachi early next morning. Twenty minutes later, while flying across the sea, his single engine suddenly cut out. Frantically turning switches, he managed to shift an airlock in his petrol line, but the experience gave him a 'bad fright'. 'I learned a lesson,' he wrote, 'and thereafter kept closer to the friendly land.'

In Karachi he was greeted by Pickthorne and Chabot, who had abandoned the race and were eager to impart news of his rivals. Matthews was in Rangoon, waiting for repairs to his plane, but the Australian flyer Cedric Hill was two days ahead. The next evening, in Allahabad, Smithy heard that the forty-year-old Hill, although exhausted, was flying by night as well by day in order to make the fastest time, and would be very hard to beat.

From Allahabad, Smithy made for Rangoon, flying over country that had troubled him even when crossing it in his three-engined *Southern Cross*. Called the Sunderbunds, or Sundarbans, it was a remote and sinister region of dense mangrove forest, intersected by hundreds of waterways belonging to the Ganges Delta where the mighty river discharges into the Bay of Bengal. It was also the home of Bengal tigers; in Smithy's words, 'a stranded airman might be lost for months – if he were not eaten by tigers meanwhile'. In his autobiography, Smithy eloquently explained the fear that this mixture of ocean and land instilled:

> When passing over regions like the Sunderbunds or over open sea, the airman experiences a curious and entirely imaginary suspicion that his engine isn't running as well as usual. One hears all kinds of noises which are really quite normal, but seem exaggerated to his ever-listening ear – a psychological reaction to the fear of a forced landing.

Smithy was not the only aviator to experience such a fear. Many pilots listen for sounds of engine failure when flying over places where, if they were forced down, they had little hope of being rescued. Happily, the rational mind is usually able to dispel these symptoms of awakening panic, but Smithy was still weak from his recent illness. He was also, in his effort to win, depriving himself of sleep and pushing himself beyond his physical and mental limits. Moreover, since adolescence he had feared being marooned in a remote place where no one could find him. If he had had companions aboard they might have laughed away his fear, but he was quite alone, without even the comfort of a radio to connect him to outside world. As he flew across the Bay of Bengal, he was consumed by a sense of dread.

The following day, as he crossed the mountainous jungle of Thailand, he felt the dread a second time: 'in this terrible country,' he wrote, 'there is no more chance in a forced landing than over the sea.' Two days later the fear revived as he flew across Sumatra. The sight of the jungle below, he wrote, 'was enough to deter the heart of the boldest airman, particularly if he be flying a single-engined plane. I could not see a patch of open ground for hundreds of miles. And a disaster here would have meant death from starvation, if one were not meanwhile killed by wild animals.'

Even so, Smithy was making an impressive journey. Hill was only one day in front, and this lifted his spirits. Indeed, Hill was in his mind when, on the tenth day of the flight, he approached the landing ground at Atamboea (now Atambua), on the island of Timor. It was nearly sunset, and his eye was caught by a shiny glint against the green background. 'My machine came to rest,' he remembered, 'and as I took off my goggles a weary, sunburned and disconsolate figure in a topee and shorts came towards me.' It was Hill. That morning he had crashed his plane during take-off.

Smithy's first reaction was to offer to carry Hill as a passenger to Darwin, for it was only half a day away, but on reflection they agreed that the Avian was too small for a passenger. Moved by Smithy's kindness, Hill then made a generous gesture. 'He asked me if I had a collapsible rubber boat,' Smithy remembered, 'and when he found that I intended to rely on inflatable tubes if forced down in the Timor Sea, he insisted that I should take his boat.' Hill could not have offered a more desirable gift, because – as Smithy recalled – 'it was the first time that I had ever faced a long sea crossing alone without the comforting roar of three engines in my ear'. This exchange of gestures between the two aviators moved a journalist on the London *Daily Mail* to comment: 'Men do not act like that after nine consecutive day of flying and at the crucial moment of a desperate struggle unless they have the right stuff in them.'

As Smithy crossed the Timor Sea the next day, he thought gratefully of Hill's rubber boat, and, since the fuel tanks in the Avian were well removed from the cockpit, he had the comfort of being able to smoke 'innumerable cigarettes'. Nevertheless, he later recalled, he 'listened anxiously to the little Gipsy engine', just in case it should suddenly fall silent.

It was with immense relief that Smithy spied Australia on the horizon. 'The long stretch of sea was astern of me,' he remembered; 'my goal was ahead. I shouted for joy.' On 19 October, in high spirits, he swept into Darwin. He had beaten Hinkler's record time by a wide margin, having completed the journey from England in nine days, twenty-one hours and forty minutes. To break another record was deeply satisfying, but more important to him was the conquest of a long-held fear, and the joy of knowing he could triumph on his own.

THE
SOUTHERN CLOUD

S mithy was still sitting in his cockpit at Mascot when two soft arms were wound around his neck and he was given a long and lingering kiss. 'Miss Powell, Miss Powell,' shouted the photographers. 'Turn the kiss on again! Just once more ... Be a sport, Mary. Another kiss. He likes it!' Mary happily obeyed, and 20,000 onlookers, who had crowded into the aerodrome to welcome the aviator home, cheered themselves hoarse.

Smithy was once again the nation's darling, and everyone seemed to want to congratulate him. Cables and telegrams poured in. There was one from King George, one from Bert Hinkler, and a special one from Nellie Stewart: 'I knew you would do it, my wonder man!' Another favourite came from the Returned Soldiers of Parramatta. 'Is he any good?' asked the Parramatta Diggers. This tickled Smithy's fancy. 'Can you beat the Diggers?' he asked delightedly. On American radio, millions heard the US air ace Eddie Rickenbacker tell Smithy over the airwaves, 'You are a wonder!' To which Smithy replied, 'I am conceited enough to enjoy hearing that.'

Despite the jubilation, there was a tragic side to his homecoming. His father was close to death, and it was a blessing to them both that he had returned in time to say goodbye. William, too ill to be nursed at home, had been moved to a private hospital in Longueville, and it was there, on 2 November 1930, that he died, surrounded by his family.

As William lay dying, he expressed a desire to be cremated, at that time not a common custom. He also asked for his ashes to be scattered from a plane over the Pacific – an even less common custom – and he entrusted Smithy with the task of organising the ceremony. After the Anglican service at St Aidan's in Longueville and the cremation at Rookwood, six Kingsford Smiths, together with Charles Ulm and two clergymen, travelled to Mascot and took off in the *Southern Sun*. Smithy circled low over Arabella Street, where his mother was watching from the verandah, and then headed out to sea. There – with Ulm flying the plane – he joined his family in a brief religious service before uncapping the urn and scattering the ashes on the breeze.

Smithy's wedding took place less than six weeks later. On 10 December he and Mary were married at the Scots' Church, Melbourne, which, being Presbyterian, allowed the marriage of divorced people. Ten thousand spectators are said to have blocked the traffic in two of the city's busiest streets as they gathered to watch the bridal party and 200 guests arrive. A squad of mounted policemen was unable to control the bystanders, most of whom were women. They jostled the guests, disarranged the bridesmaids and eventually surged into the back pews of the large church.

The unruly crowd came as a shock to Mary, who so far had seen only the beneficent side of fame. To see women clambering on top of pews to gain a better view was disturbing to the congregation, and Mary was mortified. She had so wanted her wedding to be perfect, and had given it infinite care. Her wedding dress – the pride of one of Melbourne's leading dressmakers – was of ivory georgette, her veil was pale pink tulle, and she carried a bouquet of pink waterlilies. She had four bridesmaids, each dressed in blue, and her two little nieces carried her train. Smithy and Ulm, his best man, wore their RAAF dress uniforms, and the married couple left the church under an arch of airmen's swords. A month previously Smithy had been created an honorary Air Commodore, so the Air Force theme was an appropriate one.

Hawaii had been widely predicted as the place for their honeymoon, but next day the bride and groom caught the steamer *Nairana* for the more accessible Tasmania. As they sailed across Port Phillip Bay, the planes of Smithy's friends circled above them. What was it like, asked thousands of women, to be Smithy's bride? After her wedding day, Mary would probably have said 'bewildering'. She was barely twenty, and had had a sheltered and careful upbringing. Neither her parents nor the exclusive girls' boarding school she had attended in Geelong had allowed her much experience of life. And now she was expected to cope with the vagaries of fame and the wishes of a brilliant and complex husband. Smithy, on the other hand, found her lack of experience a precious gift. He saw her as a blank page on which he could write, and he believed that, over time, he would transform her into his ideal woman.

It is also probable that Mary knew little about sex; later, she would tell the biographer Ian Mackersey how 'absurdly innocent and ingenuous' she was. This was scarcely surprising, since in polite circles it was socially taboo to mention the subject in the presence of a young lady, and well-brought-up girls were expected to stay in virginal ignorance until their bridal night. That such brides were often shocked gave credence to the common medical opinion of that era that there were many females who never felt any sexual excitement whatsoever

One imagines that Smithy had shared pleasure with too many women to believe so sweeping a medical pronouncement, and also possessed sufficient imagination to realise how disturbing her new life initially might be. Mary remembered how, on their honeymoon, he was 'very physical, but gentle and endlessly patient'. However, he found it hard to be patient with some of her prudery. Having a robust appreciation of sensual pleasure, he wanted her to share his delight.

He taught her to drink and smoke, and he told her 'highly risqué stories', which he and his flying mates found hilarious but she found embarrassing. 'A dirty mind is a perpetual solace,' he would tell her; and sometimes he made outrageous statements about sex which he found very funny but she did not. 'My highest ambition,' he would say, 'is to be hanged for rape when I am ninety-two'. Having learned to despise a lack of refinement, and to hate any sort of coarseness, Mary was worried by his bawdiness, and prayed he would not tell any of his improper stories in front of her young friends. Her prayer was usually answered. His risqué songs and stories were kept for 'smoke nights' with his male companions.

Early in January 1931 Smithy and Mary returned to Sydney and moved into an apartment at 74 Drumalbyn Road, Bellevue Hill, overlooking the charming Rose Bay. Here, Mary enjoyed playing the houseproud new wife and Smithy enjoyed indulging her. He took great pleasure in her girlish enthusiasm, although at times he was all too aware of the fourteen years' difference in their ages. While he disliked her being called his 'child bride', he contributed to this perception by giving her the pet name of 'Kiddie'. She, in turn, called him by his family nickname of 'Chilla'.

There could be no doubt that Smithy missed his old, easygoing life at Arabella Street. His mother coped with his eccentricities in a way he could never expect of a new wife. How could he ask Mary to readily accept his nervy prowling around the house, his compulsive novel-reading, his refusal to open letters – which his mother usually opened for him – or his impetuous wish to bring friends home for the night? 'In the morning I do not know how many people I will find in the house,' his mother had complained a few months previously. 'If there are too many friends,' she explained, 'they sleep on the floor.'

Smithy had no intention of loosening his ties to his mother. He made this clear when he spoke to the assembled guests at the wedding reception. 'There's a spot in my heart that even Mary may not own,' he told them, 'but Mary knows that the love I bear her is not lessened by the love that I bear my mother. She has always been my inspiration and my best friend.'

As a married man, however, he knew that he was obliged to make an effort to adapt, more or less, to Mary's view of domesticity. He also knew that he must settle down to work at the airline and become accustomed to the boredom. Certainly the airline needed him. It was starting to feel the bite of the financial depression that had enveloped the Western world since the crash of the New York stock market in October 1929. For the previous six months, passenger numbers had been quietly falling, and even prosperous businessmen began to cut out luxuries like air travel. Every Australian airline was experiencing difficulty, but Australian National Airways was particularly vulnerable since, unlike West Australian Airlines and Qantas, it received no government subsidy to cushion its losses.

Despite its falling revenues, the company had grown in size and stature over the past six months, with nine senior pilots now on its payroll, and about seventy other employees. Thanks to Ulm's industry and organising ability, and above all his emphasis on reliability, the airline was widely recognised as one of the most successful. As John Stannage observed, Smithy was 'honour-bound to help his company weather the financial crisis that had taken over the world of business'.

During his honeymoon, Smithy had lobbied the Tasmanian government to support a passenger and mail service across Bass Strait. On 16 January 1931 he captained the first Australian National Airways plane to operate the new service to Launceston, with a former RAF flyer, Jim Mollison, sitting beside him. From Mollison's autobiography comes a story that, if true, proves that Smithy was still a bachelor at heart. On the night before they were to return to Melbourne, Smithy is said to have returned, rather drunk, to the room that he and Mollison shared, bringing with him a 'slightly rumpled blonde'. Mollison was asleep but quickly awoke because 'between them they made enough clatter to bring the ceiling down'. 'We all drank,' he continued, 'and made merry together until the smallest hours.'

In addition to being an experienced pilot, Mollison was an experienced womaniser, and as flamboyant as Smithy himself. To be involved in an escapade with his celebrated captain may well have pleased his vanity. The story may be an exaggeration, or even a fabrication. On the other hand, it might be true. Smithy had been self-disciplined for much of the past year, and the strain may have been acute: no wonder, perhaps, that he broke out occasionally. But would he have attached himself to a slightly rumpled blonde just five weeks after his wedding?

Certainly his mother had faith in him. 'Every now and then he has to have these harmless flirtations,' she told an interviewer a few months before. 'It would be very difficult for a man who has done what he has done to avoid some attention from the girls, but Miss Powell understands him and knows his loyalty.' Catherine Kingsford Smith believed that Smithy was always loyal and loving to his family, even if he was unable to live by the ordinary rules of society. She tried to encourage Mary to share her view.

In the early months of 1931, Smithy took on an additional task for the airline. 'Let Smithy teach you to Fly,' ran an advertisement in the *New Sunday Times*. 'Discover for yourself the joy and thrill of handling the joy stick.' The flying school was housed at the airline's premises at Mascot, with the pilots as the instructors and a couple of Gipsy Moths as the instruction planes. Smithy's naval brother, Eric, became one of its first successful pupils. It was hoped that the school would boost the airline's business, for the loss of revenue was starting to worry the shareholders.

On the evening of Saturday, 21 March 1931, Smithy received an ominous phone call. The *Southern Cloud* had left Mascot that morning with two pilots and six passengers on the regular service to Melbourne, but had not yet reached its destination. The most obvious explanation was the weather. One hour after the plane's departure, Mascot had received a cyclone warning, but as the radios on the Australian National Airways planes could not receive messages – being solely transmitters – there had been no means of relaying the news to the pilot. Presumably the plane had landed somewhere to escape the powerful wind, which was said to have reached almost 100 miles per hour.

Since the pilot, Travis 'Shorty' Shortridge, was one of the company's best, nobody at first was unduly worried. Nevertheless, police stations along the route were alerted, and pilots and observers were warned to prepare for a possible search the following day. Next morning all regular services were suspended, and the airline's fleet set out to look for the missing plane. Other aircraft quickly joined, until at least thirty planes were searching, and would continue to search for two more weeks. Smithy led the searchers, carrying with him a doctor and a parachutist and a photographer. Numerous sightings were reported, and many hours were spent following false clues. The *Southern Cloud* would not be found until October 1958, when a bushwalker came upon its wreckage in the Toolong mountains, near Kiandra in New South Wales.

There can be no doubt that the loss of the plane devastated Smithy. For two weeks he searched for eight hours a day in the most arduous and nerve-racking circumstances over mountainous country. Believing that the plane was most likely lost in Victoria, he took rooms at the Menzies Hotel in Melbourne, where Mary and Ulm and other pilots joined him. Mary would never forget Smithy's anguish. 'He stopped smiling, he stopped joking, he couldn't sleep,' she recalled: 'The worry began to make him physically ill.' A fellow pilot remembered his 'mental distress and sorry exhaustion'. Each morning, determined to keep searching, he set out from Essendon Aerodrome, where anxious relatives of the lost crew and passengers had set up a vigil. Each evening, on landing, it was his heartbreaking duty to tell them that he had found nothing. 'It was hideous,' Mary would later say.

The loss of the *Southern Cloud* was regarded as a national tragedy, and on 4 April a civil inquiry was convened in Sydney. The first witness was to have been Smithy, but he was piloting on the Tasmanian route, so Ulm took his place at the inquiry. While searching for the lost plane, Ulm and Smithy had grown close – indeed, possibly closer than they had been at any time in the last couple of years. Watching them at Melbourne's Victoria Barracks, as they inspected maps and organised search planes, a reporter observed their united sense of purpose and quiet affinity.

Having learned his lesson from the Coffee Royal fiasco, Ulm faced the new inquiry with a calm efficiency. He stated that his airline had carried 9000 paying passengers, and had flown 671,000 miles without even one accident involving injury to a passenger. He testified to the planes' regular safety inspections, and to the company's observance of safety procedures. Stressing that he and Smithy were keen to install effective two-way radio communication in all their planes, he explained that for months he had been lobbying the government to install 'an efficient and proper system of ground wireless stations'. Lieutenant Colonel Brinsmead, the Federal Controller of Civil Aviation, endorsed Ulm's evidence. The inquiry then turned to the question of the pilots' medical fitness, and now it came to light that one pilot had dodged his regular medical examination. That pilot was Air Commodore Kingsford Smith, and his examination was long overdue. He was a national hero, but should he be exempt from the safety rules? The answer was no.

On 17 April 1931 the inquiry reconvened in Melbourne, and again Smithy wished to be excused. He claimed he felt too tired to attend, which was probably true, but there were also other reasons. One week previously, Charles William Anderson Scott – a 28-year-old English aviator who had lately worked for Qantas – had landed in Darwin. His flight from London had taken nine days, twelve hours and fifty-five minutes. He had beaten Smithy's record of the previous November by nearly half a day.

This was distressing news for Smithy, who was still suffering from the loss of the *Southern Cloud*. Nevertheless, he was at Mascot to welcome Scott, and was reported to have said: 'I have been well and truly and fairly licked, old boy. I offer you all the congratulations in the world.' Beneath his urbane exterior, however, Smithy felt pain, for Scott's achievement could threaten his own fame and professional standing.

Smithy was now at the very top of his profession. Earlier that year he had been awarded the prestigious British Segrave Trophy for 'courage, initiative, skill and a sense of adventure', and the judges' decision came as no surprise. At about the same time, the popular American magazine *Liberty* had devoted four pages to Smithy's remarkable career, with high praise coming from twenty-one of the world's greatest aviators. The British air ace William Bishop summed up their feelings: 'If all his exploits were known, the public would agree that he and his record are unsurpassed in the annals of aviation.' In America, aviators called him 'the Daddy of them all'. In terms of world fame, Smithy was now the second-most famous Australian, living or dead, being rivalled only by the opera star Dame Nellie Melba. Millions of people in nearly all corners of the globe knew his name and marvelled at his achievements.

As wise observers of human nature have testified, the upward climb in a career is often less exacting, less arduous, than remaining at the top. And once one has starred on the world's stage, to be forced to take a lesser role is a humiliation, especially for somebody who craves challenge and adventure. As he contemplated Scott's triumph, Smithy's first reaction was to long to break Scott's record. However, he was constrained by his promise to Mary to give up long-range flying. When a reporter asked if he was planning another major flight, he answered cautiously: 'Ask my wife, I am more or less an onlooker at the moment'. 'But,' he added less cautiously, 'there's life in the old dog yet.'

There was indeed energy in the old dog and, promises or no promises, he was about to act. The justification came, quite fortuitously, a day or so after his welcome to Scott. Britain's Imperial Airways, which already ran a passenger and mail service from England to Delhi, hoped to extend its service to Singapore and Australia. There was a strong feeling in Australia, however, that the section of the route between Singapore and Darwin should be operated by an Australian company. Ulm certainly believed this, and was determined to gain it for Australian National Airways.

In March 1931 the British Post Office and Imperial Airways announced that they were adopting an 'experimental plan', whereby an Imperial aircraft – a DH.66 Hercules biplane – would pick up Australia-bound letters in Delhi and convey them to Darwin. In mid-April, however, the Imperial plane carrying the mailbags crashed at the Dutch port of Koepang (today Kupang), on the island of Timor. An unlucky crash for Imperial, it was wonderfully lucky for Smithy and Ulm, since a pilot and a plane were now needed to rescue the mail and bring it on to Australia.

Smithy had the perfect plane for the rescue. The *Southern Cross* had arrived from California by ship a few weeks earlier, and 'she was flying better than ever'. Moreover, he could recruit two stalwart assistants from among his own airline pilots. As his co-pilot he chose George Urquhart Allan, usually known as 'Scotty' because of his broad Scottish accent. A former Royal Flying Corps pilot, Allan was small, tough, dogged and determined. Smithy's other assistant was Wyndham Hewitt, his company's chief engineer.

In a flurry of activity, Smithy ordered a quick refurbishment of his beloved plane. In the space of two days the fuselage was painted blue, the registration letters VH-USU were marked on the side, and the royal airmail insignia was stencilled on the cabin door. He was aglow with excitement when, at midday on 21 April, he prepared to take off from Mascot. It was 'like old times', he shouted joyfully to reporters as he donned his oil-stained flying cap and adjusted his tattered flying coat: 'We must show the British government we can be relied on.' Patting the barely dry fuselage, he declared that 'the Old Bus knows the road so well that I believe she could nose her way up the Timor Sea without a pilot at all.' Photographed as he farewelled Mary, he looked radiant but she looked resigned. 'I am not worried about him a bit,' she told reporters stoically.

A Sydney *Sun* reporter who accompanied the crew marvelled at the fun they shared during the long, cramped hours of flying. They read books, ate apples and cleaned their boots. For a few minutes they pretended to be soldiers in a dugout, this last being Smithy's idea. Smithy was the ringleader. 'I am so happy in this old thing,' he shouted above the engine noise. Remembering the anxiety he had felt flying over remote places on his previous long-distance flight, he was not looking forward to crossing the Timor Sea, but the congenial company kept his fears at bay. When they landed at Koepang, he danced a hornpipe after exiting the cockpit.

On 24 April they collected the mailbags from the crashed plane and welcomed aboard the stranded crew, who were now to travel with them to Darwin. Once again their spirits were high. In the rear cabin they played card games and wrote poems, using a bulging mailbag labelled 'Jerusalem' for a table. One of the poems ran thus:

Miles and miles of trackless sea
Below, the shark, he waits for me
Up above the sea so high
The crew play banker and poker die.

They arrived back in Darwin on Anzac Day: most fitting, said Smithy, for such an historic event. Two days later they were off again on their return flight, carrying Australian mail destined for England. They were to deliver it to the Burmese island of Akyab (today Sittwe) in the Bay of Bengal. There the mail was to be loaded onto an Imperial Airways plane bound for Delhi.

Smithy's high spirits continued to bubble throughout the flight to Akyab. Indeed, he was so euphoric that he joked that the Timor Sea was 'a mere ditch', and he was 'thoroughly contemptuous' of it. He and his crew stopped briefly at Koepang, and again at Surabaya, where the *Southern Cross* became bogged in the runway: a hundred Javanese soldiers were summoned to dig it out. Then they flew to Singapore, to Alor Star (today Alor Setar), Malaysia, and to Rangoon. Finally, on 3 May, they reached Akyab, which was 'the worst place' Scotty Allan had ever seen. There they waited in the heat for three days while the Imperial Airways plane flew to Delhi with the load of mail bound for England, and returned from Delhi with a load of mail bound for Australia. Five days later the *Southern Cross* was back in Darwin. Smithy had proved that a regular airmail service between England and Australia was viable, and that Australian National Airways was well able to operate the final sector. He and Ulm were triumphant.

Their triumph did not last because their airline was dying financially. The loss of the *Southern Cloud* had shattered public confidence, and the deepening economic depression hastened the airline's death. Revenue dwindled, creditors began to foreclose and anxious shareholders called meetings. By late June 1931 the company found itself £28,000 in debt, a huge sum. A year's loan was secured from the chief creditors, but it was considered wiser to suspend activities in the hope that services might resume the following year. From the beginning of July the fleet of planes was confined to the hangars at Mascot, and most of the pilots and staff were laid off.

Smithy and Ulm begged in vain for a subsidy from the federal government, but it, too, was seriously short of money. However, it did continue to subsidise West Australian Airways and Qantas because those two airlines flew to the more remote regions, whereas Australian National Airways flew along the east coast, where trains and ships were frequent. To Smithy and Ulm, of course, this seemed unfair. Their main hope continued to be the Singapore–Darwin airmail route, for the federal government would surely subsidise that.

Smithy himself was in financial trouble, a problem he usually bore with equanimity. However, having a wife and a marital home to support, he was more than usually disturbed by it, so he sold his Avro Avian and prepared to sell his speedboat. His nephew John, Wilfrid's son, a junior pilot who lived with his maternal aunts at 65 Arabella Street, was Smithy's frequent companion in these months, and full of unruly adolescent ideas. He torched the speedboat so that Smithy could claim the insurance, keeping the true circumstances of the fire from his uncle until after the boat was a burned-out wreck. Fifty or so years later, in describing the incident in a private letter, John claimed that Smithy, once he knew the truth, condoned the action. Certainly Smithy did not give the boy away to the police or the insurance company; if he had done so, John might have been sent to prison. Whether or not he thanked the boy – as John claimed in the letter – one cannot know. He loved young John, and since the deed was done by the time he learned of it, he may have seen no point in dealing with him harshly. And, of course, he had memories of his own unruly adolescence.

John was also involved with his uncle in a semi-secret army called the New Guard, the importance of which increased as the Depression deepened. A right-wing patriotic group favoured by ex-servicemen and dedicated to fighting communists and other radicals, it was formed in February 1931 by Eric Campbell, who was a member of the Australian National Airways board and had for some years been Smithy's and Ulm's personal solicitor.

Smithy's New Guard membership seems to have been a passing aberration. He was usually uninterested in politics unless aviation was under debate, and he accepted membership, presumably, to please Campbell. Ulm joined too, and so did Frederick Stewart, the millionaire chairman of their airline. The New Guard was strongly opposed to the New South Wales Labor premier, Jack Lang, who was Smithy's old friend and supporter.

A story seems to have circulated that a servant overheard Smithy advocating the bombing of the Lang government, but the story is unsubstantiated and seems unlikely. Smithy had a genuine regard for Lang and always appreciated his support over the Pacific flight. His opposition to Lang's premiership, if true, would not have been extreme. Nor does his role as a New Guardsman seem to have dimmed Lang's regard for him. Years later, Lang would write that his decision to back Smithy in 1927 was 'one of the most satisfying decisions' he ever made.

What Smithy needed was regular paying work, so he turned with relief to his favourite pastime, barnstorming. In the past he had given joy rides in the *Southern Cross* mainly to promote his airline, but now he set out for country towns in southern New South Wales, offering joy rides on his own behalf. The public loved the experience. From late July to mid-September he was in Albury, Buffalo, Corowa, Jerilderie, Wagga Wagga, Leeton, Griffith and West Wyalong. Wilfrid, who had abandoned his real-estate business in Albury, came with Smithy as his manager.

Emotionally, the joy flights were answers to Smithy's prayers. Domesticity with Mary was beginning to stifle him, and the financial problems of the airline bored as well as worried him. His enthusiasm for the company had never been as strong as Ulm's, and his anguished search for the *Southern Cloud* had drained much of it. Spending part of each week in the country towns allowed him to recapture the freedom and excitement he had known in earlier days.

At home, Mary realised she had been married half a year but had spent surprisingly little time with her husband. Even when he was at home, she found it difficult to engage his continued attention, for so often he seemed to be surrounded by people. The honeymoon, she recalled, 'was one of the rare times when I had him to myself'. No doubt she expostulated, and no doubt he responded kindly, for deliberate unkindness was not in his nature, and there was no doubt in his mind that he loved her and was intensely proud of her. Nevertheless, his absences and the inattention continued. Feeling lonely, she began to make frequent train journeys to her family in Melbourne. It would become the pattern of her married life.

Long-distance flights with the thrill of danger still beckoned Smithy, and new aviators appeared to public acclaim. The 26-year-old Jim Mollison, having been discharged from Australian National Airways, flew back to England, and in the course of the flight challenged Scott's record. Mollison completed the journey in eight days, twenty-one hours and twenty-five minutes: two days faster than Scott. An astonishing performance, it spurred Smithy to action. Thrusting aside his promises to Mary, he told himself that if he did not make a record-breaking flight soon, he would become a nonentity.

On 9 August 1931 – three days after Mollison reached London – Smithy announced to the press that he had ordered a new Avro Avian sports biplane with another DH Gipsy II engine, and would set out next month for England. While Mollison had taken almost nine days, Smithy hoped to complete his flight within seven. In preparation, he would undergo a 'strict and scientifically arranged training, tuning up for the terrific strain' that he believed the flight would impose. By being super-fit, he hoped to maintain a 'constant flow of energy' and escape the 'nervous apprehension' that had so disturbed him on his last solo record-breaking flight. It was a brave hope. It remained to be seen if he could make it a reality.

PANIC

S mithy's new Avian was delivered to Mascot Aerodrome for testing in September 1931. How he paid for it one does not know. John Stannage once said that if Smithy really wanted something, he ordered it on credit and gave thought to the payment much later. He was also fond of saying that debt did not worry him.

Smithy named his new plane the *Southern Cross Minor*, and about sunrise on the chilly morning of 21 September he took off from Essendon Aerodrome. He was determined to reach England in seven days and fourteen hours – a day and five hours faster than Mollison – and he vowed he would be back in Melbourne within twenty-two days. By completing such a speedy return, he intended to show that a regular air service to England was well inside the bounds of possibility.

Mary, dressed in a thick fur coat, was at the aerodrome to say goodbye. At least one gossip column was reporting her opposition to Smithy's going, and a photographer caught a picture of her young, strained face. What her parents thought is not recorded, but one assumes that they were wise enough to realise that Smithy's return to international flying was inevitable. They welcomed her warmly back into their home for the duration of the flight.

On his seat in the cockpit there was, as usual, Smithy's photograph of Nellie Stewart, despite the fact that three months previously he had been a mourner at her funeral. Later he would learn that it was bad luck to use the photograph of a dead person as a lucky charm, and would wonder if this had contributed to his ill luck, because mishaps now seemed to dog him. He was due to leave Australian soil at Wyndham, but even before he reached that tropical port an airlock had developed in his fuel line and his oil pipe fractured. Fortunately – since he was still in his native land – he could stop and receive repairs without affecting his attempt on the record.

After such a depressing start, he was half-anticipating more mishaps, but the sixteen-hour flight to Cheribon (today Cirebon), on the island of Java, was free of trouble and undue nervousness. Although he believed that his four crossings of the Timor Sea on the mail flights had banished those psychological demons, he still felt wary of crossing a stretch of sea, especially now that he was flying solo. The next day he made for Victoria Point (today Kawthaung), in southern Burma, but encountered such a deluge as he neared his goal that he was forced to put down on a beach. Unsure where he was, unable to work the emergency radio, and cold, tired and hungry, he huddled in the cockpit to await the dawn. Then, suddenly, he realised that the incoming tide might bog the plane, and he dashed into the jungle in search of wood to place under the wheels. He was returning with branches when he heard a sound that made his flesh creep. It was a tiger's roar. He 'covered two hundred yards of the beach in twenty seconds', and spent the remainder of the night marooned in the cockpit. At dawn he built wooden tracks to assist the take-off, and managed to fly to Victoria Point.

The forced landing had placed him well behind Mollison, but he pushed on, determined somehow to make up the lost time. After five hours of rest in Rangoon, he took off at three am, passing over the now familiar Akyab, and commencing to fly over the Bay of Bengal. He was about halfway across when he was stricken with a sensation that was far worse than the sense of dread that he had been half-expecting. Foreign to him, the new feeling was frightening:

It was when half way across the Bay with sixty miles of open sea on either side of me that I was suddenly stricken. I remember thinking that the midday sun on the back of my head was not very good for me. My head seemed to be getting heavy. Then suddenly I had a horrible feeling that I did not know who I was or what I was doing. I knew that I was flying a plane and that I had to reach land which was out of sight, but who I was or why I was there I didn't know. This curious attack lasted a minute or so. I was in a peculiar condition of half-consciousness, I think. The next minute I was diving at a fairly steep angle into the sea. In a sweat of apprehension I gained control of myself and then I seemed to understand that I had sunstroke.

He also understood that he needed help. Amid waves of nausea, he set his course for Calcutta's Dum Dum Aerodrome, comforting himself with the thought that, if the worst came to the worst, he could ditch the plane and try to float ashore to the Sunderbunds, which were now in sight. He arrived at Dum Dum so close to collapse that onlookers became worried, and some of them spoke to the news services. The next day – 28 September 1931 – readers around the world learned that Smithy was suffering badly from stress and seemed 'very ill'. Mary, confined to her bed in Melbourne with influenza, read the news with alarm.

At Dum Dum, Smithy borrowed a sun helmet and a bottle of brandy and rested for an hour. Although his head still ached, his nausea had eased, and the thought of his rival Mollison forced him back into the cockpit. He had arranged to spend that night at Allahabad but decided to recover lost time by flying a further 300 miles and stopping for the night at Jhansi. It was a stupid decision because, in addition to weakness, he was suffering from his old enemy, dread, and was unable to shift it. Nevertheless, in his determination to beat Mollison, he was prepared to ignore his feelings and press on. An hour or so later he was almost overcome by a 'sort of semi-panic and a desire to be safe on the ground'. Fighting his fear with willpower and swigs of brandy, he kept on flying.

Six hours of rest at Jhansi somewhat calmed him, but after take-off the weakness and dread returned. He landed for an hour at Karachi to take on more oil and petrol, and to fit a new screw on his propeller, and then he was off again to Jask, where he stayed the night. His flying time was now ahead of Mollison's and, although deadly tired, he felt easier.

He took off from Jask at about one am, groggy and dispirited, having slept for only three hours. Soon after dawn he noticed oil spots on the windshield. That called for a stop in Bushire for a quick repair, and another stop in Baghdad for a quick refuelling. Determined to reach Aleppo that night, he forced himself on with reasonable success until about nine pm, when he received a nasty scare: his oil pressure gauge began to fluctuate. Certainly he should have landed and investigated, but he was now flying over remote desert and did not dare. One suspects he was recalling the death of Anderson in the Tanami Desert, and was remembering that Anderson, having landed, was unable to take off again. 'I shall never forget this interminable stretch of desert-country over which I flew,' he wrote later, 'nor the sense of utter loneliness that came over me.' Praying the engine would not overheat, he kept on flying. At eleven pm he arrived at Aleppo, so cramped and exhausted that he had to be lifted from his plane. Four hours later he was off again.

It was now the seventh day of the flight, and Smithy was near the end of his tether. However, he knew that if he reached Rome that night, he could still beat Mollison, and that thought kept him going. Dazed and disorientated, he flew at first too far south, but by exerting every ounce of willpower he found the coast of Asia Minor, and even managed to fly over the open sea for half an hour without the expected fear. But the reprieve did not last long. On reaching the Gulf of Adalia, nausea and dizziness set in. 'I felt so bad,' he later wrote, 'that I determined to land without delay.' Outside the town of Milas, in Turkey, he saw a flat, open space on which he could safely land.

Alighting from the cockpit, he stretched out under the wing so that he could rest. A few local inhabitants gathered to look at him, and he tried to make one of them understand his needs. 'Cognac,' he asked, pressing coins into the man's hand. Then mercifully he dozed, before awakening to see a large crowd and soldiers here and there. One pointed a rifle at him; he was a prisoner. Then a car appeared, and from it came an officer, an interpreter and a bottle of brandy. He had no permission to land in Turkey, and to make matters worse he had landed in a military zone. He was being taken to Milas for interrogation, after which the authorities at Ankara would decide his fate.

The hours passed slowly in Milas, while Smithy and the local commandant waited for word from official authorities. At first Smithy was closely guarded: a soldier was ordered to stay with him, even when he was sleeping. He so disliked this intrusive surveillance that he bribed the soldier with a pair of pyjamas, thereby persuading him to stay outside the door. Since Smithy was a high-ranking air force officer, the commandant was later persuaded to release him on parole, and he was allowed to live at the house of American and English tobacco-buyers. On the third day, thanks to the efforts of the Australian government, he was given permission to leave.

Meanwhile, there was consternation in Athens and Rome when he did not arrive, and in Australia and Britain and America, where newspaper readers were nervously following the flight. London's *Daily Mail*, obviously expecting the worst, published an article that read like an obituary. 'He simply has every quality required,' it told its readers, 'brilliant technical skill, courage, good humour in adversity, and wonderful physique. He has accomplished every long distance flight worth attempting. It is a dazzling record all the more remarkable for its almost mechanical consistency.' Fortunately, Smithy lived to read the praise.

It was 3 October before Smithy reached Athens, by which time the possibility of beating Mollison's record was extinguished. Even so, he had hoped to fly on to Rome the next morning, and would have done so if his symptoms had not again appeared. It was a wise decision to rest for several more days in Athens and consult a doctor, because when eventually he reached the Adriatic Sea, he was able to cross it with little trouble. By now he was impatient to reach London and return to real life. At four-thirty pm on 7 October, he landed at Heston Aerodrome, and in little more than an hour he consumed a plate of poached eggs, spoke to Mary on the telephone and was driven to a luxury suite at the

Dorchester Hotel. Although he looked tired and had lost weight, journalists were surprised by his manner and appearance. He had spoken so frankly about his illness in interviews in Athens and Rome that they were expecting an invalid, yet Smithy now seemed 'remarkably well, considering his trying experiences'.

The following day, Smithy consulted doctors about his illness. For three hours a 'nerve' specialist and a heart specialist 'ran the rule' over him. The next day he confided their findings to reporters. Not satisfied with Smithy's self-diagnosis of sunstroke, the specialists told him that his heart, although organically sound, was 'severely strained' by nervous exhaustion and lack of sleep. Had he not been stopped in Turkey, he would, according to their view, almost certainly have perished over the next stretch of sea. They also declared that – to quote Smithy's words – his 'breakdown' was 'the climax to a series of nervous troubles', and would have occurred 'even had he not undertaken his long flight'.

These days, most doctors would have little hesitation in diagnosing Smithy as suffering from a common illness known as panic disorder. The dread, nausea, dizziness and disorientation are classic symptoms. Doctors today know more about the disorder than they did in 1931, but even then his specialists seem to have recognised its association with prolonged periods of stress, and to have known that it is sometimes induced – although not always – by an emotional 'trigger', in which case the panic is classed as phobic. For Smithy, the trigger seems to have been his long-held fear of being lost in a remote place from which he could not be rescued.

Smithy spoke openly to reporters about his breakdown, seeing no shame in it since he believed himself to be ill. Even so, he refused to disclose what those stress-inducing 'troubles' were. Anybody close to him could have guessed at them. Firstly, there was the war, which few soldiers had endured without heightened levels of stress, and after the war came the strain of returning to civilian life. The Pacific flight had been intensely stressful, and it had quickly been followed by the lawsuits, Coffee Royal and the tragic deaths of Anderson and Hitchcock. More recently, he had endured the perils of the Atlantic crossing and his illness in Holland. Added to this long list were the death of his father, the loss of the *Southern Cloud*, the dissatisfactions in his marriage and the unrelenting pressure of staying at the top. In fact, he had experienced more stress-inducing episodes in his thirty-four years than most experience in a lifetime.

Smithy's doctors grounded him until the following April. He answered them positively: 'It will be rotten to be earth-bound for the next four months but I am going to get that record.' At the Vacuum Oil offices in London, he relayed the news of his grounding to Mary over the telephone. Part of the conversation was broadcast by radio to Australia. 'Hello, Kiddie, it's all up,' Smithy told her. 'The doctors have turned me down and I will not be able to fly back. Meet me on the steamer at Colombo.' The tone of Mary's voice, on the other side of the world, conveyed her very great relief.

Before leaving England – and after – Smithy also gave interviews in which he accused his Turkish captors of unnecessarily imprisoning him. In response, the Turks began to insist that he had never been imprisoned, but simply detained for his own good. Smithy would have none of this. 'I was taken to a hut,' he told reporters firmly, 'and kept there under guard for two days. I was not even allowed to visit my plane under escort. I wonder what the Turks would describe as imprisonment if that was only looking after me. I feel that the statement impugns my character.' As future events would show, it would have been more sensible for him to have said nothing.

Soon after Smithy's phone call, Mary sailed to Colombo on the *Orsova* in company with her mother and sister. There she joined Smithy on the *Orford*, and they sailed back to Australia together. What transpired on the voyage one does not know, but clearly he did not – could not – calm her fears about his dangerous life. When he spoke to the press on his arrival, he said he had two objectives: to beat Mollison's record and to establish an aerial mail and passenger service between England and Australia. 'There will have to be some tall talking to my wife during the next four or five months,' he told his interviewers, 'but I am averse to leaving a job unfinished.'

While Smithy had been absent, Ulm was fighting to revitalise Australian National Airways. To operate the Singapore–Darwin mail route still seemed the company's prime chance, and, believing it was time to remind the prime minister of his airline's capabilities, Ulm persuaded the government to employ Australian National Airways to make a mail delivery to London in time for Christmas. Smithy arrived home to find that the *Southern Sun* – with Scotty Allan as the pilot and Lieutenant Colonel Brinsmead as a passenger – was ready to leave for England with 53,000 letters and packets of Australian mail. Smithy confessed to reporters that he longed to go with them, but this was impossible. He was still grounded.

The departure of the experimental mail flight earned welcome publicity for the airline, but on 26 November, while taking off from the swampy airfield at Alor Star, on the west coast of the Malay Peninsula, Allan crashed the plane. Brinsmead, who was travelling to England to discuss the airmail service with the British Air Ministry, then transferred to a plane of the Royal Dutch Airline, KLM. It too crashed on take-off, leaving him with serious injuries. Distressed by Brinsmead's condition and the failure of the flight, and frustrated at being grounded, Smithy itched to jump into a plane and fly to Alor Star. He now felt well, and a rescue flight would raise his spirits and boost his self-esteem. He shut his mind to his doctors' prognosis that, without adequate rest, his illness would return.

In the 1930s, psychological medicine was still in its infancy, and there were many who disbelieved that symptoms like Smithy's could spring from a purely emotional cause. They considered it more likely that his body had reacted to a poisonous substance encountered while flying. One possible culprit was carbon monoxide. It might have entered the cockpit through a leak in the exhaust pipe, although the strength of the slipstream made this unlikely. Smithy leaped upon this dubious explanation with enormous enthusiasm. It laid to rest his fears that the illness would recur, and justified the reissuing of his pilot's licence. Armed with this diagnosis, he approached a sympathetic physician, who, after careful examination, was happy to embrace it.

Within days a medical certificate was signed and Smithy's flying licence was reinstated.

The miracle continued. Although KLM and Imperial Airways offered to collect the stranded mailbags and fly them on to London, their offers were refused. Instead, the *Southern Star* was prepared for Smithy, who called a press conference to announce his departure. 'There is nothing spectacular about this flight,' he told the reporters modestly. 'We are just keeping our word ... it is my duty to Australia.' When questioned about his illness, he glibly replied that 'nervous apprehension' was a thing of the past. 'Why should I worry about this trip?' he asked, 'I know the route so well.'

By 5 December, he and his engineer, Wyndham Hewitt, were collecting Scotty Allan and the mail at Alor Star. From there they flew to Bangkok, Calcutta, Karachi, Aleppo, Athens and Rome, fighting their way across the Apennines in a blizzard, and halting for two days in Lyons, where they were fogbound. Indeed, for much of that December, fog impeded them. On what was to have been the last day of their journey, Smithy was forced to land on the beach at Le Touquet because of fog. It was 16 December when the *Southern Star* finally touched down at England's Croydon Aerodrome. A large crowd of wellwishers was waiting, and there were cries of 'Bravo, Smithy!' and 'Good old Smithy!' In reply, Smithy announced: 'I am very proud to bring the first direct airmail to England.' In Australia the headlines read 'Santa Claus Smithy' and 'ANA wins through'. His exploit had caught the public imagination. Without doubt, he was on top again.

It was to have been a quick turnaround, with Smithy and Allan flying home almost at once with the English mail destined for Australia. But before this could be, the *Southern Star* needed repairs at A.V. Roe's factory, near Southampton. Allan was sent to collect the plane on 23 December, but while returning to Croydon he lost his way in more fog. Over orchard country in Kent he ran short of fuel and was forced to land, colliding with a tree in the confusion. This compelled a further delay for more repairs, and a Christmas in London, which Smithy found cold, lonely and depressing. Back in Australia, Mary was less lonely and certainly warmer. Staying with friends at Bega, on the New South Wales south coast, she spoke to Smithy on the telephone on Christmas Day. But it was still a disappointing way to spend the second Christmas of her married life.

At last, on 7 January 1932, the *Southern Star* began its return journey. There had been more delays. Smithy had caught flu and Hewitt had left him, being replaced by Patrick Gordon Taylor, a former Australian National Airways pilot who had relocated to England. There had also been changes to the route because the British authorities had ordered Smithy not to fly over Turkey. His remarks to the press had made him an unwelcome visitor even in Turkish airspace. The journey to Darwin took fourteen and a half days; not a time record, but reportedly the longest commercial flight to that date.

The public concerns about Smithy's health were not eased by his quick and successful flight with the Christmas mail. 'All pay tribute to his marvellous spirit but what if he is disregarding his health?' asked the Sydney *Sun*. Onlookers in Darwin were keen to see if he showed signs of stress, but, snappily dressed in khaki shorts and a shirt, he looked mentally and physically alert. Yet there were other signs that he was not as fit as he appeared. As he approached the Timor Sea, he had insisted on aborting the crossing and turning back to Koepang. He blamed the torrential rain for his change of plans, and the next day, in clearer weather, he had made the crossing without incident. Nevertheless, Allan would later recall that, over the Timor Sea, Smithy appeared strained and unwell.

Smithy admitted to feeling uncharacteristically weary. Speaking to reporters, he confided that he had been invited to attend an air conference in Rome by Mussolini, Italy's air-minded dictator, who had not yet been classed as an enemy. 'I'd go like a shot,' he told them, 'but I'm just too tired.'

KNIGHT OF
THE BARNSTORMERS

O nce the applause for Santa Smithy died down, life resumed its troubled way. His money problems reappeared, his quarrels with Mary returned – for she was demanding he should give up his overseas flights – and Australian National Airways remained insolvent. Indeed, with two crashed planes, it was more heavily in debt than ever. Their only hope now was to gain the contract for the Singapore–Darwin airmail route. 'We were ready to take our part in establishing a regular air service,' wrote Smithy. 'We had demonstrated, despite accidents, floods, fogs, wretched aerodromes and poor ground organization, that the Empire mail was a practical possibility. We had the plant and the organization and the personnel.'

With so much in their favour, they could not possibly give in, yet negotiations between the Australian government and Imperial Airways seemed bogged down in official committees. As the year of 1932 progressed, however, there came a sign of hope. In midyear, the Australian government rejected an offer from KLM – which for the past two years had been flying between Holland and Java – to extend its weekly mail and passenger service to Wyndham or Darwin. The government insisted that Australia must give preference to British Empire companies.

Ulm seized his chance. For some months he had been developing a suggestion of Norman Brearley's: that, instead of acting as rivals, West Australian Airways, Australian National Airways and Qantas should combine into one large company. A pooling of their resources should give them an excellent chance of winning the Singapore–Darwin route. Frederick Stewart, the chairman of Australian National Airways, had recently been elected to parliament and was now Minister for Commerce, which made him ideally suited to present this proposal to the government. Stewart predicted that, with the aid of a suitable subsidy, this combined service could be self-supporting 'within three or four years'. Alas, the critics were not impressed. As one commentator put it: 'Both Qantas and WA Airways are actually operating but Australian National Airways is in a state of suspended animation' and has 'a deal less to put into the scheme than to take out of it'. Negotiations for the proposed amalgamation continued uneasily into the following year. Brearley, who liked and admired Ulm, was enthusiastically in favour of it, but Qantas's Hudson Fysh, who disliked Ulm, had misgivings. He saw no reason why Qantas should ally itself with a lame duck.

Meanwhile, Smithy was back at barnstorming. Having persuaded nine or so friends and relatives to help him, he planned to visit country towns across the eastern states. Foremost in his entourage was his brother Wilfrid, who, being his manager, sat in a little tent and sold the tickets. Others were mechanics, headed by a 'small, well-built man, with a charming manner and an engaging smile'. This young charmer was John Thompson Pethybridge – known as Tommy – who had been loaned by the RAAF to Doc Maidment when he came to overhaul the *Southern Cross*'s engines. Tommy so hero-worshipped Smithy that he had now left the RAAF and come to work for Smithy and Ulm. Also with the team was a pilot named Oliver Blythe Hall – always known as Pat – who gave additional joy rides in a DH.50 biplane known as the *Southern Cross Midget*. Sometimes Smithy's brother Leofric and his nephew John joined in as well. Altogether they were 'a merry crew'.

The *Southern Cross* was the most famous plane in Australia, and maybe in the world. The previous year in the Tasmanian country town of Brighton, some 3000 spectators had arrived just to gaze on it. Now, through February and early March of 1932, similar scenes were enacted in country towns in Victoria and New South Wales. When it was announced in advance that joy flights were available at ten shillings a ride, the tickets sold out immediately.

Smithy was amazed by the variety of his passengers: 'old and young, men and women, farmers, shopkeepers, girls, youths, anybody, everybody, clamoured for rides'. 'Wherever Smithy moved,' wrote one spectator, 'he was followed by a wake of small boys and girls, their eyes nearly popping out of their heads at the sight of their hero in the flesh.' Those little children begged their parents for half-price tickets, and Smithy was so touched by their artless excitement that he gave away, to those who could not afford the fare, thousands of flights and autographs. Years later, 'Des', of St Arnaud in northern Victoria, would remember how he had approached Smithy with a supplicating expression; where-upon Smithy looked at him 'like a judge', and after uttering the cryptic phrase, 'Not a word,' handed him a free ticket. Hero-worshipping young women flocked to him also, and some received free flights as well. Women were said to constitute the largest proportion of his passengers, signifying, perhaps, that his sexual attraction had increased with his fame.

Did those encounters that his mother had euphemistically termed 'harmless little flirtations' continue? Rumour says that they did. Mary did not usually accompany his barnstorming, and when his nerves played up, he needed whatever relaxation he could find. He appears to have felt no guilt, adopting the philosophy that passing encounters did not matter, provided that he still loved his wife and family. This, of course, was not how Mary would have viewed it. Smithy appears to have been confident that the press, the public and his close friends and employees would keep the secret.

His partners were not always hero-worshipping young women. He also seems to have had encounters with prostitutes. On the airmail flight to England, Scotty Allan, who was somewhat straightlaced, was shocked when a young woman knocked at his hotel door in Lyons.

She told him that Smithy had sent her, and that Smithy was already with one of her fellow-workers. Allan claimed to have rejected the offer. It must be pointed out that Allan already disliked Smithy because he believed that Smithy regarded him as an employee rather than an equal. Consequently, Allan may have felt entitled to reveal, and maybe exaggerate, his employer's indiscretions.

It is curious that Allan believed Smithy looked down on him. Smithy's egalitarianism was widely acknowledged, and one of the reasons why he was idolised. Australians were normally suspicious of 'tall poppies', and Smithy's humility and casual ways won him many fans. Indeed, in some circles he was considered too egalitarian, and was criticised for failing to uphold the dignity of his position. Those critics were appalled when he stood on his head in pubs to drink a glass of beer, or sang ribald songs and acted the fool at all-male 'smoke nights'. His latest party trick was to wear trousers he had bought in the United States which featured a novelty then unknown in Australia: a zipped fly. From time to time he would demonstrate the zip. Some thought his fooling was hilarious, others thought it was vulgar.

Possibly Smithy's seeming aloofness with Allan stemmed from his attacks of panic: in trying to control them, he set up a barrier between himself and Allan, which was misinterpreted as aloofness or even snobbery. Nor was this the only aspect of his character that Allan did not understand. He believed that Smithy's education was 'quite limited', and that he knew nothing of such great writers as Shakespeare, which was certainly not true. Allan was correct, however, when he said that Smithy's commitment to flying was absolute. As someone close to Smithy put it: 'Air, space, speed, distance were a constant, continuing challenge. No woman could compete with this; few could even live with it.'

During those weeks of barnstorming, in order to save himself extra strain, Smithy refused tiring, formal invitations. Even so, he remained surprisingly accessible to his wider public. On a local landing field, in the street or at a pub, he was always ready to shake a hand, sign an autograph, join in a drink or stop for a yarn. Wilfrid Kingsford Smith tried to keep count of the size of Smithy's public. In August 1931 he told reporters that so far Smithy had given 75,000 autographs, and he and his plane had been seen by 5 million people. Late in 1933 he claimed that Smithy had taken up 30,000 passengers. No doubt Wilfrid exaggerated, but perhaps he was not too far from the truth. A joy flight lasted about fifteen minutes and accommodated as many passengers as could be crammed into the cabin, for no safety rules curbed the numbers. The normal load was fourteen to sixteen adults, but with children sitting on adult knees the passengers could total twenty.

An escapade in Sydney would soon add to Smithy's sense of strain. The city was in festive mood, preparing for the opening of the Sydney Harbour Bridge on 19 March 1932. Smithy was doing excellent business in joy flights around the harbour. On the day of the opening, while he and Mary and her mother were celebrating at a party, he announced on impulse that he would pilot a flight over the bridge that very evening. Gathering up half a dozen of the formally dressed guests – his mother-in-law among them – he escorted them to Mascot Aerodrome and the *Southern Cross*. The flight over the floodlit bridge went splendidly until it was time to land. An unexpected wind caused the *Southern Cross* to strike the ground at such a speed that the undercarriage collapsed and the port wing broke. All were bruised and shocked.

Smithy was now in a quandary, because later that evening he was due to declare open a charity ball on an old sailing ship at Circular Quay. Should he cancel his appearance, or was he sufficiently calm and unhurt to do his duty? Shaken as he was, he managed to pull himself together. Boarding the ship to the sounds of 'For He's a Jolly Good Fellow', he charmingly explained that he had been detained by a slight accident and declared the ball open. Then he and Mary danced till dawn. 'Kingsford Smith always keeps faith with his public,' he told his friends proudly. The repairs to the plane at the naval dockyard at Cockatoo Island took many weeks, and the government did not pay the whole bill.

There were many calls on the government in these months to find him paid work. The newspapers – and Smithy himself – were not slow in pointing out that in America a flyer of his brilliance would have been regarded as a national treasure, and would have been found a remunerative post. Australian officialdom, however, did not fully trust an aviator who was both popular and unpredictable, and Smithy's cheekiness to public servants did not help him. In the words of one friend, 'He had this impish, irresistible urge to be always baiting and harassing officials.' If chided for it, he simply laughed. Officialdom at the highest level had also been responsible for barring him from a knighthood. But so spectacular were his recent successes that public opinion became difficult to ignore.

Smithy and his flying circus were in Grenfell, in New South Wales, when, on 2 June 1932, his knighthood was publicly announced. He was one of the three Australians who became Knights Bachelor in the King's Birthday Honours List. Next morning, with his flyers, he journeyed to Canberra for two days of barnstorming, and he lectured on his Atlantic flight at the crowded Capitol Theatre, where a commercial newsreel of the flight was already playing. Mary came by train from Sydney, delighted to be sharing his title, and that evening they attended a King's Birthday dinner at Government House. The following day, Lady Isaacs, the governor-general's wife, braved a joy flight in the *Southern Cross*.

In an interview with reporters, Smithy defended his barnstorming. While he was 'glad and proud to receive the honour', he did not feel that his 'mode of earning a living was beneath the dignity of a knight'. As to being called Sir Charles, he hoped that his friends would continue to call him Smithy. In country towns, where knights were not usual, his wish caused some confusion. He was frequently addressed as 'Sir Smithy', and occasionally Lord Smithy, and when Mary was with him they might be hailed as 'Sir Smithy and his good lady'. Sometimes he was affectionately called 'King Smithy', which was partly an abbreviation of Kingsford, partly an acknowledgement of his elevation, and partly a repeat of his popular nickname, 'King of the Air'.

While Smithy had been waiting in Sydney for the *Southern Cross* to be repaired, he attempted to teach Mary to fly. He was too impatient to teach her effectively, and Mary's early enthusiasm faded. 'I never enjoyed it,' she confessed, 'and after about three hours solo I gave up.' During those months, Mary became pregnant, and she was glad of the excuse to abandon her flying lessons. Smithy was overjoyed at the prospect of the baby: 'he became very protective,' she said, 'spoiling me more than ever.'

Despite Mary's pregnancy, Smithy saw no reason to linger at home. Through July and August, he and his troupe flew to Queensland, before travelling into South Australia, after which they followed the Murray River into southern New South Wales. Towards the end of the tour, he returned home briefly for his investiture as a Knight Bachelor at Government House. Splendidly dressed in his Air Vice Marshall's uniform, complete with medals and sword, Smithy became Sir Charles Edward Kingsford Smith on the morning of 8 October. 'I congratulate you,' said Sir Isaac Isaacs from his dais in the ballroom, 'on this latest mark of distinction in your great and meritorious service.'

But Sir Smithy had no intention of simply being meritorious. At the end of November, after touring Victoria, the *Southern Cross* was overhauled. New engines were fitted, the cabin was enlarged and the large tank behind the cockpit was replaced by a smaller, flatter one, thus allowing easier access between the cockpit and the cabin. Not content with barnstorming Australia, he had decided to barnstorm New Zealand.

On the afternoon of 22 December 1932, at their home at Bellevue Hill in Sydney, Mary gave birth to a son. The baby was named Charles after his father, and Arthur after his maternal grandfather. But there would be no christening in the Scots' Church for another five months. Although overjoyed, Smithy insisted on flying to New Zealand from Gerringong Beach, near Sydney, on 11 January 1933. 'I knew by then,' said Mary some years later, 'that this was going to be the pattern of our marriage – that his son and I weren't going to see very much of him, ever.'

Smithy assembled a fresh bunch of assistants for the New Zealand trip, some of whom were to fly with him across the Tasman Sea, and some of whom were to travel by steamer. His brother Wilfrid and the mechanics and pilots Tommy Pethybridge, Harold Affleck and Harry Purvis were to sail, but John Stannage was to fly, being indispensable in operating the state-of-the-art radio which had been loaned to them by Phillips Limited. Unlike the old receivers and transmitters, it dispensed with Morse code, and, like the radio telephone, it transmitted the human voice. Through the receiving station at La Perouse and the network of the Australian Broadcasting Commission, ordinary Australians, sitting beside their wireless sets at home, could hear Smithy speaking. Already familiar with his voice – for he spoke quite often on aviation matters from the ABC studio in Sydney – they were thrilled when his voice came from an aircraft high above the ocean waves.

In order to fly safely to New Zealand and back, Smithy had engaged Bill Taylor, the pilot who had accompanied him from England the previous year. In some ways it was a curious choice, because at first glance the two seemed so unalike. Fastidious, abstemious and, despite his Sydney upbringing, 'very English', Taylor seemed in many ways Smithy's opposite. His admiration for Smithy, however, was boundless, and he was able to assess his skill and character with remarkable insight.

Smithy's tough appearance, wrote Taylor, hid a sensitive, finely balanced personality, supported by 'an inner structure of fine steel'. His sensitivity enabled him to 'see, feel and predict the air, vividly and accurately'; and his 'steel' enabled him to deal capably and firmly with any situation. An unusual combination of qualities, it lay at the heart of his greatness as an airman. Although a year older than Smithy, and an experienced pilot, Taylor lacked experience as a navigator. And yet 'he never questioned my navigation', wrote Taylor admiringly; 'we found a complete mutual confidence, and I think that was one of his greatnesses'.

Smithy carried two passengers: Stan Nielson, the secretary of the New Plymouth Aero Club, and Jack Percival, a Sydney journalist who was now Smithy's publicity agent. Both felt nervous about the flight, and Smithy did little to reassure them when he explained the safety features of the plane. Prominently displayed inside the cabin were a fireman's axe, a vicious-looking hacksaw, two large cold chisels, two square-headed hammers and a coil of rope. Smithy explained that the tools were to saw off the engines if they were obliged to ditch, and the rope was so they could secure themselves to the wing, which would act as a raft. 'You can't be too careful,' he told them. 'If the worst happens, all we have to do is ride it out until someone finds us.' But in so vast an ocean there was little hope of rescue, as well he knew, and the prospect must have worried him. Determined to make the journey, his anxiety compelled him to clutch at every straw.

Fortunately, the flight to New Plymouth, which took fourteen hours, proceeded smoothly, and Stannage's presence provided Smithy with an additional source of security. Already a trusted friend and colleague, Stannage, in the previous October, had married Smithy's niece Beris, and no marriage could have pleased Smithy more. Indeed, he was in high spirits when they touched down at New Plymouth. To the cheering crowd, he said: 'Many people consider the flight across the Tasman a big feat, but it is really quite simple. I've done something far more clever recently. I've become the father of a nine pound son.' He was slightly taken aback when a man in his audience promptly shouted out that this was no reason to boast: he himself was the father of an eleven-pound son. Thinking quickly, and echoing the popular poem by Kipling, Smithy shouted back: 'You're a better man than I am, Gunga Din. But give me time!'

Across New Zealand, the barnstormers drew excited crowds. Taking advantage of specially granted half-holidays, trains brought country visitors long distances, some to ride in the *Southern Cross*, some to gaze at it. On one day in Auckland 9000 spectators were present, Smithy gave thirty-three flights and he took up more than 500 passengers, three of every four of whom were said to be women. Income from the joy flights – which cost £1 for an adult and ten shillings for a child – was reportedly £432, and to cap this off, the New Zealand government in a fit of generosity excused Smithy from paying tax on his earnings. Against this, however, they were obliged to deduct the cost of servicing the flights, which in Australia came to about £60 a day and might have been higher in New Zealand.

By the end of January 1933, Smithy's nerves were protesting and he needed to rest. It was therefore almost a blessing that on 4 February, in Palmerston North, the *Southern Cross* met with an accident. As Smithy taxied slowly toward the hangar, the wheels of the plane sank into the mud, and smashed the undercarriage and part of the wing. Affleck, Pethybridge and Percival were also aboard but nobody was hurt, and Smithy accepted the mishap with good humour. Since it was calculated that the repair work would take three weeks, Smithy snatched the chance to return to Sydney on the SS *Makura*, celebrating his thirty-sixth birthday on the way. The ship's chef presented him with a monster birthday cake, but he was in no mood for rejoicing. Age was signalling an end to his adventuring days – and, in a sense, to his life as well. For he knew that to relinquish adventure would be a form of emotional suicide.

In Sydney, the rest he craved was not forthcoming because he was obliged to confer with Ulm on the fate of Australian National Airways. Their hopes of resuscitating the airline were deteriorating, and voluntary liquidation seemed the only solution. Speaking to the press, Ulm announced that he and Smithy retained hopes of forming a new company and tendering for the Darwin–Singapore airmail route, but Smithy did not formally endorse this statement.

On 19 February 1933, still exhausted, Smithy returned to New Zealand by ship. Another month of barnstorming awaited him in cities and towns as far south as Invercargill. In addition, he was to speak at cinemas and halls that showed newsreel films of his aeronautical feats, and this task he found infinitely more taxing than flying, but somehow he summoned the strength to do it. He also faced another fear. For some time he had doubted whether he should fly the *Southern Cross* back across the Tasman Sea. Firm in his memory were the gales that had endangered the 1928 crossing. As the day of departure drew near, his anxiety reached crisis point and it appears he was briefly admitted to hospital.

In the end he resolved to fly back, choosing the Ninety Mile Beach on the west cost of the North Island for his take-off. Bill Taylor had been summoned from Australia to navigate, and Stannage and Pethybridge were aboard. Leaving before dawn on 26 March, they reached Sydney thirteen and a half hours later: an easy journey, compared with the 1928 westward crossing, which had taken twenty-three hours. When he arrived at Mascot – where his wife and son and 15,000 avid fans were waiting – Smithy jauntily proclaimed the flight had been 'a picnic'.

But was it? Taylor would recall a moment in mid-flight when Smithy, thrusting the controls into his hands, crawled back to the cabin because nausea had overwhelmed him. After an hour's rest he was fit enough to resume piloting.

Although he tried to make light of them, these panic attacks could no longer be ignored. Nobody could continue to believe that his nervous illness was caused by carbon monoxide poisoning. Garnsey Potts was the former Royal Flying Corps pilot who had replaced Smithy in the crew of the *Kangaroo* in the air race back in 1919. As a pilot, Potts knew the anxieties associated with flying, and fellow feeling moved him to submit an article about Smithy's illness to a Brisbane newspaper. 'Even the strongest nervous system,' he wrote, 'must cry out for a rest when subjected to abnormal strain over a long period.' Smithy's nervous system 'for the last eighteen years has valiantly responded to a continual and abnormal demand on its resources and reserves'. In Potts' opinion, 'the majority of pilots would have cracked up long since'. He begged Smithy to rest, and to give up trying to break records. Mary echoed this good advice. But Smithy closed his ears.

HOW LONG CAN I STICK IT?

On his doctor's orders, Smithy took a month's holiday, driving Mary and the baby and the baby's nurse on the long car journey to Melbourne. He toyed with exciting new prospects. He and Ulm may have lost Australian National Airways, but there was nothing to stop him founding his own aviation company and bidding on his own for control of the Darwin–Singapore airmail route. He was not including Ulm in his plans. He felt the need to strike out on his own.

Smithy also intended to organise a twenty-five-day flight around Australia in the *Southern Cross*, carrying ten paying passengers, seeing remote and exotic places, among which he included an aerial view of the site of Coffee Royal. After that, he proposed to fly the *Southern Cross* to England, carrying eight paying passengers; a New Zealand couple had already booked seats on that flight. Knowing the true nature of his illness, those close to him must have viewed these proposals with dismay, for by rights he should have been grounded. But nobody seems to have dared to stop him. Presumably his friends and family hoped that the presence of a co-pilot would ensure the safety of the passengers. Fortunately, neither of the flights eventuated.

Returning to Sydney in May 1933, Smithy began to promote his schemes, building a large hangar at Mascot for his planes, which had now been augmented by three Gipsy Moths. His new aviation company, which he named Kingsford Smith Air Services, offered charter flights, a flying school and an aircraft-maintenance workshop; in time, he hoped to manufacture his own planes. Stannage was his manager, Tommy Pethybridge his chief engineer, Pat Hall, Harry Purvis and Harold Durant were his pilots and instructors, while Wilfrid remained the manager of his barnstorming tours. But the backbone of the company was Marge McGrath, a secretary of legendary devotion. Four years previously, at Australian National Airways, Smithy had chosen her from 300 applicants; and despite being only eighteen and slight in build – she was so thin she was nicknamed 'Split Pin' – she had coped with his vagaries, professional and personal, as few others could have done. Very quickly she had discovered that 'he had absolutely no idea of the value of money. He'd give it or lend it to anyone, and couldn't bring himself to ask for repayment, no matter how broke we were.' When money became so scarce that salaries went unpaid, Marge took an evening job in a nightclub, and worked for Smithy during the day for nothing.

Faithful Marge was, by all accounts, one of the few women who was impervious to his amorous advances. But this did not mean that he ceased to make them, although by now his overtures had dwindled into jokey teasing, in which Stannage joined as well. As she sat working at her desk, the pair used to serenade her, Smithy providing the ukulele accompaniment. At other times, he and Stannage would discuss her in Morse code, of which she knew enough to realise that the discussions were 'flagrantly ungenteel'. According to the aviatrix Nancy Bird – who was at that time a pupil at the flying school – Marge was 'more than a match for them', and became 'the only woman Smithy would ever take at the end of the day into the bar of the Mascot hotel'. Also according to Nancy, it was Marge, aided by Stannage, who dealt with those tricky times when Smithy used his flying lessons as a cover for his assignations. His employees were sometimes hard-pressed to keep his sexual escapades secret from Mary and the newspapers.

Smithy's star in these months was high. His barnstorming tours and his lectures and radio broadcasts and newsreel films had made him almost a personal friend to thousands of Australians, and his public could not get enough of him. Journalists vied with one another to write about him. They called him 'Champion of the Air', 'Wizard of the Air', 'Ace of the Air' and 'King of Aviators'. They described him as 'the man who has flown over more of the earth than any living man', and the man who has 'made the name of Australia ring in other lands – among peoples who visualize us as a nation of kangaroos and bush'. They even tried to invest his daredevilry with an aura of respectability. Smithy was a 'stunter', wrote one reporter, 'only when occasion demands – fearless yet shrewdly calculating'.

While it was gratifying to have lived down the slurs cast by the Coffee Royal Inquiry, Smithy could still display dare-devilry, as his nephew John Kingsford Smith would testify. John remembered the night when, on sudden impulse, Smithy thrust him into the passenger seat of a Gipsy Moth and made a hair-raising flight over the Kingsford Smith home in Bellevue Hill. Wishing to 'wake Mary up' – and presumably the baby also – he began 'to do aerobatics right over the house'. 'I just kept praying,' wrote John, 'that his judgement was still intact because at the bottom of the loops he was pulling up with barely 100 feet between us and the ground'. And to cap this exploit, when they returned to Mascot, Smithy made a perfect landing on an unlit runway. John remembered the incident as 'an amazing demonstration' of his uncle's aerial prowess: it had seemed, wrote John, 'as if wings were part of his body'. Fortunately for Smithy, none of his dare-devil escapades ended so dangerously as to find their way into the newspapers. The reports that were spread by word of mouth usually ended up enhancing his reputation.

Needless to say, Mary did not approve of her husband's dare-devilry. There were many things now of which she disapproved, but which she despaired of altering, for two years of marriage had shown that no matter what she said or did, he was unlikely to change. Her reaction was to live and let live, and to enjoy the social advantages of being Lady

Kingsford Smith. Observing them together, Norman Ellison said that they enjoyed a 'happy comradeship that allowed husband and wife unconventional freedom'. That freedom was on Smithy's side, not hers.

Mary was unhappily aware that Smithy was planning another solo flight from England to Australia. The previous year, Charles Scott had flown the route a second time, setting a new record of eight days and twenty-one hours, and Smithy was itching to beat it. He was captivated by a monoplane manufactured in England but designed by an Australian flyer, Captain Edgar Percival. Called the Percival Gull, it was light and fast and indeed perfect for record breaking, and Smithy believed that it offered him an irresistible chance to beat Scott's time.

To buy such a plane he needed money – which he did not have, but which, ironically, he had possessed only few months before. His admirers had launched a testimonial fund for him, raising over £1000, but Smithy was embarrassed by their generosity and did not feel he could accept it. Instead he donated the fund to the Far West Children's Health Scheme, which provided medical services and holiday camps for children in remote, rural New South Wales. Not everyone approved of his donation; some hailed it as wonderfully unselfish, others called it irresponsible. Even so, Smithy believed that he had acted correctly and he happily bought the Percival Gull on credit. 'I'm used to debt,' he told his mother. 'In fact I'll be amazed if I'm ever out of it.'

Smithy was not alone in wishing to break records and earn glory. Ulm wished similarly, although he usually preferred to batten down his feelings and present himself as the sober man of business. Occasionally, in response to strong emotion, his suppressed ambition rose to the surface. Back in August 1930, soon after Smithy's Atlantic crossing, Ulm made an unsuccessful attempt to break the Australian endurance record. His sudden desire for public acclaim seems to have sprung from his disappointment at being left out of Smithy's Atlantic crew.

Similar feelings surfaced in Ulm when Smithy published a newspaper account of his flights in the *Southern Cross*. Written as a type of autobiography, it was ghosted under Smithy's supervision by the journalist Geoffrey Rawson, and appeared in serial form in Australian newspapers during September and October 1931. The following year it would be published as a book under the title of 'The Old Bus'.

Reading the early instalments, Ulm was appalled by the minor role assigned to him, and since he was still smarting from the criticisms directed at him by the Coffee Royal Inquiry, he was in no mood to be generous. His resentment at being passed over, his suppressed craving for glory and his envy of Smithy's spectacular success suddenly precipitated an emotional outburst. He believed that it was his own 'genius for organization' that had enabled Smithy's achievements, and he was determined to make sure this was acknowledged in future chapters. Since Smithy had left for England before the serialisation began, it fell to Leofric and Rawson to soothe Ulm's wounded feelings.

It was said by some journalists at the time, and continues to be said today, that rivalry eventually made Smithy and Ulm dislike, and even detest, one another. And yet there is no evidence of an open rupture. In the words of Scotty Allan: 'although there were lots of rumours, there wasn't to my knowledge any culminating row. They just began to go separate ways.' But was it possible that these two emotional, strong-willed men could amicably agree to part? Part of the answer may lie in an accident that befell Ulm only a few months after the final instalment of the autobiography appeared in the newspaper.

On 21 February 1932, at Laverton Aerodrome, near Melbourne, Ulm almost lost his life when, in the course of an emergency landing, his plane hit overhead electricity lines. He and his co-pilot, John Kerr, would certainly have been killed if an automatic device attached to the lines had not shut down the voltage at the moment of impact. Even so, Ulm received shocking injuries. Flung from the plane as it crashed to the ground, he was taken to the Caulfield Military Hospital, where he partially regained consciousness. However, he was so dazed by pain in his back, head and chest that he failed for a time to recognise his wife, who had rushed to his side. He suffered from shock, head injuries, severe concussion, lacerations and a broken arm. It was many months before he fully recovered.

The accident, reminding Ulm of his own mortality, seems to have provoked something akin to a psychological liberation. As he returned to health, he ceased to be the buttoned-up, and sometimes embittered, man of business. Making up for lost time, Ulm – in the words of a London journalist – 'embarked on a course of action that in the next couple of years involved him in more hazardous flights than most men would like to tackle in a lifetime.' His social outlook was transformed as well. The new Charles Ulm became less controlling, less dismissive, kinder and more in tune with the feelings of others. The aircraft designer Lawrence Wackett, who had earlier considered Ulm a difficult man to work with, now acclaimed him as a 'master of diplomacy'.

Smithy was prominent among those who benefited from Ulm's changed personality. In a speech at the Australian Flying Club six months after the accident, Ulm praised Smithy without reservation. Every country had its idol, said Ulm, but Australians could 'confidently claim that Sir Charles Kingsford Smith was the greatest airman in the world'. Smithy replied with similar generosity, acknowledging his deep debt to Ulm. The Pacific flight, he said, was a result of Ulm's 'organizing genius'. Then he made a self-deprecating 'Smithy' joke: 'It was probably the most organized flight ever accomplished. Owing to financial vicissitudes it was organized and reorganized at least 20 times!'

In April 1933, fourteen months after the accident, Ulm announced his intention of flying from Australia to England, and then on to America, whereby, like Smithy, he would have circumnavigated the world. From Australian National Airways he had bought the *Southern Moon* – soon to be renamed *Faith in Australia* – and he rejoiced as it was transformed by Lawrence Wackett into a larger and stronger machine with longer wings and multiple fuel tanks. Ulm told the press that he aimed to show the world that Australia was quite as capable as any other country of 'serious commercial aviation'.

On 21 June 1933, Ulm left for England with Bill Taylor and Scotty Allan as his crew. A crowd of fans farewelled him, prominent among them his son, John, the child of his first marriage. A Sydney *Sun* reporter watched the boy's parting from his father. 'The youngster,' wrote the reporter, 'burst into tears and clung to his father, who mingled his tears with those of the boy.' Perhaps the happiest gainer from Ulm's newfound sensitivity was this ten-year old lad, who, until a year or so previously, had been almost a stranger to his father. To forge a loving relationship with a previously neglected child cannot have been easy, and it was a testament to Ulm's skill that he succeeded.

Thereafter, fortune did not smile on Ulm. Soon after arriving in England, he flew *Faith in Australia* to Ireland's Portmarnock Beach to prepare for the Atlantic crossing. His plans were thwarted when the undercarriage collapsed and the incoming tide almost flooded the plane. He returned to England disappointed but without bitterness. Several British newspapers, pitying him, named him 'Ulm the Unlucky'; the nickname would trail him for the rest of his days.

If Smithy felt uneasy at Ulm's surge of ambition, he gave no sign of it. Presumably he was thankful that they were friends again. Also he was preparing for a flight of his own. Announced at almost the same time as Ulm revealed his plans, Smithy was intending to fly his Percival Gull – which he had named the *Miss Southern Cross* – from England to Australia. In August 1933 he sailed on the Dutch ship *Nieuw Holland* for Java, where he transferred to a plane bound for Singapore and then London.

Needless to say, Smithy's plans did not find favour with Mary. Rather than stay at home, she and her mother travelled with him as far as Singapore, and it was October before the two women returned to Australia. Meanwhile, baby Charles remained in Melbourne with Mary's sister. To leave her son, 'an engaging little chap' barely eight months old, cannot have been easy. However, the Powells seem to have been determined that Mary would accompany Smithy wherever possible, and, since she might need help, they made sure that one of them went along as well. Her parents appear to have understood that Mary's marriage had little chance of survival unless it received their emotional and financial support, and they gave both generously.

When Smithy arrived in England, Ulm was there to welcome him, and together they watched the motor racing at Brooklands and discussed the Singapore–Darwin airmail route, for which each was planning to make a tender. Smithy agreed to join Ulm's new company as an advisor if Ulm was successful. On 3 October 1933, the day of Smithy's departure, Ulm flew alongside Smithy's plane all the way from Heston, west of London, to Lympne, Kent – and, on parting, waved 'his hand in an aerial goodbye'. To reporters, Smithy was reticent about trying for a record. He hoped to reach home in seven days, but was not prepared to 'reduce myself to a shadow with successive days of sleeplessness'. 'I am a family man now,' he told them. 'The only record that really interests me is to be the oldest living airman.'

Certainly he was anxious about the flight. He had ceased to believe that he could master his nervous symptoms. He could only hope that longer periods of rest might lessen their force, and that his strong willpower might keep them at least partially at bay. His greatest fear was that he might lose consciousness in the air. He had consequently brought along a remedy: a bottle of sal volatile, a mixture of ammonium bicarbonate, water, alcohol and aromatic oils. This popular restorative against faintness was usually inhaled, although small amounts were sometimes swallowed. When taken orally, the sal volatile needed strong dilution with water, because if taken neat, the ammonia was likely to burn the throat.

On his first day out – 4 October – Smithy had no need of sal volatile, reaching his stop at Brindisi, in southern Italy, with no more than a slight queasiness as he crossed the Gulf of Genoa. But the next morning, on his way to Baghdad, his panic returned. As he flew over Athens, he wrote in his diary: 'Still getting these nervous attacks. Guess I am too old and worn out for these capers.' Approaching the Turkish coast, the attacks grew sharper and faintness began to accompany them. He prayed that he would not be forced to set down, because he had now been officially forbidden to land in Turkey. By early afternoon his faintness was worse and he feared he might pass out in the air. Reaching for the sal volatile, he took 'a couple of big doses'. These, he wrote gratefully, 'seemed to fix me up', and he was able to fly on to Baghdad.

That night in Baghdad, he 'couldn't sleep for nerves'. The next morning he 'felt pretty rotten', but nevertheless he was airborne before dawn. At about nine am panic struck again, making him feel so 'terribly sick' that he wondered if he could last the distance. An hour later he had 'a very bad turn', and became so faint that he took the plane down to 200 feet, believing he might suddenly need to land. The faintness continued all the way to Gwadar, where he landed at around four pm, still feeling 'terribly sick' and barely in charge of himself.

He went to bed immediately but could not sleep. 'I cannot sleep, that is what is killing me,' he informed his diary. Nor could he eat: 'I have had no food now for 36 hours.' On 7 October he forced himself to fly to Karachi, where he rested for four hours, craving sleep but still unable to find it. Fortunately, he managed to consult a doctor, who prescribed him bromide pills, and with their help he reached Jhodpur, where he went straight to bed.

At Jhodpur he asked himself: 'How long can I stick it?' Another sleepless night had brought him close to despair. But two hours later, as he sped over Allahabad, his despair lifted. 'Have just realized that I am more than half-way to Wyndham. Cheers,' he wrote in his diary. 'Cheers. Feel a shade better.' He even managed to keep down a bowl of cold broth. 'If only I can make Akyab tonight,' he told himself, 'the rest is easy.'

It was as well that he felt stronger, because he was about to encounter one of the toughest challenges of the flight: the Bay of Bengal. He had been dreading the crossing but by some miracle negotiated it without incident. At Akyab he had his first restful night, sleeping for six hours; the next night, at Alor Star, he slept for another six. As he hastened towards Surabaya it seemed that his troubles might be over, although there was still the Timor Sea to cross. Although he could feel panic over any isolated terrain or ocean, the most potent triggers appeared to be the Bay of Bengal and the Timor Sea. 'I suppose,' he wrote, 'that doctors would call it aquaphobia.'

It was now 11 October and he was almost home. To his diary he confided: 'Have just got to hang on somehow for another eight hours.' At eight am, as he passed Sumbawa, he felt stirrings of dread, but fortunately they soon subsided, only to return with frightening force as he came to the Timor Sea. The prospect of crossing the sea filled him with such apprehension that his stress levels were dangerously high even before he saw the water. Deep was his relief when he sighted Cape

Talbot, on the West Australian coast. A few hours later he touched down in Wyndham. He was returning to a hero's welcome, because he had performed the flight in seven days, four hours and forty-three minutes, beating Scott's time by forty hours.

On 13 October, Smithy arrived in Brisbane to scenes of rejoicing. Congratulations poured in from around the globe. In England, the 39-year-old aviator Sir Alan Cobham told the press that Smithy had proved that 'the old 'uns are as good as the young 'uns'. And Ulm, still in England, told reporters: 'Smithy is a magnificent fellow.'

From Brisbane Smithy flew almost directly to Melbourne for a very public reunion with Mary. More than 100,000 fans had gathered at Essendon Aerodrome, while another 100,000 lined the route to Mary's sister's house, in the Melbourne suburb of Caulfield. According to the newspapers, no royal visitor to Melbourne had received such a welcome, and so insistent was the crowd that Smithy was obliged to appear several times in the doorway of his sister-in-law's house with his baby son in his arms. Smithy and Mary and the baby then retreated to a holiday house in the seaside town of Mornington, where he began at last to know his son. A charming description exists of Mary and Smithy taking the little boy bathing. Afraid of the waves, little Charlie refused to go near the water, and would not be pacified until his father perched him on his shoulders and ran with him like a deer along the beach.

While Smithy was resting beside the sea, Ulm was already in the air. The repairs to his plane, which had threatened to bankrupt him, had been generously financed by a rich patron, and he was able to advance his departure date. He took off with Scotty Allan and Bill Taylor as his crew on 13 October. Six days, seventeen hours and fifty-six minutes later, they landed in Derby. Ulm had sliced almost eleven hours off Smithy's record.

Smithy was at the seaside when the press caught up with him. 'Ulm has done wonderfully well in the face of all difficulties. He is a man of great determination,' Smithy told the reporters. On 28 October he cut short his holiday and flew to Sydney to escort Ulm's plane in an aerial guard of honour from Parramatta to Mascot. It was 'a privilege', he announced, 'to welcome in true sincerity' the pilot and crew that had knocked his record 'kite high'. But at Mascot, Ulm would have none of such talk. With truth and tact, he explained to the crowd that Smithy's record was unbroken because his was a solo flight; it would remain unbroken 'for a long time', he predicted.

As a truck drew up to carry the Ulm and his crew on their ceremonial parade around the tarmac, he gestured to Smithy to join them. When Smithy modestly declined, Ulm, Taylor and Allan 'stooped down, seized him under the armpits and hoisted him onto the vehicle'. This moment – one of the proudest in Smithy's life, and in Ulm's too – seemed to heal old wounds, and rekindle hope.

THE
GLITTERING PRIZE

S mithy returned from England owing large sums of money. Fortunately, his indebtedness was somewhat relieved when the federal government awarded him a prize of £3000, but this gave only temporary relief. He needed to earn more money. Even so, he was only half-pleased when the Vacuum Oil Company offered him a position, for, as he was the first to admit, sitting day after day at a desk was not his forte. Recently the press had been urging the government to find him a post. 'How much longer will Australia be mean-spirited towards its most heroic adventurer,' asked the Sydney *Sun*. It urged the government to make him 'Director of and Consultant in Australian Aviation'. Having no desire for a government position, Smithy had answered quite sharply that he could 'battle along quite well' without one. Since the position at Vacuum Oil was only advisory and the money was vital, Smithy ended up accepting it.

Smithy's impulsive attitude to money had continued, unabated, for over a decade, and he was not about to change. He remained – to quote John Stannage – 'the prey of every cadger that happened along'. In the past, Leofric had tried to impose order on his younger brother's finances, but he had long since given up, so Stannage was placed in charge of Smithy's cheque book. At first the scheme 'appeared to be successful,' wrote Stannage, but after comparing the cheques butts with the bank statements, he changed his mind. He discovered 'certain queer withdrawls' that turned out to be counter cheques. Smithy seemed 'quite astonished that he should have been found out.'

Ulm was also in debt, and his hopes of financial security were based on winning the contract to carry the Singapore–Darwin airmail. However, by the end of 1933 neither Ulm nor Smithy seemed likely to be a winner. In fact, Smithy had publicly abandoned the idea of tendering in his own right, although he remained willing to act as Ulm's technical adviser if Ulm's bid proved successful. Over past months it had become clear that the proposed amalgamation of West Australian Airways, Qantas and Australian National Airways would never take place. Their differences were too great. When Hudson Fysh suggested that after their amalgamation, their new company should ally itself with Imperial Airways, Ulm was aghast. Unlike many Australians of that time, Ulm did not believe in giving preference to British companies and products. On the contrary, he held that the routes between Singapore and Darwin and within Australia belonged by moral right to the Australian companies that had pioneered them, and that the acquisitive Imperial Airways should be kept away from them. Undeterred, Fysh quietly forged the alliance on his own company's behalf, and in late November 1933 Qantas joined Imperial to form Qantas Empire Airways.

Disappointed by their lack of commercial success, Smithy and Ulm turned to New Zealand. An airmail service across the Tasman seemed a likely extension of the England–Australia route, and this in turn would foster valuable airmail connections between the principal New Zealand cities. Smithy also hoped that he might find a market for another project. Months earlier he had ordered a plane from Lawrence Wackett for the Singapore–Australia route. Called the Codock – after its birthplace at the Cockatoo dockyards in Sydney – it carried six passengers, ample fuel tanks and was powered by Napier Javelin engines. Its construction was underway when the government deemed it unsuitable for overseas travel because it had only two engines; Smithy's argument that it could maintain a flight of 5000 feet on one engine and so had the 'safety of a multi-engined aircraft' was rejected. Now, he realised, it might be useful as a mail carrier in New Zealand.

Hoping to create a small airline linking the New Zealand cities, and meanwhile to earn money from joy rides, Smithy sent Wilfrid to New Zealand to organise a barnstorming tour. Ulm made similar plans, although his barnstorming tour was to be carried out separately from Smithy's. In December 1933 Ulm set out for New Plymouth, taking Scotty Allan and Bob Boulton as his crew, and his wife and his secretary as passengers. Thus, Josephine Ulm and Ellen Rogers became the first women to fly across the Tasman Sea.

On 13 January 1934 Smithy took off in the *Southern Cross*, also bound for New Plymouth. Taylor, Pethybridge and Stannage formed the crew, and Jack Percival and Stanley Neilson accompanied them as passengers. It was as well that Mary had decided to cross on a ship, because the flight over the Tasman proved 'a nightmare': the radio failed and the weather was so wild that the journey took an hour and a half longer than expected. 'I would not care,' Smithy told reporters, 'to go through that experience again.' Mary was waiting for him at New Plymouth, along with a crowd of anxious fans. She felt intense relief when he jumped down from the cockpit, almost into her arms.

To further their chances of gaining future airmail contracts, both Smithy and Ulm had requested permission from the Australian government to carry airmail from Australia across with them. To their annoyance, this was refused, their flights being judged too 'experimental'. Fortunately, the postal director in New Zealand knew better. When Ulm flew back to Mascot in mid-February, he brought 44,000 letters from New Zealand with him. In the face of this, the Australian director relented, and when, soon after, Ulm crossed the Tasman again, he was entrusted with 37,000 Australian letters destined for New Zealand addresses. This, wrote Ulm, furnished ample proof that 'a Tasman airmail service can be regularly maintained'.

In February 1934 Smithy briefly abandoned his New Zealand tour and sailed back to Sydney with Mary on the SS *Lurline*. During their absence, Charles junior had taken his first steps, and his nurse had brought him down to the wharf to greet them. Reporters watched with amusement as a 'small bundle of humanity hurled himself at his mother and seemed to take pride in demonstrating several wobbly steps taken unaided'. Mary was in raptures, and even Smithy was moved. Although nowhere near as doting as Mary, he was showing an increasing devotion to his son, who was growing into a charming little boy with a mop of blond curls and a round, happy face like his mother's. Thanks to years of barnstorming, Smithy related well to small boys, especially if they liked planes. Soon after his son's first birthday, Smithy had taken him up in the Percival Gull, and had been overjoyed when the little fellow showed no sign of fear. Quite the opposite – he 'showed his nonchalance by falling asleep'. Smithy had proud visions of one day teaching his boy to fly.

Young Charles and his nurse were at Mascot to greet Smithy when, having completed his tour, he flew home a month later. Smithy bounded up to the little boy, but the child did not share his father's enthusiasm. 'This bloke doesn't seem to know me,' Smithy exclaimed as he took the protesting toddler from his nurse's arms. Mary's presence would have helped, but she was sailing back from New Zealand on the SS *Wanganella*. Indeed, a few hours earlier the *Southern Cross* had passed high over her steamship in mid ocean. From far above the ship, Smithy had rung his wife on the plane's radio telephone, and for several minutes they had engaged in affectionate chat. Did this mean that Smithy was at last coming to terms with his role as a husband and father? Perhaps it did, because at six forty-five the next morning, Smithy and his son were down at the Sydney wharf, eagerly waiting to welcome Mary home.

The tour of New Zealand had achieved less than he expected, but Ulm still had hopes of gaining the Singapore–Darwin airmail contract. Deep was his disappointment when, on 19 April 1934, the contract was awarded to Imperial Airways and Qantas. Ulm was careful in his public response. 'What's the use of squealing?' he told a reporter with a smile and a shrug. Smithy spoke more forcibly: 'I feel that the men who brought to Australia a realization of the value of air transport and have striven against tremendous odds to provide efficient service for Australia have suffered a serious injustice.' Smithy and Ulm were further dismayed when Qantas and Imperial began to eye the trans-Tasman route. Ulm was determined to nip this in the bud. A company was 'at that very moment' being formed to run that service, he told the press firmly.

It was time to seek fresh projects. While Ulm planned an air service to New Guinea, Smithy set his sights on a glittering prize. The previous year he had been approached by Sir Macpherson Robertson, a famous Melbourne confectioner who was air-minded, very rich and a generous contributor to worthy causes. Robertson planned to celebrate the centenary of Melbourne's foundation by sponsoring the longest air race in the world: a contest starting in London and finishing in Victoria's capital. A galaxy of international stars was expected to compete for the prize of £10,000 and a gold cup. Robertson wanted an Australian to win, and Smithy was his best prospect.

Knowing about Smithy's empty pockets, Robertson had even offered him £5000 towards a new plane. And since the millionaire shared the Australian bias in favour of all things British and imagined that Smithy shared it too, he took it for granted that Smithy would choose the de Havilland Comet, which had been especially designed for the race and was being sold to certain aviators, Charles Scott among them. Smithy, however, did not favour the Comet, and bravely declared that he preferred a United States plane, since the Americans specialised in 'long-range, high-speed jobs'. As this was not what Robertson wished to hear, Smithy's choice of plane was postponed.

As April 1934 progressed, Smithy felt increasing pressure to forget the American plane and order a Comet, and in the middle of that month he did so. But when he explained to the de Havilland factory that his Comet must have propellers with a variable pitch – an aid to speed and efficiency not usually found in British planes – he received a blunt response from Britain. Owing to the lateness of Smithy's order, the de Havilland factory could not guarantee any of the modifications that he was requesting. Shrugging off his annoyance, he told himself that he had never really wanted the Comet, and that he would be better off with an American plane. He sent a cable, cancelling the order.

But what was he to tell Robertson? In a flash of inspiration, he thought that Lawrence Wackett, as an aircraft designer, might intercede for him. And Wackett did intercede, visiting the multi-millionaire at his Melbourne home, and carefully explaining the technical superiority of American long-range planes. Although not wholly convinced, Robertson softened. He renewed his offer of £5000, for he truly wanted Smithy to compete and to win. But now the extent of Robertson's generosity made Smithy feel uncomfortable. How, he asked himself, could he honestly accept so large a sum in such circumstances? So they compromised, settling on a gift of £2500, and Smithy booked a passage on the next mail steamer to search for a plane in America.

As Smithy would soon discover, his decision was unpopular – even held to be unpatriotic. The British press voiced its displeasure, and in Australia the Returned Soldiers League generated so much opposition that Smithy announced: 'I'm sick to death of adverse criticism levelled at me because I'm selecting an American plane.' He added angrily: 'I probably deal more in British than most of my critics – otherwise I would have accepted the 100,000 dollar offer of a job in America five years ago.'

Smithy did not weaken in his resolve. When the SS *Monterey* sailed in May for California, he was a passenger on it, and so was Mary, and so were Arthur and Florence Powell. One doubts if Mary was happy to be making the voyage. Having been home only a few weeks, she was required to abandon her darling child yet again. Smithy had tried to mollify her by publicly announcing that the Centenary Race would definitely be his last long-distance flight, but she had heard that excuse too often to believe it. At least, on this trip, she had the prospect of her parents' support, which must have been a comfort. Also, Arthur Powell seems to have been paying the fares, which would have been a further comfort to Smithy, who was now living largely on the generosity of his in-laws.

The *Monterey* docked in Los Angeles on 19 May 1934, and Smithy's party was met by Harold Kingsford Smith, but the family reunion was brief. Within six days, Smithy had completed his search and bought his plane. A second-hand Lockheed Altair, it was a two-seater, low-winged, single-engine monoplane that had already passed through three owners, the first being an aviator named George Hutchinson, the second the actor Douglas Fairbanks, and the third the film producer Victor Fleming. Shaped like a giant winged bullet, it was the racing craft of his dreams. It possessed none of the traditional wires or struts, 'just a wing, a body and a tail of perfect form, like a beautiful blue bird poised for flight'. It was so futuristic, so fast, 'so apart from man' that he could scarcely believe it was his.

Realising that the Altair had to be partly rebuilt, Smithy sent it to the Lockheed factory in Burbank, on the outskirts of Los Angeles. There it received a new wing, four additional fuel tanks, a spate of new navigational instruments, and the very latest Pratt and Whitney Wasp supercharged engine – 550 horsepower. It already had variable-pitch propellers, metal wing flaps to assist the take-off and landing (unusual for the time), and a retractable undercarriage that folded under the wing to reduce wind resistance. The final touches were the colour and name. Smithy's new plane was painted silver and blue – like the *Southern Cross* – and it carried the name of his old mates-in-arms. It was called the *Anzac*.

So exceptional a plane was bound to be expensive, even when second-hand. Once rebuilt, its total price was more than £6000, and the expense of shipping it to Australia might add another thousand. The total cost was almost three times as much as Robertson had so generously offered. Smithy had been given a month in which to pay, but he had only six days before entries closed for the Centenary Air Race. If he could not afford the plane, there was no point in submitting an entry.

At his urgent request, Stannage lodged a public appeal through the Australian newspapers. The Melbourne *Herald* gave £500 and the *Adelaide Advertiser* gave £125. This, in turn, brought forth a donation from Smithy's former benefactor Sidney Myer of £500, with a challenge to other benefactors to give similarly. Arthur Powell responded with £500, which he is said to have later doubled. As more money was needed, Smithy sold his Codock to Newcastle's Northern Airways, and borrowed the remaining sum from his bank. By the end of the month he had paid for the plane and entered it in the race. On his entry form he stated that he had 6125 flying hours to his credit, and would be taking Captain P.G. Taylor as his co-pilot and navigator. In the Australian newspapers, he acknowledged his debt – and his hope – to his supporters: 'My backers and fellow Australians may rely that I won't let them down.'

Before leaving home, Smithy had been warned by Edgar Johnston, the Controller of Civil Aviation, that to land his plane on Australian soil he would need a certificate of airworthiness from a nation that was a signatory to the International Convention of Air Navigation. Smithy had little patience with this body, which he described as 'a lot of old women who laid down a lot of fusty regulations which were out of date before they were framed'.

Unfortunately the United States was not a signatory, so to overcome this obstacle he was advised to enter Canada, which was a signatory, and register the plane there. Or if hard-pressed, he could obtain an American certificate of airworthiness, and a declaration from the US Department of Commerce that the plane conformed substantially to international standards. He was further warned that while the US documents would satisfy Johnston – an old friend from his Royal Flying Corps days – they might not fully satisfy the Centenary Air Race's organiser, the Royal Aero Club, when he finally arrived in England to register for the race.

It is beyond doubt that Smithy received these warnings. But his attitude to officialdom was cavalier at the best of times, and in America there was no stalwart colleague – like Ulm or Stannage – to ensure that he behaved responsibly. Between 11 and 17 June, Smithy visited New York, ostensibly to obtain the documents, but he appears to have made no real effort. Much of his time seems to have vanished in social engagements. In New York he and Mary lunched with the celebrated American aviatrix Amelia Earhart. In California he visited Paramount Studios and was photographed with Mae West on the set of *Tis No Sin*, a film that was later called *I'm No Angel*. Possibly he explained to Mae, and to her co-star, Cary Grant, that, at home, he too was a film actor, having taken a leading role in a filmed history of Australian aviation. Written by Jack Percival and inspired by Smithy's book, *The Old Bus*, it was soon to be shown in Australian cinemas.

The most curious of his Californian encounters was with the celebrated New Zealand aviatrix Jean Batten, who had just flown solo from England to Australia. They conversed by radio telephone. Charles Ulm brought them together over the airwaves, and Jean could scarcely contain her excitement: Smithy had been her hero since childhood, and the first flight she ever made had been in the barnstorming *Southern Cross*. Their conversation was broadcast across Australasia by the radio station 2UE.

Of all the visits Smithy made during his ten weeks in the United States, perhaps the most heartwarming was to his old mentor, Captain Allan Hancock. Hancock had followed Smithy's career keenly from afar and was deeply concerned about his protégé's deteriorating health. Suffering himself from a speech impairment, Hancock believed that nervous maladies could be alleviated by diet and exercise. Hancock begged Smithy to eat nutritious foods, train regularly at a gymnasium, sleep more and abstain from alcohol and cigarettes. As a parting gift, he presented him with 'a book on health by way of food'. Deeply touched, Smithy resolved to try to follow Hancock's advice.

On 27 July Smithy and Mary boarded the *Mariposa* for Sydney. The previous day the *Anzac* had landed at the Matson Dock, in the port of Los Angeles, and Smithy himself had piloted its perilous descent, swooping under electricity lines and barely missing hydrants. The longshoremen were on strike at the time, but a few agreed to hoist the aircraft up by crane to the top deck, and settle it, in its entirety, onto the *Mariposa*'s tennis court.

Smithy by now was feeling unusually relaxed, thanks to Captain Hancock's health regime. 'I have gained weight,' he told reporters, 'and recovered from the jittery, nervous attacks I suffered. I sleep nine hours nightly, and have cut out drinking and smoking.' He was not even worried about his new plane's sketchy documentation. As a precaution, he had instructed the Lockheed test pilot to fly it to the Department of Commerce's aerodrome outside Los Angeles for an inspection. On returning, the pilot told Smithy that 'the machine was OK', and handed over an 'experimental' certificate. This was not, as Smithy well knew, what the Australian authorities required, but for the time being it seems to have placated his fears.

As the *Mariposa* approached Sydney, Smithy foresaw problems. How was he to convey the *Anzac* to Mascot Aerodrome? The ship was to dock at East Circular Quay, and Smithy wondered if he might move the plane into adjacent Macquarie Street and take off from there. He promised to do this at a time of minimum traffic, but even so the authorities would not hear of it. He then wondered if he might ferry the plane on a barge or lighter to Anderson Park, on the shore of nearby Neutral Bay, and use the grass as a runway. He acknowledged this would be tricky because the harbourside park was less than 150 yards long and was crossed overhead by powerlines.

Amazingly, the authorities agreed, and even removed the powerlines. But before the plane could leave the ship, customs officials demanded that the name *Anzac* be erased: the name was too sacred to be commercialised. As an Anzac himself, Smithy was not impressed, the more so when an old Anzac went out of his way to tell him, 'Yer got a good name for 'er, Smithy. The Diggers are behind yer, boy.' Even so, he complied, renaming the Altair *The Lady Southern Cross*.

As it happened, the moving of the plane to the park, and the take-off for Mascot proved relatively easy. The Altair's arrival without a certificate of airworthiness was a more difficult matter, for no plane could be registered in Australia until deemed airworthy. Documents must be summoned from America to prove its airworthiness, which was likely to take time. In the meantime, the plane – classed as a 'prohibited import' – was restricted to flights within a three-mile radius of Mascot Aerodrome. How long these restrictions would last was anyone's guess.

Smithy waited and fretted. He was buoyed by the plane's extraordinary capabilities, but since nothing was heard from the US Department of Commerce, he was beginning to doubt whether he would ever receive accreditation. To make matters worse, the ban imposed by Turkey in 1931 – forbidding Smithy to land in or fly over Turkish territory – showed no sign of being lifted. Repeated requests from the British and Australian governments had been rejected, and officials at the departments of Civil Aviation, External Affairs and Defence were beginning to believe that it would never be lifted. Those officials were also aware that Smithy had several times previously ignored the ban and flown in Turkish airspace – surreptitiously.

On 24 July, Prime Minister Joseph Lyons had written to the Premier of Victoria, warning of 'serious consequences' to Australia if Smithy was allowed to flout the ban in the Centenary Air Race. It might 'conceivably result', Lyons wrote, 'in Turkish authorities refusing subsequent applications from Australian nationals'; and as the most practicable routes between Australia and Europe involved flying over Turkish territory, this could punish all Australian planes. There was a private belief among government officials that Smithy should be grounded, and that the government should refuse to apply on his behalf to the foreign governments over whose territory he would be obliged to fly on his way to and from England.

At last some of the obstacles began to fall. After undergoing tests by the Civil Aviation Department, the Altair was granted 'provisional' registration and given the markings VH-USB. The restriction on flights outside the three-mile radius of Mascot was lifted, and Smithy and his co-pilot and navigator, Bill Taylor, sitting in cockpits one behind the other, commenced test flights to Australian capital cities. On 22 August they flew from Mascot to Melbourne in the astonishing time of two hours and twenty-five minutes. While in Melbourne, Smithy was confined to his bed with a severe attack of flu, but this did not stop him planning to fly from Laverton Aerodrome to Maylands Aerodrome, near Perth. On 8 September, before starting his flight west, Smithy dipped the Altair's wings over Box Hill cemetery, in suburban Melbourne – a tribute to his faithful benefactor, Sidney Myer, who had been buried there a few days before.

Smithy's flight time to Perth was a miraculous ten hours and nineteen minutes – less than half the time accomplished by the *Southern Cross*. With a cruising speed of about 160 miles per hour, the Altair was the fastest plane in Australia, but it still lacked an international certificate of airworthiness. Lockheed had kept no record of the multiple changes to Smithy's machine, and so could not supply the data required to the US Department of Commerce for the airworthiness certificate.

There were also difficulties over Smithy's four extra fuel tanks. When full, they added considerably to the plane's weight; some experts doubted that the Altair could bear it. Indeed, in England the organisers of the air race, the Royal Aero Club, were likely to demand that some of the tanks be sealed. Smithy by now was 'heartily sick of the delays'. The strand of egalitarianism in Australia that enjoys cutting down 'tall poppies' was once again busy, and a growing swarm of rumours suggested that he was too scared to enter the contest. 'My critics appear to forget that I have participated in one or two little flights before,' growled Smithy. 'If they think I am frightened – well, they can come with me on the race.'

For Smithy the deadline was fast approaching. He was required to be at the newly opened Mildenhall Aerodrome in Suffolk, England, for the plane to be 'weighed in' no later than 14 October 1934; and since the plane also had to be overhauled before arriving at Mildenhall, he believed that he should leave for England by 29 September. And still there was no sign of the airworthiness certificate from America, nor the requisite permission from Turkey. Fortunately, Smithy had been persuaded to send a placatory cable to the Turkish ruler, Mustafa Kemal Pasha (later Atatürk), offering to meet him in person in Ankara on the way to London, and this gesture had paved the way for a lifting of the restriction.

By 27 September – two days before his planned departure – Smithy's position had improved. The Australian Department of Civil Aviation, heeding his pleas, issued a 'provisional' certificate of airworthiness to cover the following three months. The Melbourne Centenary Air Race Committee, on the other hand, refused to give a ruling, and handed the dispute over to the Royal Aero Club in London, which had never regarded Smithy favourably. On the night of 28 September, Smithy went to bed not knowing if the club would accept the certificate. It was two in the morning when the telephone rang. Mary answered it and awakened Smithy. The Royal Aero Club was prepared to accept his entry, provided the Altair carried no more than 325 gallons of petrol. This was vexing because he had been counting on 418 gallons, which was a legal load under American regulations. With 418 gallons aboard he could fly nonstop between the race's control points, but additional stops for refuelling might impose a five-hour handicap. Nevertheless, he accepted the ruling.

Four hours after the phone call, Smithy and Taylor climbed aboard the *Lady Southern Cross*. Few spectators were present, although Mary was on hand with sandwiches and soup, and Catherine Kingsford Smith was lifted up to the cockpit to kiss her son goodbye. Smithy was eager to be away. Bill Taylor, sleepless and stressed, found the take-off rather an anticlimax. 'The spirit in which we set out,' he wrote, 'was more one of determination to see the thing through, and put up as good a show as possible, than one of real enthusiasm.'

Their first stop was to have been Darwin, but they were compelled by headwinds and the need for fuel to stop at Cloncurry, in northern Queensland. There, by a happy coincidence, they met Pat Hall and Harry Purvis and a group of government geologists who were touring the region in the *Southern Cross*. What had been planned as a brief fuel stop quickly developed into a cheerful and chatty meeting, but the atmosphere changed when Smithy asked Purvis to help him check the Altair's engine. Their routine examination revealed a dozen or so cracks in the aluminium cowling. The damage was so serious that it required immediate repair, otherwise the cowling might separate from the plane in midair. As the repairs called for highly skilled craftsmen, unprocurable in North Queensland, they had no alternative but to fly back to Mascot, with dust storms slowing their speed.

Close examination of the Altair at Mascot revealed that a new cowl must be constructed. Three skilled metalworkers toiled day and night. Meanwhile, Smithy, holding on to hope, vowed that if the cowl was ready by 4 October, he was 'still going to have a shot' at the race. News of the problem was conveyed to Charles Ulm, who was in New York. As soon as he learned of Smithy's predicament, he contacted three American manufacturers, promising to find a suitable plane and have it shipped or flown across the Atlantic for the start of the race. Smithy was deeply touched by Ulm's 'magnificent gesture', but who, he asked, would pay for the plane? He himself had no money – and Ulm had no answer.

On the afternoon of 3 October, Smithy inspected the cowl and found it far from ready. Suddenly the frustrations and disappointments of the past months became unbearable. In a rush of emotion, he decided to quit. Late that evening, he sent a telegram to the chairman of the Centenary Air Race Committee in Melbourne. He explained how deeply he regretted that circumstances were forcing him to withdraw, and sent sincerest good wishes 'for the safe carrying out of the most spectacular air race in the history of aviation'. When the committee begged him to reconsider, he remained firm. He was out of the race.

Bill Taylor remembered the shadow of depression that clouded the office at Mascot. Perhaps the only way out was to find another project for their miraculous plane. Taylor remembered Smithy staring for a time at the ceiling, then picking up the *Times Atlas* that sat on his desk. Turning to Taylor, he casually remarked, 'Let's have a look at the South Pacific.' Having studied the relevant page, he glanced at Taylor again and said, even more casually, 'We could do it all right in the Altair; how about it?'

AM I A SQUIB?

Smithy's decision to cross the Pacific was made quickly and almost desperately. He needed to perform a valiant feat to restore his reputation, because such a storm of criticism was blowing that it promised to be worse than the storm over Coffee Royal. In a way, Smithy had himself to blame. He had never attempted to hide his medical troubles from the public. Indeed, the logs of his flights, which detailed his symptoms, were published in the daily newspapers for all to read, for he saw no shame in what he was describing. He believed that faintness and fear and nausea were natural products of the strain of competing in record-breaking flights, and he took pride in the way he had managed to complete the flights – and even break records – despite his illness. He had always believed in pushing himself to his limits. That was why he was called 'Ace of the Air'.

Many people interpreted his illness differently. Some, thinking he had lost his nerve, saw his symptoms as a sign of cowardice and an excuse for failure; they alleged he had withdrawn from the Centenary Air Race because he was too scared to participate. Worse were those who claimed he was a habitual drunkard, and that his attacks of panic were cravings for alcohol. They doubted the stories of the strict regime that he maintained before a major flight: the disciplined life that Smithy summarised as 'early to bed, gym, exercises every morning, no beers and very few cigarettes'. His detractors, hearing that he liked to drink, considered that sufficient substantiation. Nor was Smithy the only 'tall poppy' of the time to be called a drunkard: Dame Nellie Melba was similarly labelled, without a scrap of evidence.

Although accustomed to being resented, he was unprepared for the heat of this resentment. Angry letters, most of them anonymous, began arriving at his Mascot office, and at the mailboxes of major newspapers. Some contained white feathers, the symbol of cowardice. Others harked back to Anderson's death. 'You surely must know he is a murderer,' one of his abusers wrote to *Smith's Weekly*. Even youthful worshippers were infected by the rumours. On the children's page of the Melbourne *Argus*, the editor gave a stern reproof to those boys and girls who were calling Smithy 'a squib'. Rumours that he was a squib – in Australian slang, a coward – were spreading fast, and he was deeply hurt.

Fortunately, powerful supporters came to Smithy's defence. Sir Macpherson Robertson, shocked by 'the many letters suggesting that Sir Charles Kingsford Smith has tricked me and taken me down', wished 'to deny this insidious suggestion'. Sometimes support came from unlikely quarters: Norman Ellison, Smithy's erstwhile critic at *Smith's Weekly*, was appalled by the unfairness of the attacks and the degree of suffering they inflicted on Smithy. 'His breezy front and casualness misled most people,' wrote Ellison, 'but from childhood he wished to be thought well of, and when he became a public figure an adverse public opinion, even though unjustified, used to sting him into white-face hurt.' With Ellison's assistance, Smithy dictated a long and frank statement concerning his entrance and withdrawal from the Centenary Air Race. Published in *Smith's Weekly*, it helped to stem some of the hostility.

Smithy's decision to fly from Brisbane to California was announced in newspapers on 5 October 1934. At almost the same hour, Charles Ulm announced his decision to fly from Vancouver to Melbourne. For two giants of Australian aviation to compete indirectly in this way pleased reporters, who were full of questions. Smithy declared that the aim of his flight was to repay his backers by selling the Altair in America, since – due to its perceived un-airworthiness – nobody in Australia would buy it. Ulm declared that his flight was preparing the way for an 'all British' airmail service across the Pacific. He had already ordered a twin-engine British Airspeed Envoy, believing it the most practical plane for the task, and he intended to carry a crew of two and a consignment of airmail with him. Might they meet in mid-Pacific, asked excited reporters. Clearly not, since Smithy proposed to leave on 20 October, while Ulm was to leave at the end of November.

With his decision announced, Smithy set about adapting the Altair for the flight. The sealed fuel tanks were reopened, and, disregarding problems of extra weight, more tanks were fitted under Smithy's seat and in the wing. Even with the extra fuel – which gave the plane a range of about 4000 miles – there were fears it might be insufficient to take Smithy from Suva to Honolulu, so, for safety's sake, he arranged for an emergency supply at Fanning Island. There was also debate about the value of a wireless. Taylor was a navigator who rejected wireless as a navigational aid, but Smithy insisted on carrying a small receiver and transmitter. Remembering Coffee Royal, he also insisted on carrying tools and a supply of emergency food, a still for producing fresh water, and a pump to empty the wing tanks. Once they were emptied, the wing of the Altair would float like a raft, thus resembling the wing of the *Southern Cross*.

At dawn on 19 October 1934, Smithy and Taylor flew the Altair to Brisbane. A few wellwishers were at Mascot to see them off, including Smithy's mother, but not his wife. As soon as Smithy left for England, Mary had moved little Charles and her maid and nanny to an apartment in Melbourne, close to the house of her sister. When Smithy abandoned the air race, she dutifully returned to Sydney. However, she was deeply unhappy about the Pacific crossing. To expose himself to untold risks simply to show he was not afraid seemed dangerously misguided, and she added her pleas to those of his friends and supporters – Robertson among them – who begged him not to go. She could not face another public farewell at Mascot. Far better, she decided, to say goodbye the day before his departure, and then return to Melbourne. 'The days will be anxious days,' she told reporters, 'but he will get there, I trust, as he always gets there.' And she added: 'I don't want to think about the flight too much.'

Officially, the Pacific crossing began at Archerfield Aerodrome, in Brisbane, at four am on 21 October. Despite the hour, about 700 were there to say goodbye and the aviators climbed aboard to strains of 'For He's a Jolly Good Fellow' and cries of 'Mr Smith! Show 'em what Australia can do'. Thereafter all went well until they passed Noumea, in New Caledonia. Rain squalls almost erased visibility, shutting down a cylinder and causing the plane to shake and thud. Taylor, in some alarm, called Smithy on the interconnecting telephone, for they were seated in separate cockpits, one behind the other, 'tandem style'. Direct speech was impossible unless they raised the Perspex covers of the cockpits and shouted above the engine noise.

Fortunately, Smithy had diagnosed the problem, and by diving sharply he restarted the ailing engine. It was a frightening episode but one he quickly solved. More troubling was the prospect of landing in small and partly wooded Albert Park in Suva. Remembering the hazards he had encountered during his famous landing there in 1928, he was thankful that they would be arriving in daylight, and so could rely on the Altair's excellent air brakes and wheel brakes. Even so, it took him two attempts to bring the plane safely down.

Poor weather, engine trouble and an aborted take-off – caused by the wind blowing the plane into the surf – delayed their departure from Fiji for eight days. Tommy Pethybridge arrived by ship to overhaul the Altair's engine, and while he worked, the aviators enjoyed their first leisure for many months. At six am on 29 October they took off from Naselai Beach for Honolulu. This promised to be the hardest section of the route, but at first it seemed easy, and their spirits rose when they saw how little fuel the aircraft was consuming. Their fears of running short were proving groundless, and at their present speed they might cut ten hours off the *Southern Cross*'s time over the same distance.

Trouble came with a burst of blinding rain. Reaching for the head-lights' switch, Smithy accidentally turned on the air brakes and the landing gear. The Altair instantly lost almost half its speed and hurtled into a violent spin, forcing him to wrestle for control. Recalling the incident, he said his wrestling seemed to have lasted a long time; in fact it was probably less than a minute. 'We thanked God,' he told the press in Honolulu, 'we had so much altitude, otherwise we would have met our doom.' As the heavy rain had also disabled the air speed indicator, he was obliged 'to wiggle the plane in order to feel the way and so guess the airspeed'. In this fashion, they safely reached Hawaii.

The landing at Wheeler Field was like old times. Five thousand Hawaiians were out to greet them, along with John and Beris Stannage and a host of expatriate Australian friends. So high were Smithy's spirits when he stepped from the plane that he announced cheekily: 'It takes the Australians to do it.' At the Royal Hawaiian Hotel, the Stannages had reserved Room 500 for him, with its view of the beach: 'My lucky room,' Smithy exclaimed delightedly, remembering his stay there in 1928. After breakfast, he and Taylor thankfully retired to their beds. After twenty-five hours in the air, they were exhausted.

To the delight of Hawaiian reporters, Smithy used his news conference next day to praise America to the skies and to settle personal scores. Perhaps it was the euphoria of a safe arrival, perhaps it was the stimulating presence of an old friend from his West Australian days, John Williams, who was now a reporter on the *Honolulu Star-Bulletin*; whatever the reason, he threw caution to the winds. His Altair – he told his eager listener – was a tremendous advertisement for the great strides made by US aircraft in recent years. If the British Empire wanted fast air services, it would need either to use American planes or beat the Americans at their own game of designing and manufacturing. He had no doubt that an American plane would have won the Centenary Air Race if it had not been ousted by ridiculous British regulations.

The British consul in Hawaii, having attended Smithy's news conference, informed the Foreign Office in London that Smithy was perceived by his audience to be 'more American in sympathies and inclinations than British'. And 'it cannot be said', he added disapprovingly, 'that Sir Charles was at any particular pains to correct this impression'.

Nonetheless, he was forced to admit that Sir Charles's flight in an American plane was 'one of the outstanding achievements of modern aviation history', and was being 'widely acclaimed as a new link in the development of a plane service between Hawaii and the coast'.

The consul enclosed in his confidential letter a curious statement, copied verbatim, that Smithy had handed to the Hawaiian press:

> Yanks are dinkum bonzer coves. What they have done beats the band and we have to crack hardy every time we appear in public. Foreigners chuck off at the Yank spirit, but they're a mob of parking coots who can't sprag us. Yank friendliness beats the band and we'll hate to do a bunk, particularly from sunny Hawaii.

Smithy's torrent of slang seems to mean, the consul dryly told the Foreign Office, that the Americans are the salt of the earth! Of course it was a joke, and highly effective. Today one can read the statement and the consul's letter at the Australian National Archives, Canberra.

Smithy and Taylor were due to do their 'bunk' the next day, but there was an unexpected delay. Holes were found in the oil tank and one of the petrol tanks: the Altair had been lucky to reach Hawaii. While the repairs were made, Hawaiian hospitality took over. One evening John Williams arrived with singers and dancers, and Smithy accompanied them on a newly purchased ukulele. It was 3 November before he and Taylor took off from Wheeler Field on a flight that Smithy anticipated would be a 'mere fleabite' compared with their previous journeys. And so it largely proved. Indeed, such was its ease that Smithy and Taylor had difficulty keeping awake – as Smithy later recalled:

> Once I took a bead on a star. I looked at it so long that it got to be the size of the sun. Then, all of a sudden, my head snapped back and I woke up. I had pulled back on the controls so far that I had almost stalled the ship. We had 10,000 feet altitude, so my little interlude had no unfortunate aftermath.

After fifteen hours they landed at Oakland Aerodrome. Their entire flying time – from Brisbane to Oakland – was fifty-two hours: astonishingly different from the eighty-four hours taken by the *Southern Cross* to fly from Oakland to Sydney in 1928. 'I'm sorry to be here so early,' Smithy joked to the waiting crowd, 'you have to blame the navigator' – whereupon he launched into generous praise of Taylor. Then he spied Harry Lyon and Jim Warner, and an excited reunion took place. Smithy seemed – wrote one observer with slight surprise – 'in excellent spirits and did not appear to be over-tired'.

No log or diary exists for this Pacific flight, but in the next couple of years Smithy and Taylor published their own accounts, both detailed, Taylor's version running to 262 pages. Smithy's nervous disorder gained no mention in either. Perhaps he was panic-free, although this does not seem likely. A more credible explanation is that the rumours about his drinking made them reluctant to mention his medical symptoms. As accusations of alcoholism cut Smithy deeply, he was eager to refute them, and later, at the airport in Oakland, he indulged in a type of charade. From the cockpit he produced a bottle of whisky, which he explained, for the benefit of reporters, was carried 'in case Taylor was sick on the way'. He made much of the fact that the bottle had not been opened. On one level he was engaging in jovial banter; on another he was sending a message to the public that he was essentially a sober man.

The formalities over, Smithy and Taylor sped off to breakfast with Harold Kingsford Smith and the Stannages, who had arrived the previous day on the *Monterey*. Police motorcyclists escorted them to Harold's house near Oakland, where more reporters waited, and Smithy was 'badgered unmercifully' about his future plans. This proved no problem: his mind was teeming with ideas. He told them that he had been consulting army experts in Hawaii about a possible 'speed flight' between California and Australia, and thought he might attempt it in a few months' time. He was considering acquiring a twin-engine Lockheed Electra, which carried ten passengers and had a cruising speed of around 200 miles per hour. If the long Pacific crossing could be performed in thirty-six hours, he believed it would become a viable proposition for commercial airlines.

The reporters also questioned Smithy about the Centenary Air Race, which had concluded while he was in Suva. The winners were C.W.A. Scott and Campbell Black; Smithy was delighted, because Scott was 'a fine fellow and an old friend'. Even so, he was sorry that he and his American plane had not taken the prize. He also said, very firmly, that the Pacific flight, almost entirely over ocean, 'is more important and difficult and demands greater skill in navigation than the Anglo-Australian race'. When Scott read this, he declared that although Smithy was undoubtedly the greatest flyer in the world, he doubted whether Smithy's flight could lead to a regular service across the Pacific.

The aviators could not dally at Harold's house. By lunchtime they were back at Oakland Aerodrome and ready to fly to Los Angeles. The crowd had swelled by about 5000, but this was small compared to the crowd of 50,000 that waited to greet them at Los Angeles, complete with city dignitaries and troupes of reporters and photographers. By evening, Smithy and Taylor were so tired that they were obliged to refuse a banquet in their honour. Over the next days the hectic pace continued. Smithy spoke at a large civic reception, visited film studios, addressed members of the motion picture industry, accepted honorary membership of the Los Angeles police force, and announced to boisterous dinner guests that he was trying to sell his marvellous Altair, the *Lady Southern Cross*. Like an auctioneer, he called for bids: 'Thirty thousand dollars for the *Lady*. She is worth all that and more.'

One incident was unwelcome. While he and Taylor were receiving congratulations in their suite at the Clark Hotel, a visitor appeared with legal papers claiming that Smithy owed a 'promoter' named Thomas Catton $2750 dollars for a debt incurred in 1928. Smithy summoned a lawyer when he heard that the Altair was about to be impounded as collateral. 'The lionised flyer is chafing at the delay,' announced one newspaper. 'He has received a flood of offers for contracts of flying stunts, lecture tours and appearances in movies. Meanwhile the Lady Southern Cross is under guard at the airport.' Fortunately, the dispute was quickly resolved.

When Taylor departed for London to join Qantas Empire Airways, Smithy – exhausted from playing the superstar – accepted Captain

Hancock's invitation to cruise on his latest yacht, the *Velero III*. Knowing Smithy's nervous problems, the captain was anxious to restore his protégé's emotional and physical health, and Smithy was equally eager to interest the captain in his latest schemes. If no buyer could be found, he would fly the Altair to England, and then on to Australia, and beat Scott's record. Or maybe he would buy a Lockheed Electra and fly a regular service between Sydney and Perth. His mind was full of exciting projects, all of which needed money.

On 10 November he returned to Oakland, where Harold had arranged for him to lead the American Legion Armistice Day Parade the following day. At first Smithy marched rather self-consciously at the head of the parade, but later he fell back to join the Australian returned soldiers. Buttonholed by reporters, he was optimistic about his plans. He told them he had been consulting financiers and would soon have an important announcement to make.

But nothing was as rosy as he hoped. Negotiations failed, his spirits sank and his nervous symptoms returned in force. No longer did the attacks of panic seem to need a 'trigger': they occurred at random. John Stannage was handling his press releases, and on 17 November he issued a sad little statement: 'Sir Charles Kingsford Smith has been harassed by autograph hunters and admirers since he landed a fortnight ago. He locked himself in his hotel suite today and sent out word that his nerves were ragged and he must get some rest.' If need be, Stannage added, 'I will take him away to some remote spot in the hills. He has been pestered and is tired of it all.'

His doctor prescribed complete rest and Smithy obeyed, confining himself to his hotel room. But by the end of a fortnight he was begging for his freedom. Charles Ulm was due to arrive at Oakland Aerodrome on 28 November. Icy weather had obliged him to cancel his departure from Vancouver, and he was now leaving on his trans-Pacific flight from Oakland. Smithy was determined to speed him on his way.

Smithy and Ulm spent 29 November 1934 in Oakland together. Details have not survived, but the newspapers described it as 'a cordial reunion'. Ulm and his crew of two were planning to fly to Hawaii the following day in his new, twin-engine Airspeed Envoy which he had named the *Stella Australis*. But storms and cloud delayed their departure until 3 December. They would have been wise to have delayed even longer, because the cloud continued. While approaching Honolulu, the *Stella Australis* became lost. At nine am Hawaiian time, Ulm radioed that he was out of fuel and about to land in the sea.

In Los Angeles that day, Smithy and Stannage were lunching at the Fox Film studios with Will Rogers, the movie star and amateur aviator. They had been chatting to actors such as Shirley Temple, and the meal had only just begun when news of Ulm's distress came through. Smithy's face turned white with shock. He was angry, wrote Stannage, 'that such a thing should have been allowed to happen'. He stood up and said, 'Let's go.' Quickly offering their apologies, the two Australians drove to their hotel, and for the next six hours tried to correlate the garbled reports from the ships and planes that were combing the ocean around Honolulu.

Powerful emotions urged Smithy to fly to Honolulu to join the searchers, but his Altair was being overhauled at the Burbank factory. Highly agitated, he begged Stannage to hurry there and collect the machine. Knowing that such a plan was 'idiotic and suicidal', and that Smithy was in no condition to fly anywhere, Stannage craftily handed him a drink containing 'knockout drops'. 'He had cracked up,' wrote Stannage, 'probably for the first time in his life.'

When Smithy awoke from his long sleep, his agitation returned. He was desperate to play his part in the rescue search, cabling John Williams in Hawaii that they must keep looking because the wing of Ulm's crashed plane would serve as a life raft, and Ulm and his crew could survive for days. As time passed and nothing was found, Smithy was forced to accept that Ulm was dead. 'It was heartbreaking', wrote Stannage, 'to see his feeble attempts to hide the quiet grief.'

The news of Ulm's death was not the only source of Smithy's pain. He felt deeply hurt when he realised that his record-breaking Pacific flight had caused no great stir in Australia or Britain. Sir Macpherson Robertson praised him warmly, but others, whose praises he had expected, were lukewarm or silent. The British magazine *Aeroplane* was frankly hostile, describing the flight as a pointless stunt. Ironically, some of the most generous praise came from two air-minded dictators, Benito Mussolini and Adolf Hitler. They realised that aircraft were shrinking the world, and transforming the nature of warfare. Seven years later, the ocean across which Smithy had so speedily flown would experience an event which even he could not have imagined – the attack on Pearl Harbor by Japanese planes.

As the year drew to a close, Smithy's emotional state worsened. On the one hand, he was depressed and longed to retreat to Sydney; on the other, he felt he must make use of the Altair while he was still its owner. He had met a rich Californian rancher named H.E. Walker, whose brother was in a hospital in St Louis. Walker engaged Smithy to fly him to St Louis, after which Smithy intended to fly on to New York. He had set his sights on beating Roscoe Turner's recent, record-breaking flight across the American continent. On 21 December he set out, but found that panic, depression and crosswinds were defeating him. Having landed Walker in St Louis, he abandoned hope of breaking the record and returned disconsolately to Los Angeles.

It was time to go home. On 8 January 1935 he boarded the SS *Monterey* in San Francisco. John Stannage was travelling with him – and possibly also Mary, who is said to have come to take care of him on the voyage. But if Mary did accompany him, the news was kept from the newspapers. The Sydney *Sun*, which prided itself on its closeness to the family, believed that she was in Sydney, preparing to welcome him home.

When the *Monterey* called in at Hawaii, so close to the scene of Ulm's death, Smithy was overwhelmed by his sense of loss. Fortunately, he found comfort in talking to John Williams, who had known both aviators almost as long as they had known each other, and could understand Smithy's grief. Williams knew how close Smithy and Ulm had grown in the last couple of years. 'Theirs was a friendship', wrote Williams, 'such as is born under fire' and not easily broken. He had learned 'from private conversations with them both that often they felt like abandoning their personal plans and partnering again'.

KING OF THE AIR

E leven planes of the Sydney Aero Club escorted the *Monterey* across Sydney Harbour, and one of them carried Smithy's mother. Down on the wharf, a bevy of reporters waited impatiently for Smithy to appear. Confronting the cameras and microphones, he spoke confidently of his future plans. He had secured the right to build Lockheed planes in Australia – 'wooden types' that would conform to Australian regulations – and he hoped to float a local company to build them in Sydney. He might also start an unsubsidised air service between Sydney and Perth; he felt sure it would be feasible because passenger planes flew across America every day. When asked if he had any more transoceanic flights in mind, his mood changed. He said he had been unwell ever since arriving in America as a result of the Pacific flight, so he was definitely out of transoceanic flying in single-engine planes. Then he thought a little and added that he might try for an England-to-Melbourne record in October.

The reception he received was reassuring, because he had wondered how the Australian people now regarded him. Any doubts about their warmth were dispelled the following evening at an Australian Broadcasting Commission community singing concert at the Sydney

Town Hall. Smithy was the guest of honour, and as he entered, the 2000 people stood and cheered and waved their handkerchiefs above their heads. Deeply moved, Smithy expressed his thanks, then asked the audience to stand in silent tribute to Ulm and his crewmen, Leon Skilling and George Littlejohn. After that, everyone sang, and the concert reached its climax with the performance of a new song called 'Smithy, the King of the Air', composed by Richard Lloyd. The words were projected onto a screen, and the tune was easy enough to allow most people to sing it. When Smithy left he received another enormous ovation.

The warmth of the welcome delighted him, nevertheless Ulm was much in his mind during the following weeks. In a ceremony of remembrance, fourteen planes flew out from Mascot and dropped wreaths into the Pacific Ocean, with Smithy leading the way in the *Southern Cross*. Of Ulm he wrote:

> He had great business capacity, 'punch' and vigour and a boundless ambition. I think his dream was to become chairman or managing director of a great world-wide aviation organization, for which he was in many respects admirably suited. He was at once both a first class air-man and a financial genius. His untimely end was one which, I think, he would have accepted with the fatalism which marked all his outlook on life.

Death and ageing were on Smithy's mind. On 9 February 1935 he turned thirty-eight. 'Many more happy landings,' exclaimed his secretary, Marge McGrath, as she handed him a ham and a bottle of pickled onions – the ingredients of his favourite meal. 'Well, I'm twenty-one today,' he announced breezily, but privately he was downcast. 'One can't be thirty-eight,' he told his friends. 'That's terribly old. I don't feel it.'

Actually, he often did feel it. He was beginning to accept that he was too old for the long-distance 'record-breaking game', and that another attempt might kill him. But how could he stop? Since boyhood his deepest satisfactions had come from confronting danger and pushing himself to his limits. That was how he had risen to the top of his profession, and kept his place there. If he denied himself those satisfactions, might it not be like an emotional suicide?

Mary must have sensed his inward struggle, but neither she nor his family and friends seemed able to help him. His worries might have seriously oppressed her, but fortunately she had a pleasant diversion. The previous year she and Smithy had moved to a rented house in Greenoaks Avenue, Darling Point, and although she loved the harbour view, she disliked the house. Luckily, an adjacent block of land with a similar outlook was for sale. Mary bought the block and was now absorbed in building a house on it. The architect she had engaged was providing her every wish, and her father was supplying most of the money. Her parents had lately moved to a house in Toorak, the most prestigious of Melbourne suburbs, and this seems to have strengthened their desire to give Mary the house of her dreams.

Projects jostled one another in Smithy's mind, but the dominant idea was an airmail service to New Zealand. Ulm and he had already paved the way, and all that was needed was government support. Smithy made certain stipulations: the service must be generously subsidised, and it must employ high-speed American planes. He favoured the twin-engine Douglas, and recommended oxygen gear because they would need to fly high to escape the Tasman Sea's storms. Early in March he abandoned a barnstorming tour of the Hunter Valley and sailed to Wellington for talks with the New Zealand government, discovering, to his dismay, that the government tended to favour his old rival, Imperial Airways. Smithy believed that the trans-Tasman route belonged by moral right to the Australasians who had pioneered it, and he refused to be put off. In the midst of the talks, he contracted what was referred to as 'flu', but which almost certainly had a psychological component.

Confined to his bed in Wellington, Smithy seemed so ill that Mary was summoned. Her presence helped and he was able to sail home with her on the *Wanganella* on 10 April. Even so, his emotional state remained low, and his doctor ordered a holiday by the sea. As word spread that Smithy was suffering from depression, his publicist, Jack Percival, wrote reassuringly in the Sydney *Sun*:

Sir Charles Kingsford Smith was feeling as fed-up, run down as a man can feel who has had a bout of the worst sort of influenza. Life was a flat spin over an endless ocean of depression with the pressure falling down, down, down. He had to have a holiday. Now, after four days, Smithy says he feels fine.

Meanwhile, Smithy was persuading the Australian government to permit a 'demonstration' airmail flight across the Tasman. With the silver jubilee of the reign of George V approaching, he foresaw excellent publicity if the flight were to carry jubilee greetings from the Australian to the New Zealand government. Moreover, the gesture would be magnified if ordinary Australians could be persuaded to send jubilee greetings to their New Zealand friends. At Smithy's insistence, Stannage travelled to Canberra and explained the plan to the acting prime minister, Earle Page. Approval was received, and Smithy, Stannage and Percival began to set it in motion.

It was Percival's suggestion to use both the *Southern Cross* and Ulm's *Faith in Australia* as the mail planes. These aircraft had carried mail across the Tasman for Smithy and Ulm the previous year, but both planes were old and in need of repair; even so, Smithy hoped to be ready by the middle of May. Meanwhile, the public was writing its jubilee greetings. Over 28,000 letters were posted, along with numerous small packages, and each item carried a special airmail stamp.

Smithy assembled two crews for the journey. A well-known naval navigator named Commander Bennett, a former employee of the Vacuum Oil Company named Beaumont Sheil and John Stannage were to accompany him in the *Southern Cross*. Ulm's former engineer Bob Boulton, the wireless operator Sid Colville and Jack Percival were to fly with Bill Taylor in *Faith in Australia*. Taylor doubted the airworthiness of the planes, but after a week of repairs they were judged by Smithy to be ready to fly. Although Stannage and Taylor expressed further doubt, Smithy was happy with the arrangements, and he remained so even after Commander Bennett fell ill and withdrew. Smithy simply decided to fly without a navigator. At this stage Stannage and Taylor considered asking Smithy to cancel the flight, but were restrained by respect for their leader, and knowledge of his admirably strong will.

Departure was scheduled for midnight on 15 May from Richmond Aerodrome, and that night the crews dined together at a hotel in nearby Windsor. After dinner, Smithy announced that the tonnage of mail did not justify the running of two aircraft. Since he – or maybe the Powell family – was bearing the cost, he decreed that only the *Southern Cross* should fly, and that Stannage and Taylor should accompany him. Taylor was perturbed. Of the two planes, *Faith in Australia* was by far the fitter, and he begged Smithy to use it. But Smithy would not listen. 'Nothing wrong with the Cross, Bill,' he told Taylor. 'Never let me down yet.' 'I believe Smithy felt,' Taylor recalled, 'that once he got his hands and feet on the controls of the Cross, she would stay in the air for him in any circumstances.'

Taylor was so worried about the safety of the plane that he considered refusing to go. And yet such was his devotion to Smithy that, close to midnight, he collected his gear and prepared to board the plane. At the last minute even Smithy had doubts. 'Here am I,' he said to Taylor, 'thirty-eight, apparently sane and sensible. And I'm going out over the ocean again in the middle of the night. Well, I'm surprised at myself. Aren't you, Bill?' Taylor, taken aback, laughed uneasily. By now Smithy's wife and mother had arrived to see him off. As he walked to the plane, he had an arm around each of them.

Notwithstanding their anxieties and forebodings, the flight went well until seven am. At that hour a metal part of the central engine's exhaust pipe – renewed only the previous week – broke loose and ripped a foot-long gash in the wooden blade of the starboard propeller. Immediately, strong vibrations shook the plane as the engine struggled in its mounting. Fearing the engine might break free and destabilise the aircraft, Smithy instantly shut it down and, as he was still closer to Australia, turned the *Southern Cross* towards home.

After such a loss of power, there was no way the heavily laden plane could maintain height. Desperate to lighten the load, the crew dropped luggage and surplus tools over the side, along with most of the fuel in the cabin tank. The mail, however, was retained, Smithy feeling honour-bound to keep it safe as long as he could. Whether he could bring the *Southern Cross* safely home was another matter. With only two engines, its speed was much reduced and it would not reach Sydney for another nine or ten hours. He doubted it could survive so long. Their best plan was to make for Port Stephens, on the north coast of New South Wales, about 600 miles away.

Stannage imparted this information by Morse code and radio telephone to Sydney, and called for immediate rescue. Since there was no regular rescue procedure, this caused confusion. A cruiser, the *Sussex*, and a steamer from Bluff called the *Fiscus* were hastily contacted, both being in the Tasman Sea; and on land, *Faith in Australia* was filled with life preservers and prepared for take-off. However, these hasty plans were plainly insufficient. If the *Southern Cross* went down in the next few hours, no rescuers could reach the crew in time.

Although Smithy had shut down the starboard engine, the propeller continued to spin irregularly, aided by the wind. In consequence, vibrations still shook the plane, and would continue to do so until the blades were trimmed to equal length. Leaning out the cockpit window, Taylor tried to trim the broken blade with a hacksaw, and Smithy tried to help him by placing the plane in the most favourable position. But the hacksaw could not help. The shuddering went on and simply had to be endured.

The *Southern Cross*, now at 500 feet, was travelling at the slow speed of around sixty miles per hour. Labouring against a headwind, it was close at times to stalling. It stayed aloft only because of the skill with which Smithy was flying. The Old Bus, wrote Taylor, 'responds to his touch like a horse to its master': 'she leans on the air, staggering, but staying on her feet', she begs for help and receives exactly what she needs. Smithy is 'literally holding her in the air with his hands and feet, juggling, coaxing her to do it, and getting the extra response nobody else could get from the machine'.

In this masterly way, he piloted the plane until noon, by which time the port engine showed signs of running out of oil. With the engine gasping its last breaths, Smithy took off his heavy flying suit, arranged his demolition tools methodically under the escape hatch and prepared to ditch. Although facing the situation he had dreaded since boyhood, he remained admirably calm and focused. To Stannage he said: 'Looks like we're for it this time, old son'. He and Stannage broke open the emergency bottle of whisky and lit a last cigarette. It almost ceased to matter if they got drunk or blew up the plane.

Taylor, who had temporarily taken over the controls, had a better idea. Handing the plane back to Smithy, he removed his shoes, belted his coat and grabbed a length of line that was tied around the mail bags. This he fastened to a steel strut, then wound the rest of the rope around his waist. Then he opened the cockpit window and climbed out. He intended to transfer oil from the dead starboard motor to the port motor.

Standing on the strut, he could barely keep upright against the force of the air. Nevertheless, he edged along the strut towards the dead motor, bracing his shoulders and neck against the wing. He would later explain that he managed to perform this extremely perilous action by emptying his mind of rational thought, reminding himself that this was 'do or die', and concentrating intensely on the job at hand.

Stannage now became Taylor's assistant, anticipating all his needs. He watched Taylor remove the cowling, screw by screw, in order to reach the oil tank, and then he himself leaned out of the cockpit window to pass a spanner so that Taylor could open the drain plug. Taylor managed to open it by sitting astride the strut and hooking his left arm around a steel tube, thus leaving his hands free to work. As Stannage moved to hand him a thermos flask in which to catch the oil, Taylor, in order to receive it, stood up and stretched his arms out wide. Smithy, watching from the cockpit, never forgot this sight of Taylor. Later, he would write: 'With arms outstretched from cabin to motor mount, almost on tip toe, and his head as far back as he could bend it over the wing edge, he was a man crucified in the air.' Smithy is said to have had no real Christian belief, but at that perilous moment his thoughts sped back to the faith of his boyhood.

When part of the oil was drained, Taylor passed the thermos back to Stannage, who emptied it into his briefcase; Stannage then passed the thermos back to be refilled. It was a frustrating exercise, because whenever Taylor started to pour oil into the thermos, the slipstream sucked away a few precious drops. After performing the manoeuvre often enough to extract about a gallon of oil, Taylor edged back along the strut and entered the cockpit. He was close to collapse but had no time to rest. Someone now had to climb out of the opposite window and edge along a strut to the port engine; and since Taylor was exhausted, Stannage volunteered to try. Small in stature and physically weak, he was defeated by the blast of air that roared towards him, and was forced to crawl back. Fortunately, Taylor was sufficiently rested to try again. His first two attempts failed because he, too, was defeated by the 'howling blast like all the devils of hell let loose'.

This stalemate might have continued if Smithy had not fathomed the problem. Since the port engine was still firing when Stannage and Taylor tried to edge towards it, they were facing the combined force of the propeller's slipstream and a headwind. According to Smithy's later calculation, the air pressure against a human body must have equalled a gale of about 100 miles per hour.

The only solution was to briefly shut down the port engine. Pushing open the throttles, Smithy coaxed the plane up to 700 feet and then cut the motor. Without the additional blast from the propeller, Taylor found that he could 'force a passage against the air'. Along the strut he edged, and then set about removing the cowling and uncovering the oil tank. At this juncture, he was obliged to pause because, with only one engine to support it, the plane had dropped to within twenty feet of the sea. Smithy was obliged to restart the motor and climb again. At 700 feet he again cut the motor, and Taylor resumed his task. Having unscrewed the cap, he took the oil-filled thermos from Stannage and emptied it into the tank – and kept repeating this action until the oil in Stannage's briefcase was gone.

Would this cure the engine? Taylor remembered 'a moment of awful suspense', until shouts and 'a thumbs-up' sign from the cockpit confirmed that the oil pressure was rising. As Taylor lay jammed against the engine mounting, he experienced a surge of 'magnificent exhilaration'. Even the thought of many more trips for oil did not disturb him. He remembered thinking: 'It works – it works – it works!'

The oil that Taylor had so perilously transferred lasted less than an hour, and four more hours of flying lay ahead of them. In all, Taylor would perform his precarious journey five more times, and each time Smithy would cut the port engine as required until the task was done. Each time the *Southern Cross* responded like a faithful horse obeying its master. 'The Old Bus knows me,' Smithy declared triumphantly, 'she always comes through for me.' It was an astonishing piece of piloting.

Nor was the port engine their only worry. As the afternoon wore on, the centre engine began to falter. If it ran out of oil there was no remedy, for Taylor was unable to reach its tank. Their main hope was to lighten their load. Having been instructed by the Australian postal director to do what he must, Smithy ordered Stannage to dump the mail. The fourteen big bags went over the side – about 24,000 letters. Stannage and Taylor rejoiced because they had long wished to jettison the mail, although neither was prepared to defy Smithy and do it without his consent.

At around four pm the *Southern Cross* limped into Mascot Aerodrome. Thanks to Stannage's frequent bulletins, which had been passed to radio stations and newspapers around Australia, the public had followed the flight with high emotion. Taylor's extraordinary heroism was headline news. A crowd of supporters had gathered at Mascot and broke into cheers as Smithy's oil-stained team appeared. Smithy was barefoot, haggard and expressionless: 'like a man from the dead'. Only Stannage showed animation. 'God, what a time we've had,' he gasped as he hugged his wife. To Mary, Smithy said, 'It's all right, darling, they can't kill me,' and fell on her neck in a long embrace. Mary was accompanied by Smithy's physician, Dr Matthew Banks, who fended off officials and hustled Smithy home.

Although he looked like a ghost, Smithy was stronger than one might expect. Admittedly he was exhausted, and in some pain. The failure of the starboard motor had compelled him to keep his left foot continually on the rudder in order to offset the thrust from the port motor; otherwise, the plane would have drifted off-course. This relentless pressure had set up neuritis in his maimed left foot. However, it can be deduced from the written reports that he had suffered no attack of panic during the remarkable ordeal. Indeed, Stannage and Taylor had been impressed by his reassuring calm. 'Gee,' Stannage told the wireless operator in Sydney at the height of the crisis, 'I had vertical breeze up at first, but feel much better after seeing Smithy's smiling fizzog smile back at me.' As many sufferers seem to discover, when their backs are to the wall they have no time for panic; all feelings and thoughts are directed towards action.

When he arrived home, Smithy soaked himself in a hot bath. Indeed, he stayed in the water for such a stretch of time that Mary became worried and went to find him. He had fallen asleep with the water lapping his chin. 'It's funny,' said Smithy, 'that after beating the Tasman that seemed so hungry to get us, I should go home safely and then nearly be drowned in my bath.'

Not all the airmail had been dumped in the ocean. Stannage had overlooked seven bags, and these Smithy offered to take across the Tasman the following day in *Faith in Australia*. Wisely, the postal director refused. Indeed, on doctor's orders Smithy and Stannage were each confined to bed. Three days later both were well enough to escort their wives to a performance of the musical *High Jinks* at Sydney's Theatre Royal. When they entered their box, looking fit and relaxed, the audience clapped and cheered with a sense of relief and pride.

The heroic flight fired the public's imagination as few other exploits could have. Newspapers in both hemispheres reported the story with awe, especially praising Taylor's extraordinary courage. While it was true that a few critics – like the one writing for the British magazine *Aeroplane* – called the flight a useless stunt, the public did not agree, and the mean-spirited comments brought forth angry refutations. A writer in London's *Daily Mail* summed up the prevailing view: Smithy and his companions, he wrote, had added a 'new epic to the tales of the air'. Indeed, they had 'thrilled the Empire'. The king took the unusual step of bestowing on Taylor the Medal of the Order of the British Empire. It was a medal given for conspicuous bravery, and seldom awarded; later it came to be called the George Cross.

Reluctant as Smithy was to admit it, his Old Bus was no longer airworthy. He had no choice but to find the 'old girl' a final home. He wished he could afford to donate her to the nation, but he needed the money from her sale. However, he was ready to accept a low offer, if only the nation could be persuaded to make one. The Returned Soldiers League and the Australian Flying Corps Association seconded his request, for they believed that the *Southern Cross*, if placed in a museum, 'would be an inspiration for generations of aviators as yet unborn'. Eventually, the federal government offered Smithy £3000 and agreed to exhibit the plane in a national museum in Canberra.

For his old plane, Smithy planned a grand farewell. 'This aeroplane has been a living thing,' he told the press. 'No Arab could think more of his thoroughbred horse, no sailor regard his tall ship with deeper affection ... she has never really let me down.' At ten-fifteen on the morning of 18 July 1935, in front a large and reverential crowd, he tenderly patted the fuselage and prepared to take off. Beside him in the cockpit sat fourteen-year-old John Ulm. Smithy believed that 'Charles Ulm should be with me on the last journey', and he had invited young John to represent his father. On the short flight to Richmond, where the plane was to be temporarily housed, Smithy dipped the wings in a salute over Keith Anderson's grave, and made another salute over his mother's home in Arabella Street. Travelling in the rear cabin were Mary, Stannage, Taylor, Beau Sheil and a charismatic Methodist preacher named Colin Scrimgeour, who was president of the Auckland Aero Club.

At Richmond, a ceremony and speeches awaited them. The Minister for Defence, Archdale Parkhill, proclaiming that Smithy was 'the greatest aviator of all time', declared: 'The *Southern Cross* now becomes the property of every Australian.' Tight-lipped and close to tears, Smithy responded with a thank you on the plane's behalf. Turning towards his beloved Old Bus, he spoke moving words: 'To you, old friend, I bid a fond and affectionate farewell.' More distressed than he was prepared to admit, Smithy drove his car back to Sydney in time to board the *Aorangi*, bound for New Zealand. He and Beau Sheil were due to discuss their latest trans-Tasman plans with the New Zealand government.

Smithy's hopes for an air service to New Zealand were stronger than ever. Early in June he had formed the Trans-Tasman Air Service Development Company, of which he, Taylor, Stannage and their wives were directors, along with the now indispensable Beau Sheil. With a capital of £2000, its stated aim was 'to initiate, promote and carry out research' on the Tasman route, but Smithy already knew what he wanted. He was requesting a subsidy from both the Australian and New Zealand governments to be able to conduct a biweekly service across the Tasman Sea. So determined was he to launch the airline within the next twelve months that he was already calling for applications from pilots and technical staff. When discussions with the New Zealand government were completed, he planned to board a ship for California, where he would buy aircraft for the trans-Tasman service. He also planned to reclaim his Altair, the *Lady Southern Cross*, and ship it to England.

Mary and his mother were on the wharf to see him off. It was a wretched farewell. Previously at times of parting, Catherine had stoically placed her boy in God's hands, believing that divine will and his own skill would protect him. On this occasion she felt no such confidence. Like Mary, she believed that he was making a foolishly dangerous decision.

Both women knew that he intended to fly his Altair from England to Australia, following the route of the Melbourne Centenary Air Race. Determined to restore his reputation among those who had called him a squib, he was intent on breaking the record of Scott and Black, the winners of the race, who had completed the flight from Mildenhall to Melbourne in seventy-one hours. Given the state of his nerves, it seemed a recipe for disaster, but neither woman felt capable of stopping him. 'Oh I do wish, if he comes back safely, that he will never do these long flights again,' Catherine told a reporter in an unusual burst of candour. Mary agreed, but exhibited greater resignation: 'I hope he will give up these flights after he completes this one, but I will not insist. A wife should not direct a husband as to what he should do or not do.'

Mary and his mother had a further reason to fear. At the end of June, Smithy had undergone a minor operation, the nature of which was not publicly disclosed, but was probably an investigative gastroscopy, a new procedure in Australia. He appears to have been suffering from a stomach ulcer, but refused to allow illness to deflect him from his purpose. 'I don't call this rashness or foolishness,' wrote Beau Sheil. 'There was something magnificent about this willingness of the spirit scorning the weakness of the flesh. It was the spirit of Smithy.'

On 25 July, in Wellington, Smithy and Sheil discussed the trans-Tasman air service with New Zealand's acting prime minister and cabinet. Thanks to an unofficial government agreement with Imperial Airways, Smithy was almost defeated before he started. Moreover, there was a patriotic cry of horror – even greater than the one he had received from the Australian government – when he proposed using American Sikorsky S-42 flying boats, similar to those used regularly by Pan-American Airlines in the North Pacific. Smithy defended himself by pointing out that 'both Australia and New Zealand would become more valuable units of the Empire when connected by efficient air services', but the accusations of anti-patriotism spread as far as London.

Negotiations with the government broke off after only one day, and Smithy once again succumbed to 'a chill'. However, he was not too ill to visit a Japanese naval ship that was then in port. He had been invited by the Japanese government to make a goodwill flight to Japan the following year, and was trying to learn the language. Hearing of Smithy's studies, the captain invited him on board 'to try out his Japanese'.

Two days later, Smithy boarded the *Monterey* for San Francisco. Once there, he visited Harold in Oakland, attended parties in Hollywood and held discussions with Igor Sikorsky in Los Angeles about his flying boats. There were also meetings with Tommy Pethybridge and Ulm's former engineer Bob Boulton, who had sailed for California the previous month so as to be on hand when Smithy arrived. Their task was to overhaul the Altair. Smithy tested the plane in person on 24 August, and at the end of the month flew it to Cleveland for the National Air Races. From there he flew on to Chicago. Two weeks later, in New York, he arranged to ship the plane across the Atlantic, with Pethybridge and Boulton as its accompanying mechanics. Meanwhile, he booked a passage for himself on the liner *Britannic*.

Arriving in England on 23 September 1935, Smithy arranged, at some expense, for the Altair to be lifted by crane from the freighter to the dock, and then from the dock to a barge. The barge carried the plane to a mud beach at All Hallows, near the mouth of the Thames. Smithy piloted the Altair from there to Croydon Aerodrome. 'That flight off the Thames mud-bank,' wrote Beau Sheil, 'was a feat of airmanship requiring Smithy's own brand of skill' – it made the take-off from Anderson Park seem like child's play.

Now began arguments about the plane's registration and the amount of fuel it was permitted to carry. Sheil had arrived in London to help Smithy, and they could scarcely credit the ensuing rigmarole. The registration at first seemed relatively easy because Smithy had come armed with an airworthiness certificate from America. However, the certificate was for the plane in its original state, and did not cover the subsequent modifications and the extra fuel tanks. Here they struck obdurate officialdom. With its original tanks full the plane was deemed airworthy; with the additional tanks full it was not. The additional tanks would have to be sealed, even though the plane had crossed the Pacific in the most taxing of circumstances with every tank full and come to no harm. Smithy had no patience with the stupidity – for such did he label it. Eventually, he saw no alternative but to agree, but he did not give in without a struggle. Less than a week before he was due to leave for Melbourne, he was still demanding permission for sufficient fuel to take him from England to Baghdad. When he did not get it, he confessed to Sheil that he would surreptitiously fill up en route.

The New Zealand and Australian governments had by now officially turned down his trans-Tasman proposition, so he decided to continue without a government subsidy. He could not proceed without a wealthy business partner, though, and fortunately he seemed to find one in the

London-based British Pacific Trust, one of the most powerful aviation companies in the world. A formidable rival to Imperial Airways, it was eager to extend into Australia. It liked Smithy's 'well-matured plans', and recognised Smithy's moral right to the trans-Tasman route – a response that greatly pleased him. Moreover, it had no objections to Sikorsky flying boats; indeed, it was arranging for them to be manufactured in England, and was expecting Smithy to buy two of them. When it proposed to inject £125,000 into his enterprise, he felt all his dreams were coming true. But there was a serious drawback: the British Sikorsky flying boats would not be ready for at least two years. Smithy could not wait that long and was obliged to cancel the negotiations. There followed a succession of frustrating interviews with bankers and financiers as he searched for another business partner.

The wrangles over the fuel tanks and the fruitless business discussions were sapping Smithy's strength and wasting his time. He had planned to leave on 15 October – at which time he would have had the advantage of a full moon – but four days passed before he could get away. Mary was waiting for him in Melbourne. She had recently had a short spell in hospital; maybe, as the papers said, for a tonsillectomy, or maybe for a curettage after a miscarriage. Now she was convalescing at her parents' home in Toorak. Afraid, perhaps, that Smithy might arrive back ill, she had invited his physician, Dr Banks, to be on hand for his arrival.

Smithy was relieved to have a co-pilot on the flight back to Australia. At first the overzealous Air Ministry had forbidden a second occupant in the Altair, but the ban had been lifted and he was able to take the eager young Tommy Pethybridge. The previous June, Tommy had married his Tasmanian sweetheart, Jessie Locket, and she did not want him to go. But if Smithy needed him, there was no question of him refusing. It was said of Tommy that he 'would have flown to hell with the Boss he idolized'.

To be ready for an early departure, Smithy, Pethybridge and Sheil spent the night before the flight at the hotel at Croydon Aerodrome. Smithy seemed unwell during dinner and went to bed as soon as it was over. At about two in the morning, Sheil was dismayed to be awakened by a night porter, who said Smithy seemed very ill. Hastening to his room, Sheil found him restless and feverish, but a dose of aspirin and a hot drink somewhat restored him, and presently he fell asleep and slept until the following afternoon. In the meantime, Sheil cancelled the flight, and when Smithy awoke, Sheil summoned a doctor, who diagnosed a 'severe chill' and ordered rest for the next two days. Sheil then put it to Smithy that he was in no condition to attempt to break a record. He and the Altair should sail home on a ship, for which Sheil would book a passage the next day.

Smithy was too ill to argue, but he reminded Sheil that shipment home would require more money that he possessed. Fortunately, he then remembered that the Australian government still owed him £3000 for the *Southern Cross*, and he cabled Stannage to approach the government on his behalf. Stannage obeyed, but Smithy's request for payment was denied. He was informed that until the government had adequate proof that the mortgages raised on the *Southern Cross* in 1927 had been discharged, no money could be released. The government's best offer was £500, loaned against the surety of Smithy's household furniture. At this, Smithy flew into a rage and told them where they could put their money.

In a heartbreaking conversation on the telephone, Mary begged him to return by ship. Her father would willingly pay for Smithy's passage, and the shipping of the Altair. Smithy, suddenly proud, would not hear of it. As Mary would later recall, he kept saying, 'If only they'd pay up for the Old Bus I wouldn't need to do this.' That the meanness of the Australian government involved his beloved *Southern Cross* seemed the final insult.

The berth that Sheil had booked was cancelled on Smithy's orders. At seven am on the frosty morning of 23 October, Smithy and Pethybridge flew out from Croydon Aerodrome, farewelled by a large crowd. Smithy's itinerary was Croydon, Marseilles, Baghdad, Allahabad, Singapore, Surabaya, Wyndham and Melbourne, a journey which he hoped would take a miraculous sixty-seven hours. To overcome the restrictions on his fuel, he planned to defy officialdom and, at his first stop, load as much fuel as he could carry.

The flight onward from Marseilles turned into a nightmare. A devastating hail and snow storm over mountains near the Gulf of Corinth limited visibility to ten yards, and shredded the edges of the wings. This, plus formations of ice on the body of the plane, rendered the Altair almost uncontrollable. Fifteen minutes of icing and buffeting, and freezing cold – how he blessed the Altair's electrically heated seats – convinced Smithy that he must land at Brindisi, even though he had no permit to set down in Italy. The aerodrome was in darkness, but hastily summoned motorists lit the runway with their headlights. When he arrived back at Croydon, mechanics told him they had never seen a plane more battered and still flying.

Smithy was still determined to try for the record, and two weeks later he and Tommy Pethybridge took off from Lympne Aerodrome, in Kent. Climbing into the cockpit at six-thirty am on Wednesday, 6 November, he announced that from now on there would be no turning back – whatever befell them, he was determined to 'stick it'. It so happened that the Altair was not the only plane travelling towards Australia that day. Two young flyers, one Australian-born, the other Australian by adoption, were attempting record-breaking solo flights from England to Australia. One was Jimmy Melrose, who, at twenty, had been the youngest competitor in the Centenary Air Race; the other was Jimmy Broadbent, aged twenty-five, and he was making the faster time. Both were flying Percival Gulls, and both were attempting to break Smithy's solo record of seven days, four hours and forty-seven minutes. Having departed on 2 November, they were now approaching India.

Fortunately, the weather was clear, and Smithy made excellent time, arriving at his first stop, in Athens, in eight hours and two minutes. His next stop was to be Baghdad, and the Altair sped across the Middle East like a big blue bullet, flying at 10,000 to 15,000 feet and cruising at more than 200 miles per hour, covering the distance in seven hours – one of the fastest flights Smithy ever made. After the quickest of stops – for he allowed himself little time off for rest on this flight – he and Pethybridge were off to Allahabad, where they stopped for just an hour to refuel, wash and eat. Smithy's hopes were high because their flying time from

England was only a couple of hours behind Scott and Black's record; he believed they could make up the difference once they reached Australia. Meanwhile, their next stopping place was to be Singapore, which entailed flying over the dreaded Bay of Bengal.

No log has survived for this next section of the flight, nor was there an official flight plan, so it is impossible to know Smithy's exact route on leaving Allahabad. However, by flying over three aerodromes he signalled his progress, and thus it is possible to trace much of his journey and the time he took to fly it. Wishing to avoid a long crossing of the Bay of Bengal – a sensible precaution, given the strain on his nerves – he left Allahabad at 10.58 pm on Thursday, 7 November, Sydney time, and flew 470 miles to Calcutta, passing over Dum Dum Aerodrome at 1.36 am Sydney time on Friday, 8 November. From Calcutta, he seems to have cut across a corner of the Bay of Bengal and flown 420 miles to Akyab, passing over its aerodrome at a time that was not officially recorded but was said to have been about midnight, local time. From Akyab he seems to have continued along the coast to Rangoon – 330 miles – passing over its aerodrome at five am Sydney time, or about one-thirty am local time. It was expected that he would then fly south over the ocean but close to the Malay Peninsula, reaching Singapore about daybreak. He had already announced that he was hoping to reach Darwin by midnight on Friday, 8 November, Sydney time, and to reach Melbourne by noon the next day, Saturday, 9 November 1935. If he managed to keep to this schedule, he would have covered the distance in sixty-seven hours and broken Scott and Black's record.

It so happened that Jimmy Melrose was flying his Percival Gull between Rangoon and Victoria Point in the early hours of Friday, 8 November, local time. He was over the Bay of Bengal, about 150 miles from the west coast of the peninsula, when – to use his own words – he 'saw the exhaust flame of the Altair go past me at 2.30 am'. The night was dark, and the flame was about 200 feet above him, so some have doubted if he really saw it, but Melrose was convinced that he did. On reaching the aerodrome at Victoria Point, he fully expected to hear that the Altair had passed overhead. Instead, he found Broadbent, who, when questioned, declared that he had seen no sign of the plane. Melrose continued on to Singapore and did not find the Altair there either. Melrose and the world were beginning to realise that Smithy and Pethybridge were missing.

Melrose concluded that the Altair had gone down somewhere between Rangoon and Victoria Point. With remarkable selflessness, he abandoned his attempt on the solo record and joined the searchers, who had begun assembling almost at once, under the command of Air Commodore Sydney Smith of the RAF in Singapore. Squadrons of Hawker Fury fighters, Vickers Vildebeest torpedo bombers, Audax reconnaissance planes and flying boats prepared to lead the search. At dawn on Saturday, 9 November, Melrose accompanied two flying boats up the west coast of the peninsula, looking among the 800 or so islands in the Andaman Sea that make up the Mergui Archipelago. Lying close to the coast, the islands, of all shapes and sizes, stretch for about 300 miles and cover an area of about 10,000 square miles. Not content with examining the coast and the islands, Melrose also flew inland, towards the mountains, searching the jungle as well.

The sweep along the west coast of the peninsula was as thorough as could be managed. Even so, it found nothing, and by 15 November the search was winding down. This caused consternation in Australia, for Smithy was a national hero, and the public believed it was far too soon to give up hope. As one observer put it, 'He is Smithy and we have to find him.' Forgetting past differences, the federal and New South Wales governments donated £2000, and Scotty Allan was dispatched to Singapore in a Qantas DH 86 to begin an Australian search. Stannage, Taylor and Purvis were determined to go also, and set out in a plane called the Gannet, stopping in Cloncurry to refuel. There, Taylor fell seriously ill, and since neither Stannage nor Purvis was capable of flying the Gannet, their expedition was abruptly curtailed.

It was now that Smithy's brother Leofric tardily announced that Smithy had intended to fly down the east coast of the peninsula, not the west. Suddenly revitalised, the remaining searchers diverted their attention to the east coast. At the same time, flares were reported being seen at an island off the west coast, Sayer Island, south of Victoria Point. In neither place was anything found. Hope of finding Smithy and Pethybridge alive was fast disappearing.

Mary remained in Melbourne until mid-November, as did Jessie Pethybridge, and the two women comforted each other. Then Mary returned to Sydney, accompanied by her mother and young Charles. Neither she nor Catherine Kingsford Smith could accept that Smithy was almost certainly dead. When, early in December, the Anglican archbishop of Sydney spoke of a memorial service at St Andrew's Cathedral, they told him firmly that Smithy was alive and would be found.

Smithy's brothers – with the exception of Harold – did accept his death. Wilfrid was convinced that Smithy had passed out in the air. He believed that Tommy, sitting in a separate cockpit behind his boss, had been unaware of the danger until it was too late. In view of Smithy's state of health, this was a reasonable theory. The danger of having pilots in separate cockpits, unable to communicate easily during a crisis, had been suspected during the Pacific flight. Beau Sheil agreed with Wilfrid, but few others seem to have favoured this explanation. Most people assumed that a failing engine had forced the Altair down. Years later, when more was known about the problems of fatigue in Pratt and Whitney Wasp supercharged engines, Lawrence Wackett offered the theory that the supercharger impeller on the *Lady Southern Cross* had failed.

On 7 December, Prime Minister Lyons announced in federal parliament that the search must end. A motion was passed, expressing the parliament's regret at Smithy's death, and members stood in silent tribute to his memory. Lyons, John Curtin, who led the Labor Party, and Jack Beasley, who led the Lang Labor Party, gave heartfelt eulogies, which were echoed in newspapers across Australia and across the world. Obituarists proclaimed that Smithy's aerial exploration had earned him a firm place in history, alongside such great explorers as Magellan, and that, having helped shape the future of the world, he could justly be named the 'King of the Air'.

There were some, however, who kept searching, and one was Jack Hodder, a pilot who had been taught to fly by Tommy Petheridge at the Kingsford Smith Flying School. He now worked for the Tavoy Tin Dredging Corporation, which was based at the port of Tavoy, close to the middle of the Mergui Archipelago. In May 1937, Hodder heard from one of the native Salones, or 'sea gypsies', whose boats sailed the archipelago, that the wheel of a plane had been found in the sea below a cliff on what was then called Aye Island, but which today is known as Kokunye Kyun. Travelling north, Hodder located the island, which lay a mile or so offshore. Resembling a rock rising from the sea, it was about a mile long, a quarter of a mile wide, 460 feet high and covered with jungle. It had no beach and was surrounded by reefs.

As Hodder soon discovered, the wheel was 'complete with a fork member, and attached to the fork was a bar, which was part of the retractable gear'. The bar bore the words 'Aerol and Cleveland Pneumatic Tool Company, Cleveland, Ohio no. 2621'. A careful examination at the Lockheed Aircraft Factory at Burbank confirmed its authenticity. The Lockheed experts believed that the plane must have been planning to come down on the land, because the landing gear, which was folded into the wing during flight, appeared to have been deliberately lowered.

On examining the cliff below which the wheel had lain, Hodder formed the conclusion that Smithy had been attempting to land on the island. A swathe appeared to have been shorn from the tops of the trees close to cliff's edge, and part of the cliff had fallen away. Hodder presumed that the undercarriage had hit the clifftop, broken off and tumbled into the sea, while the remainder of the plane had 'pancaked' into the treetops. His search, however, did not reveal wreckage, which would have confirmed his theory.

Others put forward different explanations. Some, like Stannage, wondered if Smithy had struck the island accidentally in the darkness. Others thought that he had been searching for a beach on which to land, and believing, mistakenly, that one lay below the cliff, had landed on a reef. The island was searched three times in the following eighty years, and divers inspected the sea beneath the cliff – indeed, the sea all around the island – but no sign of a plane was found.

As we have learned in recent years during the long search for Malaysia Airlines Flight 370 in the Indian Ocean, currents can carry wreckage long distances, and Smithy's wheel may have floated far from its original landing place. One day, unexpectedly, the Andaman Sea may give up its secret. Then the world will know where, and possibly how, Smithy and Pethybridge died.

Few more beautiful places could be imagined for a last resting place. A traveller who visited the islands in 1939 has left this description: 'As far as the horizon stretched were hundreds of islands of every fantastic shape and size, covered with green forests from which I could hear the shrill cries of monkeys as we passed close to them.' In between the islands 'were channels of the clearest blue water and you could see down to a depth of 160 feet. The islands were all shapes and sizes, from great limestone rocks, hundreds of feet high, to small boulders like beehives.' In this maze of islands, a man 'could live for months on end without ever seeing another human being or the handiwork of man or civilisation'.

One likes to think that Smithy and Tommy Pethybridge sleep peacefully in this water of clearest blue.

NOTES AND SOURCES

ABBREVIATIONS

Anderson court case: Transcript of evidence given in *Keith Vincent Anderson v. Charles Edward Kingsford Smith and Charles Thomas Philippe Ulm, Supreme Court of NSW*, in equity, 5633/1928, held in the archives of the Supreme Court of New South Wales, Sydney.

CKS: Charles Kingsford Smith

EKS: Elsie Kingsford Smith (Charles' sister)

FM: Norman Ellison, *Flying Matilda: Early Days in Australian Aviation* (Sydney, 1957). This biography of Smithy was commissioned by his family in 1937, but never finished. It was published in its unfinished form in 1977 along with other aeronautical essays.

EC NLA: The Ellison Collection, National Library of Australia, Canberra.

Hitchcock court case: Transcript of evidence given in *Henry Smith Hitchcock v. Charles Edward Kingsford Smith and Others in the Supreme Court of New South Wales*, 1929 (6/1458), located in transcripts of evidence of various courts and commissions and boards of inquiry 1899–1985, held in the State Archives of New South Wales, Kingswood.

MFL: *My Flying Life: An Authentic Biography*, prepared under the personal supervision of and from the diaries of the late Sir Charles Kingsford Smith, with a preface by Geoffrey Rawson (London, 1939). (Written in the first person as Smithy's autobiography, and first published in 1937.)

Mitchell: Manuscript collection, Mitchell Library, State Library of New South Wales, Sydney.

NAA: National Archives of Australia, Canberra.

SC: Kingsford Smith's plane, the *Southern Cross*.

SC *Story*: C.E. Kingsford Smith, *The* Southern Cross *Story* (Sydney, 1995). First published under the title of *The Old Bus* in 1932.

SC *Trans-Pacific*: C.E. Kingsford Smith and C.T.P Ulm, *The Story of the* Southern Cross *Trans-Pacific Flight, 1928* (Sydney, 1928).

SMH: *Sydney Morning Herald*

Smithy: Ian Mackersey, *Smithy: The Life of Charles Kingsford Smith* (London, 1998).

TNA: The National Archives (United Kingdom), Richmond, Surrey.

CHAPTER 1: SUNNY FACED, SUNNY HEARTED, SUNNY TEMPERED

CKS arrives in Brisbane *Brisbane Courier*, 11 June 1928, pp. 13–4; Melbourne *Age*, 11 June 1928, p. 11; Brisbane *Truth*, 10 June 1928, p. 1. **Early life of William Charles and Edward Eldridge Smith** Mackay *Daily Mercury*, 6 June 1934. William, born 26 March 1852, came to Brisbane in 1876 as a clerk in the Bank of New South Wales and worked in branches at Mt Perry and St George – see *Brisbane Courier*, 21 August 1876 and Brisbane *Queenslander*, 6 May 1876. In 1877 he moved to the newly opened branch in Cairns, was wounded in a skirmish with Aborigines and returned to Brisbane in 1878, where he married, and resumed work at the St George branch – see J.W. Collinson to Winifred Sealby, 11 July 1951, MLMSS2571, Mitchell. In 1882 he transferred to the Queensland National Bank and became manager of its Cairns branch. **Marriage of CKS's parents** *Brisbane Courier*, 24 April 1878, p. 2. **Wreck of the *Ranelagh*** *Maitland Mercury*, 6 May 1882; Winifred Sealby to J.W. Collinson, 22 October 1944, MLMSS2571, Mitchell. **Early history of Cairns** J.W. Collinson, *Early Days of Cairns* (Brisbane, 1939), pp. 62, 150; 'What Queensland Meant to Kingsford Smith', *Brisbane Courier Mail*, 4 January 1936, p. 19 and 'Early Days of Cairns', *Cairns Post*, 14 July 1939, p. 12 – both articles by CKS's sister Winifred Sealby. *Aladdin Cairns Post*, 14 May 1885. **William's public spirit** *Cairns Post*, 23 April 1885. **R.A. Kingsford** Obituary, Cairns *Morning Post*, 7 January 1902. **William transfers to Mackay** *Cairns Post*, 5 November 1885. **William founds Smith Brothers** *Cairns Post*, 14 January 1886 and *Brisbane Courier*, 10 February 1886. **Prevalence of malaria in Cairns** Report on malaria, *Cairns Post*, 4 November 1897; advertisements for malaria cures, *Cairns Post*, 10 June 1886. **William's illness and holidays in Tasmania** *Cairns Post*, 25 May and 28 September 1889, 19 March and 19 July 1890, 25 May 1895. **Fairview** *Launceston Examiner*, 8 June 1889 and 16 January 1899. **William joins Bank of North Queensland** *Cairns Post*, 3 September 1892. **William leaves Cairns** *Cairns Post*, 7 May 1896. **Bank policy concerning malarial employees** William's successor in Cairns, J.P. Canny, caught malaria and was transferred to the Sydney office; see 'Early Day Banking in North Queensland', *Townsville Daily Bulletin*, 28 January 1937. **William in Bank of North Queensland** Manager in Cairns from 10 August 1892; assistant manager in Brisbane from 1 June 1896;

manager in Rockhampton from 13 to 30 December 1897; manager in Sydney from 17 February 1898 to 5 January 1899. Records of the Bank of North Queensland are now held by National Australia Bank, and supplied courtesy of the Manager of Historical Services. **Family awaits CKS's birth** From Catherine's diary, quoted in *Smithy*, p. 2. Mr Mackersey gained access to the diary, which is now owned by one of Catherine's descendants. Thanks to Mr Mackersey's kind assistance, I was able to contact the descendant but was unable to gain access to the diary. So I am sourcing the diary through Mr Mackersey's book. **CKS's birth notice** *Brisbane Courier*, 13 February 1897. **Address and appearance of CKS's birth place** A menu for a luncheon given in CKS's honour by Queensland residents in Sydney on 27 June 1928 gives his birthplace as the Smith family home on the corner of Riverview Terrace and Hamilton Road; see MS 1882/3/43, EC NLA. Mackersey, *Smithy*, p. 1, describes the birth house as a cottage called *Corbea* on the corner of Whyenbah Road and Riverview Terrace. However, recent historical research by the Brisbane City Council endorses the Hamilton Road address, and a plaque has been placed on the site. **Kingsdown** Advertisement for Kingsdown, *Brisbane Courier*, 13 March 1897 – the family moved there soon after CKS's birth. **CKS in childhood** Winifred Kingsford Sealby, *Recollections Personal and Scenic: Our Dumb Friends* (privately printed, n.d.), MS1882/3/44 EC NLA. **CKS's siblings** Richard Harold Kingsford Smith, born 7 May 1879; Winifred Kingsford Smith, born 15 November 1880; Wilfrid Kingsford Smith, born 12 March 1882; Leofric Kingsford Smith, born 15 September 1884; Eric Kingsford Smith, born 27 February 1887. **The family arrives in Sydney** Catherine's diary, see Mackersey, *Smithy*, p. 7. **William's departure from Bank of North Queensland** Catherine's diary, see Mackersey, *Smithy*, p. 8. Records, now held by the National Australia Bank, do not explain his abrupt departure. **Catherine's journey to Cairns** *SMH*, 16 Jan 1899 and Winifred Sealby, *Recollections Personal and Scenic*. **Catherine in Cairns** *North Queensland Register*, 30 Jan 1889. **Family life and entertainment** Winifred Sealby, *Recollections Personal and Scenic*. **Walter Bentley** *SMH*, 20 September 1927. **Harold in Bank of North Queensland** *Cairns Post*, 16 Oct 1900; **William's visit to Cairns** *SMH*, 28 March 1900. **Death and will of R.A. Kingsford** Brisbane *Queenslander*, 11 January 1902; *Brisbane Courier*, 31 March 1902; will of Richard Ash Kingsford, ID 742270, file number 84/1902, State Archives of Queensland.

CHAPTER 2: VANCOUVER BECKONS

Move to Longueville See death notice of William's mother, Elizabeth Eldridge Smith, *SMH*, 8 May 1901. William is described as living at Longueville. **CKS and Sir Toby** Winfred Sealby to J.W. Collinson, 25 March 1950, MLMSS2571, Mitchell; *Brisbane Courier*, 16 July 1897; *SMH*, 2 June 1898. **CKS jumping from the family shed** Sydney *Sun*, 3 June 1928. **CKS's early school days** Winifred Sealby, *Recollections Personal and Scenic* and *FM*, pp. 196–97. **Enticement to Canada**

West Australian, 12 August 1905 and *Brisbane Courier*, 24 April 1904. **Harold in Cairns** Cairns *Morning Post*, 16 October 1900 and 8 May 1903. **Adoption of the prefix 'Kingsford'** Ellison in *FM*, p. 197, claims 'Kingsford' was added to avoid 'postal mis-identification' in Vancouver. Winfred Sealby in *Cairns Post*, 14 July 1939, claims that the prefix was to 'differentiate them from other paternal relatives'. By January 1901 Harold was calling himself 'R.H. Kingsford Smith' and offering acting lessons – see Cairns *Morning Post*, 15 January 1901. **William sails to Vancouver** Passenger list of the *Aorangi*, *SMH*, 28 December 1902. **Kingsford Smiths sail to Vancouver** Passenger list of the *Aorangi*, Vancouver *Weekly News Advertiser*, 11 August 1903. Catherine is listed as Mrs Kingsford Smith. **CKS and his family on board ship** *Daily Telegraph Pictorial*, 6 June 1928 and Ellison, *FM*, p. 197; Mackersey, *Smithy*, p. 12, believes Catherine's brother Arthur accompanied them but he is not listed as a passenger, nor can I find mention of him in Vancouver. A relative of Catherine's mother, a Miss Southerden, travelled with them. **Winifred sailed on the *Moana*** See passenger list, *SMH*, 10 August 1903. **Family residences, occupations and businesses in Vancouver** *Henderson's City of Vancouver Directories*, 1904 to 1908 editions. Winifred was a stenographer, working with her father in the claims department of the CPR; Elsie was a stenographer in a business on 'Cordova West'. The family lived at 1334 7th Avenue West from late 1903 to late 1906. Stories that 'they moved about like gypsies', living restless, aimless lives seem to come from a descendant, Catherine Robinson, and are quoted in Mackersey, *Smithy*, pp. 12–13. Evidence does not support them. **Catherine, a feminist** Catherine's obituary, *SMH*, 23 March 1938. **CKS's schools** *Vancouver Daily Province*, 9 June 1928. **Storm in the sound** *FM*, p. 198. **Harold arrives in Vancouver** Passenger list for *Aorangi*, *SMH*, 5 September 1904; interview with Harold in *Victoria Daily Colonist*, 2 October 1904; Harold worked as an accountant in the booksellers Clarke and Stuart Co., 441 Hastings West – see *Henderson's Vancouver Directory 1905*, p. 435. **Australian Club** Inaugural meeting, *Victoria Daily Colonist*, 25 August 1905; *Adelaide Advertiser*, 13 October 1905; *Launceston Examiner*, 15 November 1906. **Masonic lodge** *History of the Grand Lodge of British Columbia*, Victoria, B.C., 1971, pp. 199–222. **Smith Brothers** *Henderson's Directory for 1906*, pp. 585–86. Smith Bros' first premises were 'on Lonsdale between 3rd and 4th', North Vancouver. **Canadian economy 1904–07** Harold A. Innis, *A History of the Canadian Pacific Railway* (London, 1923), pp. 159–62. **CKS loves trains** *FM*, p. 198. **Some family members return to Sydney** Winfred sailed on *Miowera* – see passenger list, *SMH*, 17 July 1906, and Winfred Sealby to Ellison, MS1882/1/8 EC NLA. Catherine, CKS, EKS and Harold's wife and children returned on *Moana* – see passenger list, *Brisbane Courier*, 3 December 1906. **CKS's social skills beyond his years** *FM*, p. 200. **CKS's cousins** Rupert Swallow, born 29 June 1897; Robin Swallow, born 25 February 1893; Godfrey Kingsford, born 13 February 1894; Raymond Kingsford, born 6 April 1898; Philip Kingsford, born 22 March 1895. **Near drowning** *SMH*, 3 and 5 January 1907 and Sydney *Sun*, 26 October 1930;

for Catherine's state of mind, see her diary entries, Mackersey, *Smithy*, p. 14. **Back to Vancouver** Harold's wife, Elsie, and children, Beris and Basil, arrive back on *Moana* on 10 April 1907. **Catherine, CKS and Eric return on *Aorangi*** See passenger list, *Vancouver World*, 31 August 1907. **Kingsford Smith and Co.** *Henderson's Vancouver Directory 1908.* **Home to Australia** William, Catherine and CKS sail on *Aorangi*, passenger list, *SMH*, 24 February 1909.

CHAPTER 3: CHARLES THE CHORISTER

William, weary and disillusioned Winfred Sealby, 'What Queensland Meant to Kingsford Smith', Brisbane *Courier Mail*, 4 January 1936, p. 19. *Kintore* I am indebted to Mr Eliot Ball, local historian at Stanton Library, North Sydney, for identification of *Kintore* and for locating William's occupation in the municipal rate book for 1911. **Musicality of the Swallows and Kingsford Smiths** For Marjorie Swallow in JCW Company, Winfred Sealby, *Recollections Personal and Scenic* and Perth *Daily News*, 28 April 1909; for Rupert Swallow's singing career, see *Rockhampton Morning Bulletin*, 25 May 1931. **CKS's appearance at his interview** Recollections of Canon Melville Newth given to Ian Mackersey – see Mackersey, *Smithy*, p. 17; for CKS's enthusiasm for baseball, see *Vancouver Daily Province*, 5 June 1928. **At St Andrew's choir school** *FM*, p. 199; M. Deasy, *Conversations with Choristers: The Story of St Andrew's School* (Sydney, 2007), pp. 16–17; St Andrew's Choir School brochure, MS1882/3/45 EC NLA; Winfred Sealby to J.W. Collinson, 8 January 1945, MLMSS2571, Mitchell. For Edward VII's service, *SMH*, 21 May 1910. For Joseph Massey, *SMH*, 'Jubilee of Church Organist', 29 July 1921; CKS's school reports are from St Andrew's Cathedral's archives, courtesy of Dr Allan Beavis. **CKS's album** Sydney *Sun*, 4 April 1937, p. 14. **CKS's popularity** Winifred Sealby, *Recollections Personal and Scenic* and *FM*, p. 201. **CKS, prankster and speedster** *FM*, p. 119. For CKS speeding in Cremorne, Sydney *Sunday Times*, 10 June 1928. **Expertise with motorbikes** From a letter from Mr Robb, Sydney *Sun*, 5 June 1928. **Love of machines** CKS's fascination with trams was reported to Mackersey by Canon Newth – see Mackersey, *Smithy*, p. 17. **Sydney Technical High School, visits to the docks, apprenticeship and camping** For visits to the docks and vocal imitations, Sydney *Sunday Times*, 20 July 1930. For reunion with old schoolmates and use of nicknames, *Brisbane Courier*, 1928, p. 13. For camping, fisticuffs and winding armatures, Beau Sheil, 'An Australian Immortal Named Smith', Brisbane *Courier Mail*, 17 April 1937.

CHAPTER 4: SAPPER SMITH

CKS's physical characteristics See 'Australian Imperial Force Attestation Papers of Persons Enlisted for Service Abroad', Barcode 8334261, NAA and *FM*, p. 200. Ellison gives CKS an inch more in height that the Army doctor. For CKS's staccato speech, see Sydney *Sunday Times*, 10 June 1928. **CKS's first girlfriend** This was Betty Tuckwell. I'm indebted to her daughter Katherine for this information. **Military cadets** See Australian Imperial Force Attestation Papers, Barcode 8334261, NAA. **Eric on HMAS** *Sydney Burnie Advocate*, 26 April 1934, p. 7. **First weeks in the Army** Certificate of Attesting Officer, 50/474, Barcode 8334261, NAA; J.E. Fraser, 'Kingsford Smith. His start as an Airman', *SMH*, 9 June 1928, p. 11; *FM*, pp. 203–06. **Nellie Stewart** *FM*, p. 204; Melbourne *Age*, 15 June 1928; playbill of *Madame Du Barry*, 1914: Marjorie Swallow played Mlle Guinard, but Elsie does not seem to have been an actress – she possibly worked in the company office. **Voyage on *Ajana* and arrival in Egypt** *FM*, pp. 205–09; two letters from CKS to his parents: 3 July 1915, headed 'Heliopolis', and 27 July 1915, MS 1882 /3/31, EC NLA; Theo Barker, *A History of the Royal Australian Corps of Signals, 1788–1947* (Canberra, 1987), pp. 56–57.

CHAPTER 5: GALLIPOLI – NOTHING TO RAVE ABOUT

Life in the dugouts D. and M. Anthony (eds), *Letters Home: To Mother from Gallipoli and Beyond* (Sydney, 2009), pp. 144–47; J.E. Fraser, 'Kingsford Smith. His start as an Airman', *SMH*, 9 June 1928, p. 11; *FM*, pp. 210–11 – see this also for CKS's account of the evacuation from Gallipoli; CKS to his parents, 25 January 1916, MS 1882/3/31 EC NLA; L.A. Robb, Sydney *Sun*, 5 June 1928 (Robb lived in the next dugout). **Back in Egypt** Barker, *A History of the Royal Australian Corps of Signals*, pp. 57, 84–85. **Tel el Kebir, Moascar, Serapeum and promotion to corporal** See 'Casualty Form – Active Service for Charles Edward Kingsford Smith, no. 1017', Barcode 8334261, NAA; C.E.W. Bean, *Anzac to Amiens* (Canberra, 1968), pp. 218–19; for march to Moascar, P. Lindsay, *Fromelles* (Melbourne, 2008), pp. 37–38. **France** 'The War Diary of the 4th Signals', WO95-3475, TNA. CKS departed Serapeum on 1 June, embarked Alexandria on 2 June, disembarked Marseilles on 8 June, went north by train to Bailleul, near the Belgian border, on 8 June and marched to Merris on 11 June 1916; *FM*, p. 212. **CKS joins RFC** *Official History of Australia in the War of 1914–18*, vol. VIII (Australian War Memorial, Canberra), see introduction; *FM*, pp. 214–16; J.E. Fraser, *SMH*, 9 June 1928; 'Application for Employment as a Flight Lieutenant in the Royal Flying Corps', 16 September 1916, WO 339 89023A, TNA.

CHAPTER 6: ONE OF THE VERY BEST FIGHTERS

At Denham *FM*, pp. 218–21. Ellison quotes freely from CKS's letters home, the originals of which are now mostly lost. **At Oxford** 'Casualty Form – Active Service', Barcode 8334261, NAA. It states that CKS entered No. 3 School of Aeronautics, Oxford, on 26 January 1917, was in hospital from 22 February to 1 March 1917 and arrived at Netheravon on 20 March 1917; A. Revell, *Brief Glory: The Life of Arthur*

Rhys Davids (London, 1984), pp. 67–68. Davids was a fellow recruit at Exeter College and also at Netheravon – see p. 70. **Netheravon** *FM*, p. 225; CKS's log book, MS 209, folder 9, item 24, NLA. **Learning on a Maurice Farman Shorthorn** Stannage, *Smithy* (London, 1951), pp. 2–5. **Fear of heights**, *Cairns Post*, 8 June 1928. **Upavon** *FM*, pp. 225–27. For Spad, see J.M. Bruce, *Aeroplanes of the Royal Flying Corps (Military Wing)* (London, 1982), pp. 554–55; CKS became a Second Lieutenant in the Special Reserve of Officers on 17 March 1917 – see *London Gazette*, 5 April 1917. He gained his wings and became a flying officer on 13 May 1917 – see *London Gazette*, 1 June 1917, p. 5401. He had qualified in artillery, observation, bomb-dropping, formation flying, patrols, and Vickers and Lewis guns. **St Omer** *FM*, pp. 227–28. **La Lovie** M. O'Connor, *Airfields and Airmen of the Channel Coast*, p. 132; J. Rawlings, *Fighter Squadrons of the RAF and Their Aircraft* (London, 1993), p. 58; *FM*, pp. 229–33. **Aerial combat** *No 23 Fighter Squadron (Short History)*, pp. 28–32, AIRI/690, Air Historical Branch, Air Ministry, TNA; 'Combats in the Air, Offensive Patrol, 10 August and 14 August 1917', AIR-1221-204-5-2634, TNA; Stannage, *High Adventure* (Christchurch, 1944), pp. 78–86. Stannage reports that CKS told him it was the famous ace von Richthofen who shot CKS down, but there is no indication of this in official records.

CHAPTER 7: A BIG THING TO HOPE FOR

Sir John Ellerman's Hospital *FM*, pp. 231–38. Ellison quotes from CKS's letters home, now lost; O.C. Wilkinson to CKS's parents, *Brisbane Courier*, 26 November 1917, p. 7. **Military Cross** *London Gazette*, 9 January 1918, p. 643; for investiture, *FM*, p. 237. **Before Medical Board** 'Application for Leave to Proceed Abroad', WO339-89023B, TNA. **Voyage to Australia** On the *Star of Lapland*, see New York passenger lists 1820–1957, ancestry.com. **In Sydney** *FM*, pp. 238–39; interview with CKS regarding German pilots, *Maitland Daily Mercury*, 6 March 1918, p. 8; photo of family under fig tree, see *Australian Women's Weekly*, 2 September 1944. **Sydney Medical Board** Copy of telegram granting three months' extension, WO339-89023B, TNA. **Voyage to England** *FM*, pp. 240–41; **Eastchurch** CKS goes to Training Depot Station 204 on 26 August 1918, to Chatham Hospital on 10 November 1918, and returns to TDS 204 on 19 December 1918 – see AIR/76/469, TNA; for 'nervy as usual', see CKS to parents, 23 Sept 1918, from Shoreham, MS1881/3/31 EC NLA. **Need for excitement** *MFL*, p. 18. **Aerial jousting, pheasant shooting and possible job with A.V. Roe** CKS to parents, 11 February 1919, MS 1882/3/31, EC NLA; for offer to demonstrate the Avro in Australia, see CKS to parents, 12 January 1919, MS1882/3/31 EC NLA. **Plentiful girlfriends** Interview with James Cross, who remembered CKS at Eastchurch – see Mackersey, *Smithy*, pp. 33–34. **Pneumonia** *FM*, pp. 243–44. **Maddocks, Blackburn and the air race** Hull, *Daily Mail*, 6 June 1919, p. 5 and 13 June 1919, p. 3; *Lancashire Evening Post*, 31 May 1919, p. 4; *Yorkshire Post*, 11 June 1919, p. 8;

CKS to parents, 17 April 1919, MS1882/3/31 EC NLA; *FM*, pp. 246–55. **Monash lends support** *Maitland Weekly Mercury*, 21 June 1919, p. 12. **Billy Hughes's objections** *MFL*, pp. 18–19. **Kingsford Smith Maddocks Aeros Ltd** *Burnley News*, 3 September 1919, p. 5 and 6 September 1919, p. 12. **Over-insuring of planes** Interview with Stannage in Mackersey, *Smithy*, pp. 9–11; R. Williams, *These Are the Facts* (Canberra, 1977), pp. 111–15. As well as condemning the over-insuring, Williams declared that he knew of 'no foundation for the assertion' that Hughes prevented CKS from participating in the air race.

CHAPTER 8: THERE'S A LOT OF FIGHT LEFT IN ME

Black days *FM*, pp. 256–57. **Stony broke** *London Gazette*, 28 July 1922, p. 5649, for liquidation of Kingsford Smith Maddocks Aeros Ltd. **Scarlet fever** *FM*, p. 255 and EKS to parents, 6 December 1919, MS 1882/3/31 EC NLA. **Ince's offer** *FM*, pp. 256–57; for announcement in the United Kingdom, *Dundee Evening Telegraph*, 17 June 1919. **Search for a sponsor** *FM*, pp. 258–59; EKS to parents, 15 February and 14 March 1920, MS 1882/3/31 EC NLA. **Aerial stuntman** For Locklear's death, A. Ronnie, *Locklear: The Man Who Walked on Wings* (London, 1973), pp. 272–73. CKS could not have witnessed Locklear's death plunge because EKS wrote to parents on 23 August 1920 (MS 1882/3/31 EC NLA) that CKS had worked for Moffett in Nevada for the last thirty-five days. Stannage, *Smithy*, p. 13, says that CKS performed movie stunt work; Beau Sheil in *Caesar of the Skies: The Life Story of Sir Charles Kingsford Smith* (London, 1937), p. 33, says that CKS flew for Art Wilson, a stunter at Universal Studios. Sheil also describes on pp. 33–34 how CKS managed to hang by his toes from a plane's undercarriage. **Moffet's Flying Circus** Advertisement, MS 1882/3/31 EC NLA; for CKS's progress in Moffett's employment, see EKS to parents, 14 July, 23 August, 3 November and 10 November 1920, MS1882/3/31 EC NLA. **Rice patrol** CKS to EKS from Willow, 22 September 1920 and EKS to parents, 14 March 1920, MS 1882/3/31 EC NLA. CKS seems to have flown the rice patrol in both March and September; *FM*, pp. 267–68. **Painting Shell signs** EKS to Leofric, n.d., MS 1882/3/31 EC NLA. **Homeward voyage** *FM*, pp. 267–69 and EKS to Leofric, n.d., MS 1882/3/31 EC NLA. **CKS's siblings' careers** For Winfred's 'Kookaburra Tea and Luncheon Rooms', advertised as the 'business man's rendezvous', see *Hebrew Standard*, 2 January 1920; for Wilfrid's import–export business, Kingsford Smith and Co., see *Daily Commercial News and Shipping List*, 22 November 1920; for Eric's naval career, see *SMH*, 7 April 1920.

CHAPTER 9: HARUM SCARUM ANTICS

The Kingsford Smith family house Heritage Report for 73 Arabella Street, Longueville, compiled for the owners, Amelia and Charles Slack-Smith, in 2014 by Colin Brady and kindly supplied by Mr and Mrs Slack-Smith. **CKS broods**

FM, p. 270. **Bad war memories** Stannage, *High Adventure*, pp. 79–83. **Setting up of Diggers Company** *Wellington Times*, 4 November 1920 and 20 January 1921. **CKS joins Diggers Aviation** *FM*, pp. 270–72; Ellison interviewed Lionel Lee, and most of Chapter XIX in *FM* comes from that interview. Beau Sheil, in *Caesar of the Skies*, Chapter VII, has a similar account. Although their biographies of CKS were published twenty years apart, both Ellison and Sheil wrote their books in the late 1930s and relied on Lionel Lees' testimony. **Oberon** Bathurst, *National Advocate*, 16 March 1921; *Wellington Times*, 14 March 1921; *Bathurst Times*, 14 March 1921 – see also advertisement for joy rides, 6 April 1921. **CKS said to be first pilot into Oberon** *Barrier Miner*, 6 June 1928. **Harum scarum antics** Stannage to Ellison, 23 April 1937, MS1882/1/6 EC NLA. **Dubbo crash** *Dubbo Liberal and Macquarie Advocate*, 15 March 1921 and 22 March 1921; *Wellington Times*, 14 March and 17 March 1921; interview with Dulcie and Oliver Cook, Mackersey, *Smithy*, p. 4. **Riverslea crash** *SMH*, 22 July 1921 and *Wellington Times*, 21 July 1921. **Early version of the Cowra crash** *Cowra Free Press*, 23 July 1921. **Later version of Cowra crash** *Cootamundra Herald*, 17 July 1950. **Pilot's licence** CKS applies to Civil Aviation Department of Defence for civil pilots' licence, 28 June 1921, MS 1822/4/43, John Oxley Library, State Library of Queensland. **Letter home** CKS to his mother, 4 August 1921, MS 1882/3/31 EC NLA. **Diggers Aviation wound up** *Wellington Times*, 19 December 1921.

CHAPTER 10: HAPPY AND CONTENTED

Western Australian Airways The airline changed its name to West Australian Airways in 1926. **Length of route** The longest route in Europe was Nimes in France to Casablanca in Morocco – 8000 miles. Western Australian Airways' route was 12,000 miles, according to *Geraldton Guardian*, 23 Oct 1923: the air route began at Geraldton so as not to compete with the rail service between Perth and Geraldton. **Brearley secures contract** *SMH*, 1 August 1921. **CKS applies for the pilot's job** Diggers Aviation supplied a glowing reference and the rail fare to Point Cook – see *FM*, p. 284. CKS's OC at Eastchurch supplied a list of planes CKS had flown: Maurice Farman, BE2C, BE2E, BE12, Martinsyde, Sopwith 1 and a half Strutter, Sopwith Scott (Pup), Sopwith Camel, Sopwith Snipe, Sopwith Dolphin, Avro, Spad, SE5, Handley Page, DH6, DH4, Bristol Fighter, Small AW, FE2B, Newport Scout, RE8, Bristol Monoplane, Albatross D5, Fokker Biplanes – see MS 822/4/73, John Oxley Library, State Library of Queensland. **Brearley accepts CKS** Brearley to CKS, 7 September 1921, MS 1882/3/31 EC NLA. **Interview at Point Cook** CKS to mother, n.d., MS 1882/3/31 EC NLA; Norman Brearley, *Australian Aviator* (Rigby, 1971), pp. 81–82. **Pilots arrive in Perth** *Geraldton Guardian*, 19 November 1921; advertisement for stunting by Brearley and his 'world famed birdmen', see *West Australian*, 3 December 1921. **Official Airways opening** CKS to parents, 20 November and 3 December 1921, MS 1882/3/31, EC NLA.

Crash at Murchison House Perth *Sunday Times*, 11 December 1921. **Survey flight** *West Australian*, 28 December 1921; CKS to parents, 30 December 1921, MS 1882/3/31 EC NLA; *FM*, p. 286; Brearley, *Australian Aviator*, pp. 96–99. **CKS flies doctor to Carnarvon** CKS to parents, 11 February n.y., MS 1882/3/31, EC NLA; Perth *Daily News*, 9 February 1922; *West Australian*, 9 February 1922. **Cargo** 'Pilots of the Purple Twilight', Perth *Daily News*, 23 July 1923. **Excellent reports of CKS** Brearley to CKS, 23 May 1922, MS 1882/3/31 EC NLA. **Pilots as pioneers** *West Australian*, 26 October 1921. **CKS's conviviality** Interview with June Dupre, Mackersey, *Smithy*, pp. 55–56. **Dengue fever** *West Australian*, 7 July 1922; CKS to parents, 25 June n.y. (but 1922 on internal evidence), MS 1882/3/31 EC NLA; **Verona** Interview with Verona's son, Mackersey, *Smithy*, pp. 57–58; for CKS's dive to oyster beds, CKS to parents, 25 June 1922, MS 1882/3/31 EC NLA. **Verona's husband** 'Death of Frederick Graves', *Western Mail*, 27 March 1924. **Graves divorce case** *West Australian*, 12 May 1917, p. 8. **Verona's breach of promise case** *West Australian*, 12 December 1916, p. 8. **Smithy convalesces in Perth** CKS to parents, 25 June 1922, MS 1882/3/31 EC NLA. **Forced landings** *FM*, pp. 290–91; CKS to parents, 30 July and 1 September 1922, MS 1882/3/31 EC NLA. **CKS's finances** CKS to parents, 11 February, 14 March, 25 April, 3 May, 1 September, 19 November, 31 December 1922, MS 1882/3/31 EC NLA. **Tired of itinerant life** CKS to parents, 31 December 1922, MS 1882/3/31 EC NLA. **Race Week** *Northern Times*, 9 September 1922. **Thelma Corby's first sight of CKS** Interview with Thelma, in *Smithy*, pp. 66–68. **Solar eclipse** John L. Robins, 'Wallal: The 1922 Solar Eclipse Expedition to Test Einstein's Theory', History of the Department of Physics at UWA, no. 9. **CKS to parents about eclipse** 1 October 1922, MS 1882/3/31 EC NLA; Brearley, *Australian Aviator*, pp. 104–05. **CKS at Meentheena** CKS to parents, 1 October and 26 October 1922, MS 1882/3/31 EC NLA; Graham J. Wilson, *Pilbarra Bushman: The Life and Experiences of W. Dunn* (Hesperian Press, 2002), pp. 8–12. **Description of Meentheena homestead** See advertisement for its sale, Perth *Sunday Times*, 9 October 1921. **Thelma's background** Thelma went to WA School 74 in the Perth suburb of Leederville (see exam results in *West Australian*, 26 December 1914) and to Perth City Commercial College (see Pitman exam results, Perth *Daily News*, 29 November 1918 and *West Australian*, 15 March 1919). Thelma was musical from childhood – see report of children's cantata, *West Australian*, 18 August 1913. **CKS's sheet music** CKS to parents, 1 October 1922, MS 1882/3/31 EC NLA. **Christmas week in Port Hedland** Carnarvon *Northern Times*, 6 January 1923. **CKS feels he must marry** CKS to parents, 31 December 1922, MS 1882/3/31 EC NLA. **Thelma's former engagement** *Pilbarra Goldfield News*, 20 Jan 1920 and Perth *Daily News*, 5 May 1921. **Marriage** Wilson, *Pilbarra Bushman*, p. 11; CKS to his mother, 9 June 1923, MS 1882/3/31 EC NLA. **Rail trolley** Travel writer Ernestine Hill rode on the rail trolley a few years later. She described it as a raft of about six feet by three feet that ran along the rails, with a motor attached – see Ernestine Hill, *The Great Australian Loneliness* (London, 1937), p. 51.

CHAPTER 11: A LAST FLUTTER AT THE FLYING GAME

Anderson's background Perth *Daily News*, 16 April 1929; **Keith Mackay** For Mackay selling part of his land, *Great Southern Herald*, 5 December 1923; for Mackay favouring air travel, *Geraldton Guardian*, 23 October 1923. **Thelma at Arabella Street** Interviews with Thelma and John Kingsford Smith, Mackersey, *Smithy*, pp. 69–70. **Lebbeus Hordern** *SMH*, 4 June 1914 and 5 April 1921; *FM*, p. 295. **CKS buys block next to Meentheena Station** CKS to parents, and Thelma to parents-in-law, 2 September 1923, MS 1882/3/31 EC NLA; CKS to parents, 21 Nov 1923, MS1882/3/31 EC NLA; see also Agricultural Bank to Officer in Charge of Records, Military Headquarters, Sydney, 21 November 1923, 1523 /23 SSS WAW MB, NAA. CKS was hoping for a monetary grant under the government's Soldier Settlement Scheme. **Sale of Meentheena** McDonald and McKenna were owners of Meentheena. In 1921 the first mortgagee foreclosed and the property was put up for sale, but no buyer could be found. The property was re-mortgaged but in 1924 was put up for sale again, this time successfully; see adverts in Perth *Sunday Times*, 9 October 1921 and 10 February 1924. Smithy's and Mrs McKenna's blocks were on separate titles so they retained their leases after the sale. **Race Week** CKS gives joy rides, see Carnarvon *Northern Times*, 15 September 1923. **Sacked by Brearley** *Australian Aviator*, pp. 109–111; CKS to parents, 15 February 1924, MS 1882/3/ 31 EC NLA. **Thelma not philosophical** Interview with Thelma, Mackersey, *Smithy*, p. 73. **CKS buys share in Carlin Transport Co.** CKS to parents, 27 March and 2 June 1924, MS 1882/3/31 EC NLA. **Gascoyne Transport Co.** *FM*, pp. 298–301. In June 1924, CKS and Anderson, with the financial help of Anderson's mother, bought out Carlin and formed their own company – see Carnarvon *Northern Times*, 5 September 1924, and advertisement, 13 June 1924. CKS persuaded EKS, her husband, Bert Pike, and Phil Kingsford to work for the company – see CKS to parents, 4 December 1924, MS 1882/3/31 EC NLA. **CKS buys speed trucks** *Geraldton Guardian*, 6 March 1924. **CKS's flying and business career** Typewritten account of CKS's career, sent by EKS to Leofric, n.d., MS 1882/3/31 EC NLA. It was written after CKS's death to help Norman Ellison write CKS's biography. **Mackay intends to buy Hordern's flying boat** CKS to parents, 27 April 1924 and 2 June 1924, MS 1882/3/31 EC NLA. **Mackay dies** CKS to parents, 23 July 1924, MS 1882/3/31 EC NLA; *Kalgoorlie Miner*, 17 July 1924; *Great Southern Herald*, 19 July 1924. **McKenna's part-Aboriginal son** Wilson, *Pilbara Bushman*, p. 8. **McKenna's cattle stealing** For his first offence, *Pilbara Goldfield News*, 18 September 1902; for his second offence, Wilson, *Pilbara Bushman*, pp. 13–14 and *Northern Times*, 6 March 1925. **McKenna's trial** *West Australian*, 6, 10 and 11 June 1925. **Henry Mosely** Perth *Mirror*, 2 January 1924; Carnarvon *Northern Times*, 25 January and 21 November 1924. **June Dupre names Mosely as Thelma's admirer** Interview with June Dupre, Mackersey, *Smithy*, p. 77. **CKS accuses Mosely** See divorce petition, Sydney *Truth*, 28 October 1928. **Thelma's concert** Carnarvon *Northern Times*, 21 November 1934. **CKS returns to Sydney** CKS to mother, 26 July 1925, MS 1882/3/31 EC NLA.

CHAPTER 12: A STEP NEARER HIS DREAM

CKS and prime minister 'Documents relating to proposed flight across the Pacific', A 458, M 314/4 NAA. Includes material on Hordern's flying boat, the Widgeon, Billy Hughes, and the proposed itinerary. **Letter to Thelma** Sydney *Truth*, 28 Oct 1928. **Bon Hilliard** For birth notice, *SMH*, 7 November 1896; for engagement to Laycock, Sydney *Sunday Times*, 5 September 1920; for broadcasting work, *SMH*, 22 June 1935; interview with Bon's relatives, Mackersey, *Smithy*, pp. 82–83. **Sale of Gascoyne Co.** EKS's typescript to Leofric, MS 1882/3/31 EC NLA. **CKS applies to be Aero Club instructor** 'Application for a Position as Instructor to the Australian Aero Club, NSW Section', MLDOC909, Mitchell. **Hitchcock and New Guinea** For Hitchcock's career, *Brisbane Courier*, 23 April 1929; for New Guinea charter flights and setting up of Interstate Flying Services, see advertisement for Interstate Flying Services and EKS's typescript, MS 1882/3/44 EC NLA. **Progress of Perth–Sydney flight** Perth *Daily News*, 24 and 28 January 1927 and 16 February 1927; *Kalgoorlie Miner*, 29 January 1927; Dick Smith and Pedr Davis, *Kookaburra: The Most Compelling Story in Australia's Aviation History* (Sydney, 1980), Chapter 3. **CKS turns thirty** *MFL*, pp. 21–22. **Anderson's engagement to Bon** Sydney *Sun*, 17 April 1927, p. 26. **Thelma's answer to CKS's letter** Sydney *Truth*, 28 October 1928. **Ulm older than his years** Brearley, *Australian Aviator*, p. 180. **CKS's meeting with Ulm** Transcript of evidence given during Anderson court case.

CHAPTER 13: THE BIG FEAT

Ulm's early life Emerald Hill *Record*, 16 June 1928 and Burnie *Advocate*, 7 June 1928. According to electoral rolls, the Ulm family lived first in Hawthorn Road, Caulfield, and then at 30 Dundas Place, Albert Park. By 1914 the family was living at 27 Keston Ave, Mosman; Ulm attended the Albert Park State School from 1906 to 1909; for his father's work as a photographer, see Sydney *Sun*, 12 April 1929; see also Charles Ulm, 'Ulm's Life Story', Brisbane *Courier Mail*, 4 and 5 March 1935, where Ulm describes his schooldays, going to war, becoming air-minded, founding Aviation Services Company and his first meetings with Campbell Jones and CKS. **Earlier meeting of Ulm and CKS** SC *Trans-Pacific*, pp. 13–15. **Ulm's tender** 'Proposals for Air Services Between Perth and Adelaide' and letter to Sir Charles Rosenthall, MS1882/3/34, EC NLA; *FM*, pp. 302–03. **Ulm's view of CKS and Anderson and Ulm's ambition** 'Ulm's Life Story', Brisbane *Courier Mail*, 5 March 1935. **CKS and Ulm as soulmates** *MFL*, pp. 21–22. **Ousting of**

Pike and Anderson Ulm's evidence, pp. 67–70, and CKS's evidence, p. 38, in Hitchcock court case. **Ulm's divorce** *SMH*, 16 December 1926. **Josephine Callaghan** *Mullumbimby Star*, 14 June 1928. **Lindbergh New York–Paris flight** *SMH*, 23 May 1927. **'I'm all for it'** Ulm's evidence, Hitchcock court case, p. 69. **Flight around Australia** SC *Trans-Pacific*, pp. 14–19; Dick Smith and Pedr Davis, *Kookaburra* (Lansdowne Publishing, 1980), Chapter 4; *Northern Territory Times*, 24 June 1927; Perth *Daily News*, 25 June 1927; *Geraldton Guardian*, 28 June 1927; *West Australian*, 30 June 1927; Carnarvon *Northern Times*, 2 and 9 July 1927; Melbourne *Herald*, 7 July 1927. On the Camooweal to Darwin section, CKS beat Bert Hinkler's record (Brisbane to Bundaberg) for the longest non-stop flight so far made in Australia. **'If any man can do it'** Perth *Daily News*, 25 June 1927. **Anderson learns he is being ousted** Adelaide *News*, 7 July 1927; Melbourne *Herald*, 7 July 1927; Kalgoorlie *Western Argus*, 12 July 1927.

CHAPTER 14: A COUPLE OF SPENDTHRIFT DREAMERS

Finance 'Pacific Flight', A 458, M 314/4 NAA, pp. 154–57; for Le Maistre Walker's subscription fund, see 'Proposal from Captain Kingsford Smith for Flight from San Francisco to Sydney', MS1882/3/32 EC NLA. **Lang's help** 'Charles Kingsford Smith and the Gamble of the Pacific', Sydney *Truth*, 21 March 1954, p. 40. **Confrontation with Anderson at Collaroy and Carlton Hotel** CKS's evidence, pp. 39–44, 56, and Ulm's evidence, pp. 2–5, Hitchcock court case. **Departure** *Newcastle Sun*, 15 July 1927. **Voyage to America** R.E. Haylett, 'The Men Who Met the Challenge', *Union Oil Bulletin*, 1928, MLMSS3359/3 X/1-13, Mitchell. **Dole Air Race and the lessons learned** Sydney *Truth*, 21 March 1964, p. 40; SC *Trans-Pacific*, pp. 23–28. **Buying the Fokker** *SMH*, 19 August 1927; *MFL*, pp. 24–31; SC *Trans-Pacific*, pp. 29–32. **CKS approaches US Navy for help in buying the engine** SC *Trans-Pacific*, pp. 35–38. The name of the admiral who helped them is mistakenly given as PEABLES in SC *Trans-Pacific* – in fact, it was Rear Admiral Christian Joy PEOPLES, General Inspector Supply Corps for the Pacific Coast. **Lang gives further money** *SMH*, 19 August 1927. **CKS's skill during test flight** Letter to Ellison from Atlantic Oil Company executives R.A. Pope and Walter B. Phillips, 22 May 1959, MS1882/3/32 EC NLA; for CKS's instinctive handling of the large Fokker, see P.G. Taylor, *The Sky Beyond* (London, 1963), pp. 31–33. **Todd, wireless operator** Melbourne *Argus*, 7 October 1927. **Bavin backs out** *SMH*, 24 Oct 1927. **Ulm tramps the streets** 'Ulm's Life Story', Brisbane *Courier Mail*, 5 March 1935. **Endurance record** *MFL*, pp. 33–38; Melbourne *Argus*, 5 Dec 1927; Warwick *Daily News*, 5 Dec 1927; Adelaide *Register*, 21 Dec 1927; Adelaide *News*, 9 January 1928. For the endurance flights, *Southern Cross* was briefly renamed *Spirit of California*. **Expiry of guarantee** Ulm's evidence, Anderson court case, pp. 72–73.

CHAPTER 15: ROCK BOTTOM – BUT NOT FOR LONG

Poverty *MFL*, p. 38. **Todd's misdeeds** 'Coffee Royal Inquiry', Lismore *Northern Star*, 13 June 1929; for Todd's arrival back in Australia, Melbourne *Argus*, 2 January 1928. **Legal agreement** Agreement dated 15 December 1927 and affidavit by Ulm, pp. 1–4, Anderson court case. **Hinkler's flight** Grantlee Kieza, *Bert Hinkler* (Australia, 2012), Chapter 20. **Anderson's departure** Anderson's evidence, pp. 35–40 and Ulm's evidence, pp. 53–71, Anderson court case. **Reasons for Ulm's antagonism to Pond** Brisbane *Daily Standard*, 7 April 1928 and *Barrier Miner*, 16 April 1928. **At Rogers Field** R.E. Haylett, 'The Men Who Met the Challenge', *Union Oil Magazine*, July 1928, MLMSS3359/3/1-12, Mitchell; letter to Ellison by Atlantic Oil executives R.A. Pape and Walter B. Phillips, 22 May 1959, MS 1882/3 EC NLA. Haylett's account is especially valuable because it was written so close to the event. **Andrew Chaffey** See George Chaffey, *Australian Dictionary of Biography* vol. 7, (Melbourne, 1979), p. 600; **Allan Hancock** Sally Capon, 'Allan Hancock Rose from Tar Pits to the Wild Blue Yonder', *Santa Maria Times*, 13 February 2005; SC *Trans-Pacific*, p. 44–49; 'Ulm's Life Story', Brisbane *Courier Mail*, 5 March 1935. **Preparations** *MFL*, pp. 40–44; SC *Trans-Pacific Flight*, Chapters IV and V; *SC Story*, Chapters 1, 2 and 3. **Cables to and from Anderson** Anderson's affidavit, presented during Anderson court case. **Harry Lyon** 'Interview with Lyon', Sydney *Sunday Telegraph*, 22 June 1958; Sydney *Sun*, 11 June 1928; *SMH*, 2 June, 11 June and 20 June 1928; 'Aviation: The Pacific Ocean Crossed', Brisbane *Queenslander*, 14 June 1928. **James Warner** *SMH*, 2 June 1928; *Los Angeles Examiner*, 6 June 1928; Rockhampton *Morning Bulletin*, 7 June 1928; *Honolulu Star Bulletin*, 6 June 1928. **Contract with Lyon and Warner** 'The Trans-Pacific Flight, Being the Saga of the *Southern Cross* by James W. Warner, as Told to John Robert Johnson', *Liberty* magazine, 19 April 1930. **Ready for departure** *MFL*, p. 45.

CHAPTER 16: WE ABSOLUTELY WON'T FAIL

Departure SC *Story*, pp. 69–71; SC *Trans-Pacific*, pp. 66–72; cutting from *San Francisco Bulletin*, 7 June 1928, ML SS3359-16. **Women give CKS farewell kisses** Sydney *Sun*, 1 June 1928. **CKS feels elated** *MFL*, p. 46. **Conditions on board and progress of flight** SC *Trans-Pacific*, Chapters VI–X; SC *Story*, Chapter IV; Michael Molkentin, *Flying the Southern Cross* (Canberra, 2012), pp. 56–59, 72. **Plasticine ear plugs** *Vancouver Daily Province*, 9 June 1928. **Lyons and Warner's clothes** Cutting from *Los Angeles Daily Herald*, n.d., in Harry Lyon's scrapbook, MS5312, NLA. **Radio broadcasts** *Wellington Times*, 21 June 1928 and *Brisbane Courier*, 7 June 1928. **William listens to broadcasts with next-door neighbour** *Daily Telegraph Pictorial*, 6 June 1928 and Sydney *Sun*, 6 June 1928 – Mrs Minich was his next-door neighbour. **Catherine KS's belief that CKS was saved from drowning for a purpose** See Brisbane *Telegraph*, 16 June 1931. On later flights

Southern Cross was registered as a telegraph office and only designated people could access the Morse messages – see Sydney *Sun*, 2 August 1928. **Baseball scores and card games** Hobart *Mercury*, 4 June 1928. **Sailing on the Milky Way** SC *Trans-Pacific*, pp. 91–93.

CHAPTER 17: A SHOT AT A DOT

CKS's tribute to Ulm *Newcastle Sun*, 4 June 1928. **Conditions on board and progress of the flight** SC *Trans-Pacific*, pp. 110–53; SC *Story*, pp. 35–47; *MFL*, Chapter VI; Molenkin, *Flying the Southern Cross*, pp. 80–108. **Landing in Suva** London *Times*, 6 June 1928. **Warner's nakedness** Sydney *Sunday Telegraph*, 22 June 1958.

CHAPTER 18: KINGSFORD SMITH, AUSSIE IS PROUD OF YOU

Difficulty over clothes SC *Trans-Pacific*, pp. 158–59. **Ulm and Warner and Lyon** Ulm vented his dissatisfaction with the Americans in notes at the back of his log book – see Molenkin, *Flying the Southern Cross*, pp. 166–69. The Americans vented their dissatisfaction with Ulm in Warner's 'The Trans-Pacific Flight', *Liberty* magazine, 19 April 1930. **Ulm praises Lyon and Warner** Sydney *Sun*, 6 June 1928 and *Los Angeles Examiner*, 6 June 1928. **British Consul's questions** *Honolulu Star Bulletin*, 6 June 1928. **Grand Pacific Ball** *San Francisco Bulletin*, 7 June 1928. **At Naselai beach** Sydney *Sun*, 8 June 1928. **Lyon and Warner praise CKS** *Brisbane Courier*, 6 June 1928. **Kava drinking and bonfire** *MFL*, p. 65 and *Christchurch Star-Sun*, 20 August 1958. **Progress of flight** Molenkin, *Flying the Southern Cross*, pp. 115–53; SC *Trans-Pacific*, Chapters XVII; SC *Story*, Chapter VI. **Wireless bulletins** *Mullumbimby Star*, 14 June 1928; for holding down the key, *Brisbane Courier*, 7 June 1928; for Basil Kirke, Perth *Sunday Times*, 17 January 1954. **Hancock thanked by wireless** *Vancouver Daily Province*, 9 June 1928. Ulm also thanked Sidney Myer. **Arrival and reception at Brisbane** Brisbane *Daily Standard*, 11 June 1928; for 'Kingy, you darling' and 'my kingdom for a smoke', *Vancouver Daily Province*, 11 June 1928. **Interviews at Lennon's Hotel** *Brisbane Courier*, 11 June 1928 and *Honolulu Star Bulletin*, 11 June 1928. **Lyon, Warner and Ellison** *FM*, pp. 132–34. **Accolades from overseas** Sydney *Sun*, 11 June 1928; Melbourne *Argus*, 11 June 1928.

CHAPTER 19: EXCEEDINGLY FAMOUS

Arrival in Sydney Melbourne *Age*, 11 June 1928 and *SMH*, 2 June 1928. **Catherine's nerves** *Daily Telegraph Pictorial*, 6 June 1928. **Longueville celebration** *Dubbo Liberal and Macquarie Advocate*, 6 July 1928 and *SMH*, 11 June 1928.

Technical School Old Boys Dinner *Brisbane Courier*, 18 June 1928. **Arrival in Melbourne and Canberra** Hobart *Mercury*, 14 June 1928 and Townsville *Daily Bulletin*, 18 June 1928. **Australian–American friendship and distribution of prize money** Documents relating to the Trans-Pacific flight, esp. p. 48 and p. 110, A458, M314/4, NA of A. **Lyon and Warner speak to the press** Brisbane *Daily Standard*, 18 June 1928. **Ulm's statement re: Lyon and Warner** *Newcastle Morning Herald*, 13 June 1928. **Hearst money for Warner and Lyon** Lismore *Northern Star*, 12 June 1928. **CKS's and Ulm's prize money** *MFL*, p. 72. **CKS's gifts to parents** *SMH*, 11 June 1928. **CKS buys Studebaker car** *Newcastle Sun*, 12 June 1928 and Perth *Daily News*, 14 July 1928. **CKS reads novels** Interview with Catherine KS, Perth *Sunday Times*, 20 July 1930. **William's gratitude** Sydney *Sun*, 12 April 1929. **CKS's many roles** Sydney *Daily Guardian*, 10 April 1929. *Rio Rita* **and** *The Girl Friend* *Newcastle Sun*, 12 June 1928; Melbourne *Table Talk*, 21 June 1928; Melbourne *Age*, 15 June 1928. **Nellie Stewart** Sydney *Sun*, 10 June 1928. **Female fan mail** Sydney *Sun*, 17 June 1928. **Calls for knighthood** Sydney *News*, 28 November 1928. **Air Force Cross** See discussion in documents A2926 – A6, NAA; Richard Williams, *These Are the Facts*, p. 202. **Lyon and Warner agree to join flight to New Zealand** Cairns *Northern Herald*, 27 June 1928. **Lyon and Warner sail home** Sydney *Evening News*, 23 June 1928. **Flight to Perth** Broken Hill *Barrier Miner*, 14 August 1928; Mackay *Daily Mercury*, 27 August 1928; *MFL*, pp. 7, 75–78. **McWilliams** *Brisbane Courier*, 17 July 1928. **Request for New Zealand radio operator** Letter, 23 July 1928, AD 314/4, NAA. **CKS's visit to Carnarvon and CKS's career as a Mason** CKS was eager to attend a Masonic Lodge meeting in Carnarvon. On 15 February and 26 July 1925 he wrote to his parents (MS 1882/3/31 EC NLA) describing how he had joined the Masonic Lodge in Carnarvon, being initiated on 9 April and passing the second degree on 1 July. He passed his third degree at his father's lodge in Sydney on 3 September 1925 and remained a keen Mason for the rest of his life; see *New South Wales Masonic Magazine*, March 2007. **Take-off from Perth** To reduce weight for take-off at Perth, McWilliams, Litchfield and a passenger, William Broadhurst, boarded the plane at Tammin; see *MFL*, p. 78. **Trans-Tasman flight** The aviators' families farewelled them at Richmond; see *Daily Telegraph Pictorial*, 11 September 1928; *MFL*, pp. 79–86; Broken Hill *Barrier Miner*, 13 September 1928; Adelaide *News*, 12 September 1928. Hood and Moncrieff had been lost trying to cross the Tasman earlier that year. Smithy dropped two wreaths into the sea in their memory. **In New Zealand** For female fans, see New Zealand *Evening Post*, 12 September 1928; for 'own kith and kin', see 17 September 1928; for the long wait in Blenheim and makeshift jazz band, see Canterbury *Press*, 29 September and 15 and 16 October 1928; for Ulm gaining his wings, see Sydney *Daily Guardian*, 10 October 1928. **Non-observance of Sabbath** *Daily Telegraph Pictorial*, 11 September 1928. **Flight back to Richmond** Brisbane *Queenslander*, 18 October 1928; Sydney *Sun*, 14 and 15 October 1928. *MFL*, p. 88, states that they left New Zealand on 7 October. In fact, they left a week later, on 18 October 1928.

CHAPTER 20: THE ORDEAL OF COFFEE ROYAL

Founding of Australian National Airways 'Ulm's Life Story', Brisbane *Courier Mail*, 7 March 1935; Ellen Rogers, *Faith in Australia: Charles Ulm and Australian Aviation* (Sydney, 1987), pp. 29, 36–37: ANA was registered as a company on 11 December 1928 and underwritten prior to registration. **Ives divorce case** Perth *Daily News*, 21 June and 27 June 1928; *West Australian*, 4 December 1928 and 26 September 1929; Maitland *Weekly Mercury*, 26 September 1929. **CKS's divorce** Sydney *Truth*, 28 October 1928. **Anderson and Hitchcock sue CKS and Ulm** *Kalgoorlie Miner*, 20 July 1928; Adelaide *Advertiser*, 10 August 1928 – see also for writ served at RSL dinner. **Anderson tries to fly with CKS and Ulm** Sydney *Sun*, 2 August 1928. **Anderson's failed flight to England** Davis and Smith, *Kookaburra*, p. 40 – see also p. 42 for purchase of the plane *Kookaburra*. **Settlement of Anderson's case** *West Australian*, 23 February 1929 **and Hitchcock's case** *SMH*, 21 March 1929. **Purpose of Australia–England flight** *Newcastle Sun*, 8 June 1928; *SMH*, 21 May 1929; for publicity regarding their plans for ANA, see Brisbane *Queenslander*, 7 February 1929. **Invitation to Party** Ulm Papers, MLMSS3359/3X/1-12, Mitchell. **CKS feeling ill** *SMH*, 21 May 1929; Sydney *Sun*, 25 and 28 March 1929; Sydney *Evening News*, 31 March 1929; Sydney *Daily Guardian*, 2 April 1929. **Ulm disregards Chateau's advice** Sydney *Sun*, 16 May 1929. **Copy of Chateau's telegram** Ulm papers, MLM223359/5, Mitchell. **Catherine's farewell to CKS** Sydney *Sun*, 31 March 1929 and Broken Hill *Barrier Miner*, 7 November 1935. **Forced landing and life at Coffee Royal** *MFL*, pp. 91–118 and SC *Story*, pp. 81–112; for an early and concise account by CKS, see Adelaide *Register News Pictorial*, 20 April 1929; Brearley, *Australian Aviator*, pp. 143–44; 'Coffee Royal file', Ulm Papers, MLMSS3359/4/1-6, Mitchell, which includes the report of the Committee of the Southern Cross Air Inquiry and a report of the Citizens' Southern Cross Rescue Fund. **Ulm's diary** Sydney *Sun*, 14 and 15 April 1929, has the simple, early version; Rockhampton *Capricornian*, 18 April 1929, gives a later version. See also 'Ulm's Log Up to Date', Brisbane *Courier*, 15 June 1929. **Bruce's uninterest and the public's protests** Sydney *Sun*, 4 April 1929; for poem, see *Daily Telegraph Pictorial*, 9 April 1929; letter from Lismore Sub Branch of RSSILA to prime minister Bruce, 12 April 1929, AH458 AH 314/4, NAA; Adelaide *Chronicle*, 13 April 1939. **Rescue efforts** Brearley, *Australian Aviator*, pp. 144–48; 'Citizens Rescue Fund', A 458 AH 314/4 NAA. **Anderson and Hitchcock** Davis and Smith, *Kookaburra*, pp. 46–51, and p. 98 for CKS having saved Anderson's life years earlier.

CHAPTER 21: A MOST PERPLEXING PARADOX

Found Adelaide *Register News Pictorial*, 20 April 1929; Stannage, *High Adventure*, Chapter 2. **Rejoicing across the nation** Sydney *Sun*, 12 and 13 April 1929; *Canberra Times*, 13 April 1929; *Daily Guardian*, 13 April 1929; for Darlington

Children's Home, see Sunday *Sun*, 14 April 1929. Station 2BL repeated the news at intervals and supplied extra telephonists to deal with the calls. St Andrew's Cathedral held a thanksgiving service. **Food drops** Sydney *Sun*, 14 and 15 April 1929. **Anderson joins the search** Smith and Davis, *Kookaburra*, pp. 54–55. **Tonkin and Marshall visit Coffee Royal** Sydney *Sun*, 15 April 1929; *Daily Guardian*, 15 and 16 April 1929; Rockhampton *Morning Bulletin*, 20 April 1929. **Rumours of publicity stunt** Adelaide *News*, 17 May 1929; *Brisbane Courier*, 15 June 1929; C.E. Kingsford Smith, SC *Story*, p. 112. **Take-off and flight to Derby** Melbourne *Argus*, 6 June 1929; Hobart *Mercury*, 19 April 1929. **CKS searches for Anderson** Sydney *Sun*, 20 April 1929. **Finding Anderson's body** *Brisbane Courier*, 22 April 1929. **Smith newspapers out for blood** *Daily Guardian*, 15 April 1929, writes of a 'paradox', and on 16 April 1929 enlarges on the theme with an article entitled 'Flight Flops in Own Backyard'; see also *Daily Guardian*, 22 and 27 April 1929; *Smith's Weekly*, 27 April 1929. **CKS flies over Anderson's plane** SC *Story*, pp. 109–110. **Arrival in Richmond** *SMH*, 29 April 1929. **Edward Hart condemns Coffee Royal** Cross-examination during SC Air Inquiry in Sydney *Sun*, 27 May 1929. **Anderson's death is reminiscent of Capt. Scott's** SC *Story*, p. 108. *SC* **Air Inquiry** See file labelled 'Southern Cross Air Inquiry', 14 May 1929 to 16 June 1929, MLMSS3359/4/1-6, Mitchell. **Air Inquiry is announced to the public** Brisbane *Queenslander*, 2 May 1929. **Women in jury box** Sydney *Sun*, 16 May 1929. **'All I have is a bill'** Adelaide *News*, 17 May 1929. **Investiture** *SMH*, 4 June 1929. **CKS in witness box** Perth *Daily News*, 17 May 1929. **Ulm's quick temper** 'Ulm's Life Story', Brisbane *Courier Mail*, 12 March 1935. **Porteous's evidence** Melbourne *Argus*, 6 June 1929. **Lush's evidence** *Brisbane Courier*, 15 June 1929. **Todd's evidence and Ulm's response** Sydney *Sun*, 12 June 1929. **Findings of SC Air Inquiry** See report of committee, MLMSS3359/4/1-6, Mitchell. **Ulm's reaction** 'Ulm's Life Story', Brisbane *Courier Mail*, 12 March 1935.

CHAPTER 22: HE PURSUED ME QUITE RELENTLESSLY

Ulm's talent for organisation The book that displayed his talent for organisation was C.E. Kingsford Smith and C.T.P. Ulm, *The Story of* Southern Cross *Trans-Pacific Flight*, published October 1928. **Review of SC *Trans-Pacific* book** Hobart *Mercury*, 13 October 1928 and *The Bulletin*, 'The Red Page', 3 October 1928. **Ulm's reaction to *SC* Air Inquiry** 'Ulm's Life Story', Brisbane *Courier Mail*, 12 March 1935. **CKS's reaction to *SC* Air Inquiry** SC *Story*, p. 110. **CKS's reaction to scurrilous letters** 'How I Regard Public Fame', quoted in Beau Sheil, *Caesar of the Skies*, p. 105. **CKS's illness** Catherine's diary, quoted in Mackersey, *Smithy*, p. 218. **Anderson's funeral** *West Australian*, 8 July 1929. Hitchcock was buried quietly in Perth at the request of his family. **Hinkler's trophy** Perth *Daily News*, 21 June 1929. **CKS's quiet departure** *Daily Telegraph Pictorial*, 26 June 1929. **Flight to England and arrival at Croydon aerodrome** SC *Story*, Chapter XII; 'Log of the

Southern Cross by Co-Commander CTP Ulm, 25/6/1929', MLMSS3359/4/1-6, Mitchell Library; Sydney *Evening News*, 5 and 7 July 1929; Sydney *Sun*, 5 and 7 July 1929; *Daily Telegraph Pictorial*, 5 and 7 July 1929; Hobart *Mercury*, 12 July 1929. Smithy met the Prince of Wales at the RAF pageant at Hendon; see *MFL*, p. 31. **Ulm buys Avros** Brisbane *Queenslander*, 24 October 1929. **CKS flies *SC* to Amsterdam** *MFL*, pp. 31–32; **CKS meets Aussie, the kangaroo** *Smith's Weekly*, 7 December 1935. Aussie was attracted to aviators and was almost killed by James Mollison's propellor in 1931. **CKS sails to New York, meets Fokker, goes to California** CKS arrives in New York on 28 September 1929, *MFL*, pp. 132–34. **CKS visits Vancouver** *MFL*, p. 134. **Shipboard courtship of Mary Powell** Interview with Mary Powell, Mackersey, *Smithy*, pp. 229–31; interview with Mary Powell, *Australian Women's Weekly*, 16 June 1976, pp. 10–11. **Powell family background** CKS describes Mary's father as a 'rich rancher' in *Honolulu Advertiser*, 30 March 1930; for Powells' address in Footscray, see electoral rolls, 1919 and 1921; for Arthur Powell's death notice, see Melbourne *Argus*, 4 September 1952. Mary's older siblings were Thelma and Reginald: obituary of William Hamilton Powell, *Footscray Independent*, 5 May 1894; obituary of Robert Gustavus Powell, Melbourne *Australasian*, 3 February 1934 and Melbourne *Argus*, 2 February 1934; for Mosstrooper, see Brisbane *Truth*, 3 August 1930; for Gus's ability as a rough rider, see Brisbane *The Week*, 14 January 1888. **CKS becomes engaged to Mary** Melbourne *Argus*, 30 November 1929; Adelaide *Register News Pictorial*, 29 November 1929; *Cairns Post*, 20 December 1929. **Flight to Sydney** 'Jolly Good Trip', Sydney *Evening News*, 3 December 1929; interview with Mary Powell, Mackersey, *Smithy*, p. 232.

CHAPTER 23: RIGHT AROUND THE WORLD

Preparations for the inaugural flight Brisbane *Queenslander*, 24 October and 14 November 1929. **Inaugural flight** Brisbane *Telegraph*, 2 January 1930 and Lismore *Northern Star*, 2 January 1930. **Forced landing of *Southern Sky*** *Cootamundra Herald*, 23 January 1930 and article by Ron Gibson, *Northern Star*, 2 January 1980. **Satisfied customers** Cornell's letter, *Urana Shire Advocate*, 1 May 1930; clipping, n.d., from *Smith's Weekly* for Mary Anne's letter, MS 1882/3/40 EC NLA. **ANA statistics** Sydney *Sun*, 19 September 1930. **CKS's attitude to Ulm** CKS's remarks to Robert Thomas, see news cutting, n.d., MS 1883/3/40 EC NLA. **CKS's half-hearted regrets** *The Worker*, 18 September 1929. CKS makes a show of cancelling the Atlantic flight when Ulm is refused leave, but soon after announces the flight will go ahead – see Scotty Allan's interview, Mackersey, *Smithy*, p. 263. **CKS sails to America** Sydney *Evening News*, 15 March 1930. CKS's mother is reported to be distressed as he boards the ship. **CKS meets van Loon** *MFL*, p. 135; for reunion with *SC*, see *MFL*, p. 136. **CKS chooses van Dijk and Stannage** 'Personal memories of the *Southern Cross* and the Brave Ocean Pilots

Kingsford Smith and van Dijk', a manuscript account by K. Zimmerman, John Oxley Library, State Library of Queensland; Stannage, *High Adventure*, Chapter 6. **Preparations for the flight** Interview with Stannage, Mackersey, *Smithy*, p. 76; *MFL*, pp. 137–39; Launceston *Examiner*, 23 June 1930 and 25 June 1930 – includes CKS's statement of the odds being twenty-five to one. **Progress of the flight** Stannage, *High Adventure*, Chapter 7; for 'the big stiff', Sydney *Sun*, 25 June 1930; interview with Stannage, Mackersey, *Smithy*, pp. 81–83; *MFL*, pp. 140–52; 'Wireless Log for Crossing the Atlantic', *Flight Magazine*, 4 July 1930, MS1882/37 EC NLA; 'Smith's Story', Melbourne *Age*, 30 June 1930; *New York Times*, 25 June 1930.

CHAPTER 24: A HIT AND NO QUESTION OF IT

Congratulatory cables Adelaide *Register News Pictorial*, 27 June 1930. 'Donnerwetter' is an exclamation like 'My word!' **Clothes** Hobart *Mercury*, 28 June 1930. **Welcome at Roosevelt Field** *Cairns Post*, 28 June 1930. **Tickertape parade** *New York Times*, 28 and 29 June 1930; *New York Sun*, 27 June 1930. The headline in the *New York Sun* is 'Airmen Decline Triumphant Broadway Parade', but this is misleading because the article then explains that the parade will go ahead but over a shorter route, leaving from the Roosevelt Hotel instead of the docks. **Phone calls** *New York Times*, 28 and 29 June 1930; Sydney *Sun*, 29 June 1930. **Fokker's yacht party and reception in Washington** *New York Times*, 30 June 1930; Stannage, *High Adventure*, pp. 59–56: an aquaplane is 'a platform attached by ropes astern of a powerboat'. **CKS meets Lindbergh** *Newcastle Morning Herald*, 3 July 1930. **CKS is a hit** 'Life and Thought in America', Melbourne *Age*, 11 August 1930; Herbert Brookes to P.M. Scullin, 2 July 1930, encloses news cuttings from *New York Times* and *Gazette Montreal*, A458 AQ314/4 NAA. **Disapproval of CKS's promotion** R. Williams, *These Are the Facts*, p. 202. **Journey's end** *MFL*, pp. 157–58; Sydney *Evening News*, 7 July 1930.

CHAPTER 25: THE RIGHT STUFF IN HIM

Offers for SC Telegrams: CKS to V.C. Anderson, 9 July, and CKS to Robert Kloeppel, 7 July, MS 1882/3/32, EC NLA. **CKS visits Capt. Hancock** *MFL*, pp. 158–59; Stannage, *High Adventure*, pp. 66–76. **CKS and Stannage develop empathy** 'Real Smithy', *Newcastle Sun*, 10 December 1935; interview with Stannage, Mackersey, *Smithy*, p. 76. **CKS confides his wartime experiences** Stannage, *High Adventure*, Chapter 10. **CKS and Stannage arrive in England** Melbourne *Argus*, 31 July 1930. **CKS falls ill** Stannage, *High Adventure*, pp. 89–90; *MFL*, pp. 159–60. **CKS's symptoms of nervous collapse** Rockhampton *Evening News*, 23 October 1930. **Flight to Australia** For itineraries of CKS's rivals, *SMH*, 20 October 1930; for progress of the flight, *Daily Telegraph Pictorial*, 20, 21 and 23 October 1930 – see article on 20 October for CKS's abstinence from alcohol;

Sydney *Sun*, 18 and 20 October 1930; *MFL*, pp. 160–81. **CKS's fear of going down over jungle or sea** Log of the flight, published in SC *Story*, pp. 166–69; telephone conversation with Eddie Rickenbacker, quoted in *Daily Telegraph Pictorial*, 24 October 1930. **Encounter with Hill** London *Daily Mail* quoted in Hugh Buggy, 'Smithy', Melbourne *Argus*, 3 March 1956. CKS greatly admired Hill and publicly praised him; see *MFL*, pp. 184–85.

CHAPTER 26: THE *SOUTHERN CLOUD*

Mary's kiss *Daily Telegraph Pictorial*, 24 October 1930. **Congratulatory telegrams** Sydney *Sun*, 20 October 1930. **Father's death and funeral** *SMH*, 4 November 1930; *Cairns Post*, 5 November 1930. **CKS's wedding** Adelaide *News*, 10 December 1930; Melbourne *Herald*, 10 December 1930; Melbourne *Sun*, 11 December 1930. **Departure on honeymoon** Melbourne *Sun*, 13 December 1930. **Young girls' ignorance of sex** Anne de Courcy, *The Fishing Fleet* (London, 2012), pp. 14–15, 42, 140; interview with Mary, Mackersey, *Smithy*, pp. 230, 248. **CKS's child bride** Interview with Mary, *Australian Women's Weekly*, 16 June 1976. **CKS's eccentricities** Interview with Catherine KS, Perth *Sunday Times*, 20 July 1930. **CKS expresses love for his mother at the wedding** Melbourne *Herald*, 11 December 1930. **ANA's size and strength** Brisbane *Daily Standard*, 9 January 1931; for numbers of pilots and staff, Murwillumbah *Tweed Daily*, 11 April 1931; I.R. Carter, *The Southern Cloud* (Melbourne, 1963), pp. 84–85. **CKS's duty to help ANA** Interview with Stannage, Mackersey, *Smithy*, p. 90. **CKS's possible philandering** James Mollison, *Playboy of the Air* (London, 1937), p. 59; interview with Catherine KS, Perth *Sunday Times*, 20 July 1930. **Flying school** For advertisement for flying school, *New Sunday Times*, 22 February 1931. **Loss of *Southern Cloud*** I.R. Carter, *The Southern Cloud*, gives detailed account of the loss of the plane and the subsequent search; *MFL*, pp. 189–91. **CKS's anguished searching** 'With Kingsford Smith, Ten Hours Intensive Searching', Melbourne *Age*, 26 March 1931; Garnsey Potts, Brisbane *Sports and Radio*, 14 October 1933; Hobart *Mercury*, 11 April 1931; SC *Story*, pp. 184–90; interview with Mary, Mackersey, *Smithy*, pp. 251–2. **CKS's and Ulm's unity of purpose** Sydney *Sun*, 25 March 1931. **Brinsmead's evidence** Adelaide *Advertiser and Register*, 18 April 1931. **C.W.A. Scott arrives** *Daily Telegraph Pictorial*, 11 April 1931; for dinner with Scott, *Sunday Guardian*, 19 April 1931; Hugh Buggy, 'The Loss of the *Southern Cloud*', Melbourne *Argus*, 10 March 1956, quotes CKS's words to Scott. **CKS awarded Segrave Trophy** Sydney *Sun*, 20 March 1930. **CKS is voted top aviator by the magazine *Liberty*** Richmond River Herald, 12 December 1930. **CKS constrained by his promise to Mary** Sydney *Evening News*, 20 March 1931. **ANA mail flight to the rescue** SC *Story*, pp. 191–95; *MFL*, pp. 195–98; Sydney *Sun*, 20, 21, 24, 25 and 26 April; see also news cuttings in Ulm papers, MLMSS3359-26, Mitchell. **ANA's financial situation** Brisbane *Daily Standard*, 9 July 1931; Sydney *Sun*, 24 November 1931.

John KS torches speedboat *SMH*, 7 April 1931. **John KS's confession** Interview with John KS, Mackersey, *Smithy*, p. 254; John Kingsford Smith, *My Life Story*, pp. 30–32. **The New Guard** Keith Amos, *The New Guard Movement* (Melbourne, 1976), p. 40. **CKS's supposed plotting against Lang** Mackersey, *Smithy*, p. 253–54. **Lang's admiration for CKS** Sydney *Truth*, 21 March 1924, p. 42. **Joy riding** *MFL*, pp. 198–99; interview with Mary, Mackersey, *Smithy*, p. 248. **CKS tries never to be unkind** Stannage, 'Real Smithy', *Newcastle Sun*, 10 December 1935. **CKS orders new Avro Avian** *Albury Banner*, 28 August 1931. **CKS aims to be super-fit** Sydney *Sun*, 24 September 1931.

CHAPTER 27: PANIC

Mary's photograph Kalgoorlie *Western Argus*, 2 October 1931. **Gossip about Mary's opposition** 'Random Jottings', Sydney *Referee*, 19 August 1931. **Nellie Stewart's funeral** *SMH*, 24 June 1931. **Progress of the flight** Log of the flight, 209/5, Folder 5, items 5–9, NLA; *MFL*, Chapter XXII; SC *Story*, pp. 208–17; Brisbane *Telegraph*, 1 October 1931; for change of schedule, see Adelaide *News*, 28 September 1931; for Mary's reaction, see Perth *News*, 28 September 1931 and Sydney *Sun*, 2 October 1931. **CKS's attacks of panic** For CKS's statement to reporters, see Dubbo *Western Age*, 14 October 1931; for imprisonment in Turkey, arrival in England, phone call to Mary and doctors' diagnosis, see Adelaide *Chronicle*, 15 October 1931. **Mary sails on** *Orsova* Sydney *Sun*, 12 October 1931. **CKS's interview on** *Orford* Adelaide *Advertiser*, 16 November 1931; see also for complaints against his Turkish captors. For reaction to these complaints, see documents refusing permission for CKS to land in Turkey or fly in its airspace, D314/1/7 NAA and *SMH*, 21 December 1931. **Sydney doctor diagnoses carbon monoxide poisoning** *MFL*, p. 221; Adelaide *Chronicle*, 3 December 1931. **CKS's licence restored** Sydney *Sun*, 28 November 1931. **Experimental mail flight** *MFL*, pp. 222–27; Sydney *Sun*, 16, 17, 22 and 23 December 1931 and 19, 22 and 24 January 1932; *Daily Guardian*, 28 November 1931; Brisbane *Telegraph*, 17 December 1931; Burnie *Advocate*, 23 December 1931; Sydney *Daily Telegraph*, 22 January 1932. See also news cuttings, many unsourced, in Ulm papers, MLMSS3359-27, Mitchell. **Panic symptoms over Timor Sea** Interview with Allan, Mackersey, *Smithy*, pp. 266–67.

CHAPTER 28: KNIGHT OF THE BARNSTORMERS

Hopes for Singapore to Darwin airmail route *MFL*, p. 227; Brisbane *Truth*, 16 October 1932; E. Rogers, *Faith in Australia*, pp. 77–78; Lismore *Northern Star*, 25 January 1932. **Proposal for ANA, WAA and Qantas to combine** John Gunn, *The Defeat of Distance* (Brisbane, 1985), Chapter 9. **Barnstorming** *MFL*, pp. 227–29; Wagga *Daily Advertiser*, 6 November 1933; for Wilfrid as a ticket-seller in a tent, *Bombala Times*, 28 October 1932; for Tom Pethybridge, Sydney *Sun*, 14 November 1935; anecdotes re: barnstorming, see *Smith's Weekly*, 21 December 1935. **Rumours of CKS's infidelity** Interview with Nancy Bird, Mackersey, *Smithy*, p. 272; interview with Scotty Allan, Mackersey, *Smithy*, p. 267 – Allan speaks of encounters with prostitutes and CKS's seeming aloofness so at odds with the egalitarianism described in *FM*, p. 324. **CKS demonstrates zip fastener** Mackersey, *Smithy*, p. 274. **Figures for CKS's barnstorming** *Australasian*, 18 April 1931, says his plane was viewed by 3000 in Brighton, Tasmania; *SMH*, 24 October 1933, gives number of passengers to date; *Goulburn Evening Penny Post*, 26 August 1931, gives Wilfrid's figures; *Walcha News*, 30 June 1933, gives number of passengers in *SC* cabin and claims they took up to 200 passengers a day, sixteen adults at a time in the cabin; log book for flights – including joy flights – for 1931–35, MS 209-207-212 EC NLA. **Detailed accounts of a joy-riding visit** 'Kingsford Smith's Visit', *Grenfell Record and Lachlan District Advertiser*, 6 June 1932 and 'Flying with Kingsford Smith by Milton Ryder, Aged 16', *Australasian*, 18 April 1931. **Joy ride over Sydney Harbour Bridge and shipboard ball** *Smith's Weekly*, 7 December 1935; *FM*, p. 321; *Smith's Weekly*, 21 December 1935; R. Russel, *Great Deeds, Great Lives* (Sydney, 1966), p. 21, quoting Marge McGrath. **Knighthood** 'Sir Smithy', *Canberra Times*, 4 June 1932. **Mary's flying lessons** Interview with Mary, Mackersey, *Smithy*, pp. 271–72. Catherine KS, while she never learned to fly, was in favour of women pilots. **Investiture** *SMH*, 10 October 1932. **CKS's son's birth** Sydney *Sun*, 22 December 1932. **'The pattern of our marriage'** Interview with Mary, Mackersey, *Smithy*, p. 274. **Flight to New Plymouth** New Zealand *Evening Post*, 11 January 1933. **Radio transmitter** Stannage, *High Adventure*, p. 98–99. The transmitter was already in use on KLM's Amsterdam–Java flight. On flight to New Zealand it was more successful on the homeward journey than the outward. **Taylor's insight into CKS's character** P.G. Taylor, *The Sky Beyond* (Melbourne, 1963), p. 33. **Safety features of SC** R. Clarke, *Great Deeds, Great Lives*, p. 22. The safety features had been reinstated after their abandonment on the Coffee Royal flight. **Stannage marries Beris** Melbourne *Argus*, 8 October 1932. **CKS announces his son's birth in New Plymouth** Melbourne *Sun News Pictorial*, 12 January 1933. There are several versions of this anecdote – see also interview with Mary, *Australian Women's Weekly*, 16 June 1976. **Statistics of the tour** For attendance figures, Leeton *Murrumbidgee Irrigator*, 3 February 1933; for no tax on earnings, *FM*, p. 322; for servicing costs, *Bombala Times*, 28 October 1932. *SC* **crashes** Adelaide *Mail*, 4 February 1933. **CKS sails on** *Makura* Sydney *Sun*, 11 February 1933. **ANA goes into voluntary liquidation** Adelaide *Advertiser*, 2 March 1933. **CKS in hospital** Mackersey, *Smithy*, p. 276. **Flight home** Adelaide *Chronicle*, 30 March 1933. **Warning to CKS** Garnsey Henry Mead St Clair Potts (later publicity manager for Qantas) writes to Brisbane *Sports and Radio*, 14 October 1933.

CHAPTER 29: HOW LONG CAN I STICK IT?

CKS and family motor to Melbourne New Zealand *Evening News*, 4 April 1933. **CKS plans passenger flights to UK and around Australia** Sydney *Sun*, 29 March 1933; itinerary of flight around Australia, A458 CL314/4 NAA. **Staff of Kingsford Smith Air Services** Nancy Bird, *My God It's a Woman* (Melbourne, 1988), pp. 22–26; *FM*, pp. 322–24. **CKS superlatively praised by journalists** Port Pirie *Recorder*, 27 March 1933. **CKS's dare-devilry** Interview with John KS, Mackersey, *Smithy*, p. 281. **CKS and Mary's happy comradeship** *FM*, p. 318; Ellison described their relationship differently in an early draft of his book, saying they enjoyed 'unconventional freedom' – see box 1, file 3, EC NLA. **Percival Gull** *Cairns Post*, 30 May 1933. **CKS's gift to Far West Children's Health Scheme** *Brisbane Courier*, 27 April 1933; *Canberra Times*, 22 July 1933. **Ulm attempts Australian endurance record** Perth *Sunday Times*, 3 August 1930. CKS and Ulm had set the previous record during their Melbourne to Perth flight in 1928. **CKS and Ulm supposedly split up** Interview with Scotty Allan, Mackersey, *Smithy*, pp. 263–64; see p. 264 also for Ulm's resentment over the serial publication of *The Old Bus*. **Ulm almost dies in crash** Adelaide *Advertiser*, 22 February 1932; Burnie *Advocate*, 22 and 24 February 1932. **Ulm's newfound sensitivity** Lawrence Wackett writing in Sydney *Sun*, 16 December 1934, speaks of Ulm's intolerance in early life of those who did not share his enthusiasms, but says that in later life Ulm learned 'the value of personality and tact and soon became the master of diplomacy'. Wackett also remarks on Ulm's sudden passion to circumnavigate the world. **Ulm's emotional farewell to his son** Sydney *Sun*, 21 June 1933. **Ulm calls CKS 'the greatest airman in the world'** *SMH*, 26 August 1932. **Ulm aims to circumnavigate the world** E. Rogers, *Faith in Australia*, pp. 80–93. **CKS and Mary sail to Java** Brisbane *Daily Standard*, 23 August 1933; Sydney *Sun*, 6 October 1933; Melbourne *Sun*, 12 October 1933. **CKS's departure from Lympne** Brisbane *Courier Mail*, 5 October 1933. **Progress of the flight** *MFL*, pp. 239–40; Brisbane *Courier Mail*, 12 October and 14 October 1933; Brisbane *Daily Standard*, 11 October 1933, for Cobham's comment; *SMH*, 2 November 1933, for CKS's speech to businesswomen's lunch in which he admits to fear of passing out in the air. **Melbourne welcome** Melbourne *Argus*, 19 October 1933. **CKS and Mary and baby Charles at Mornington** Melbourne *Argus*, 25 October 1933, says they were at Mornington; however, Mary, in a later interview, thought they were in Sorrento – see Mackersey, *Smithy*, p. 286. **CKS's tribute to Ulm and Ulm's tribute to CKS** For CKS exclaiming that Ulm has sent his record 'kite high', Brisbane *Courier Mail*, 30 October 1933; for Ulm explaining that CKS's solo record stands, Townsville *Daily Bulletin*, 30 October 1933; for CKS being hoisted onto the truck, Melbourne *Age*, 30 October 1933.

CHAPTER 30: THE GLITTERING PRIZE

Job for CKS Sunday *Sun*, 15 October 1933; Brisbane *Courier*, 27 October 1933, reports that CKS is to be an advisor to Vacuum Oil Company. **Stannage on CKS and money** 'Real Smithy', *Newcastle Sun*, 10 December 1935. **Failure of amalgamation of WAA, ANA and Qantas** J. Gunn, *The Defeat of Distance*, p. 188. **Ulm's refusal to give preference to British airlines** *Queensland Central Herald*, 16 November 1933. **The Codock** *Newcastle Sun*, 28 May 1934; for Codock being safe on one engine, see Melbourne *Argus*, 2 March 1934. **Ulm flies to New Zealand** E. Rogers, *Faith in Australia*, pp. 96–103. **CKS flies to New Zealand** Sydney *Sun*, 4 January 1934; Brisbane *Courier Mail*, 15 January 1934; New Zealand *Evening Post*, 15 January 1934. **Figures for airmail carried** Melbourne *Age*, 6 September 1934. **Charles junior learns to walk and takes to the air** Sydney *Sun*, 10 and 15 January 1934. **CKS and Mary alternate between Sydney and New Zealand** For their arrival back in Sydney, see Sydney *Sun*, 14 February 1934. On 10 March they return to New Zealand on *Monterey*. CKS completes his tour and returns home on 29 March – see Sydney *Sun*, 29 March 1924. Mary returns on *Wanganella*, Sydney *Sun*, 29 and 31 March 1934. **Ulm's response when Qantas wins the contract** *Dubbo Liberal and Macquarie Advocate*, 21 April 1934. **CKS's response** Melbourne *Argus*, 21 April 1934. **Hands off the New Zealand route** Brisbane *Telegraph*, 21 April 1934. **Centenary Air Race** Perth *Daily News*, 13 September 1934. **CKS prefers US plane** Katanning, *Great Southern Herald*, 14 April 1934. **CKS orders a special Comet but is refused** Sydney *Sun*, 22 April 1934. **CKS appeals to Robertson** *SMH*, 4 May 1934. **CKS is criticised as unpatriotic** Launceston *Examiner*, 8 May 1934; Sydney *Truth*, 6 May 1934; *FM*, pp. 311–12; Sydney *Sun*, 6 May 1934. **CKS sails to California** Sydney *Sun*, 30 April 1934. **Description of *Altair*** P.G. Taylor, *Pacific Flight: The Story of the Lady Southern Cross* (Sydney, 1937), pp. 119–20; *SMH*, 20 July 1934. **CKS begs money to pay for *Altair*** Adelaide *Advertiser*, 22 June 1934. **CKS enters the Air Race** Brisbane *Courier Mail*, 31 May 1934. **Legal tangles** *Brisbane Telegraph*, 30 June 1934. **CKS's opinion of ICAN** Sydney *Sun*, 7 February 1935. **CKS visits Mae West** *Newcastle Sun*, 12 June 1934. **CKS lunches with Amelia Earhart** Melbourne *Argus*, 18 June 1934. **CKS speaks by phone with Jean Batten** Brisbane *Telegraph*, 11 May 1934. **CKS meets Captain Hancock** Article by John Williams, Honolulu *Star Bulletin*, 5 July 1934 – see news cuttings MLM SS3359-2071, Ulm Papers, Mitchell. **CKS lands *Altair* on the deck of *Mariposa*** Sydney *Sun*, 9 July 1934. **CKS flies *Altair* from Sydney docks to Anderson Park** Melbourne *Age*, 18 July 1924; *FM*, pp. 309–10; P.G. Taylor, *Pacific Flight*, pp. 5–8. **Further legal tangles** Sydney *Sun*, 18 and 25 July 1934; *SMH*, 26 July 1934. **Trouble with Turkey** 'Centenary Air Race', A461 G314/1/7 Part 1, NAA. The file includes PM Lyons' letter to the Premier of Victoria. For CKS's respectful cable to Mustapha Kemal Pasha, *Canberra Times*, 22 September 1934. **Test flights allowed** For Mascot to Melbourne, Hobart *Mercury*, 8 September 1934; for Melbourne to Perth, Melbourne *Argus*, 10 September 1934; for Perth to Adelaide to Sydney, Sydney *Sun*, 11 September 1934; Taylor, *Pacific Flight*, Chapter 11. **Even more legal tangles** Melbourne *Age*, 22 September 1934; *SMH*, 21 and 28 September 1934; Sydney *Sun*, 28 September 1934. **Permission to enter the race** Sydney *Sun*, 30 September 1934; P.G. Taylor, *Pacific Flight*, pp. 30–34. **Cracks found in the cowling** *SMH*, 1 October 1934. **CKS withdraws from race** Launceston *Examiner*, 4 October 1934. **Ulm offers CKS a plane from America** Sydney *Sun*, 4 October 1934. **CKS and Taylor decide on Pacific flight** Taylor, *Pacific Flight*, p. 36.

CHAPTER 31: AM I A SQUIB?

Reason for Pacific flight 'Kingsford Smith Was Not Afraid to Fly', Brisbane *Courier Mail*, 4 January 1936. **White feathers** *FM*, p. 310; see also a collection of scurrilous letters and a white feather sent to *Smith's Weekly*, MS1883/3/32 EC NLA. **Children reproved for calling CKS a squib** Melbourne *Argus*, 8 November 1934. **CKS's distress** *FM*, pp. 310, 317. **CKS's health regime** Sydney *Sun*, 16 October 1934. **Robertson defends CKS** Brisbane *Courier Mail*, 6 November 1934. **CKS's statement in *Smith's Weekly*** *FM*, pp. 310–31. **Preparations for Pacific flight** Taylor, *Pacific Flight*, Chapter IV. **Take-off from Brisbane** Brisbane *Worker*, 24 October 1934; Taylor, *Pacific Flight*, pp. 74–75. **Mary leaves for Melbourne** Sydney *Sun*, 16 and 18 October 1934. **Flight to Fiji and Hawaii** Taylor, *Pacific Flight*, Chapter V – Chapter XII; *MFL*, pp. 250–54; Sydney *Sun*, 29 October 1934; Brisbane *Courier Mail*, 22 October 1934. **CKS accidentally hits airbrakes** Brisbane *Courier Mail*, 4 January 1936. **Arrival in Honolulu** Launceston *Examiner*, 31 October 1934; Honolulu *Star Bulletin*, 29 October 1934. **British Consul's letter** 4 November 1934, AV61 B314/1/7, NAA. **In Honolulu** Taylor, *Pacific Flight*, Chapter XIII. **Flight to Oakland** Taylor, *Pacific Flight*, Chapter XIV. **CKS falls asleep at controls** Brisbane *Courier Mail*, 4 January 1936. **Arrival in Oakland** Taylor, *Pacific Flight*, Chapter XV; Sydney *Sun*, 5 November 1934; Melbourne *Argus*, 6 November 1934 – see also for arrival in Los Angeles. **Bottle of whisky** Hobart *Mercury*, 6 November 1934. **CKS answers reporters' questions** For speed flight, *SMH*, 7 November 1934; for congratulations to Scott and Scott's reply, Brisbane *Courier Mail*, 6 November 1934. **CKS lionised in Los Angeles** Sydney *Sun*, 9 and 10 November 1934; for CKS's address to motion picture industry, *SMH*, 10 November 1934 – see also for *Altair* being impounded for debt and CKS's visit to Hancock's yacht; for CKS made honorary policeman, Brisbane *Telegraph*, 8 November 1934. **CKS visits Harold KS and leads Armistice Day parade** Melbourne *Argus*, 14 November 1934; letter from Noel Wisdon, *Smith's Weekly*, 10 December 1935. **Nebulous plans** Canberra *Times*, 17 November 1934. **CKS ordered to rest** Adelaide *Advertiser*, 19 November 1934. **Reunion with Ulm** Sydney *Sun*, 30 November 1934. **Stannage recalls Ulm's death** 'Loss of Ulm', Brisbane *Courier Mail*, 13 December 1935. **CKS longs to search for Ulm** *West Australian*, 6 December 1934; Brisbane *Telegraph*, 7 December 1934.

Praise from Mussolini and Hitler Melbourne *Argus*, 1 December 1934. **CKS visits St Louis and abandons record flight across America** Brisbane *Telegraph*, 22 December 1934; *Newcastle Morning Herald*, 24 December 1934. **Williams writes of CKS and Ulm** Honolulu *Star Bulletin*, 7 December 1934, MLMSS3359, add on 2071 Ulm Papers, Mitchell.

CHAPTER 32: KING OF THE AIR

Interview on Monterey Sydney *Sun*, 28 January 1935. **Concert** *SMH*, 30 January 1935. **Memorial ceremony for Ulm** Adelaide *Chronicle*, 21 February 1935. **CKS praises Ulm** *MFL*, p. 260. **CKS's birthday** Brisbane *Sunday Mail*, 10 February 1935. **Mary builds a new house** Sydney *Sun*, 'Social notes', 22 March 1936; Adelaide *Advertiser*, 9 November 1937. **CKS catches flu and Mary nurses him** New Zealand *Evening Post*, 10 April 1935. **CKS is depressed** Sydney *Sun*, 27 April 1935. **Proposal for Jubilee airmail flight to New Zealand** Stannage – writing on CKS's behalf – to Earle Page, 12 April 1935, AD 314/4 NAA. **Near disaster on the Jubilee airmail flight** *MFL*, pp. 260–78; P.G. Taylor, *Call to the Winds* (Sydney, 1939), pp. 1–67; Stannage, *High Adventure*, Chapter 17; 'The Old Bus Has Done It Again', Sydney *Sun*, 15 May 1935. **Arrival at Mascot** Sydney *Sun*, 15 May 1935. **Praise from London *Daily Mail*** Quoted by Sydney *Sun*, 16 May 1935. ***Aeroplane* magazine calls the flight a stunt** Brisbane *Telegraph*, 23 May 1935. **CKS offers to deliver remaining mail** Sydney *Sun*, 16 May 1935 – see also for CKS asleep in bath. **High jinks** Sydney *Sun*, 19 May 1935. **Taylor awarded medal of order of British Empire** Adelaide *Chronicle*, 15 July 1937. ***Southern Cross* given to the nation** Sydney *Sun*, 17 and 18 July 1935; *SMH*, 18 July 1935. **Trans-Tasman Development Company** *West Australian*, 7 June 1935. **CKS recruits pilots** Sydney *Sun*, 19 July 1935. **Mary and Catherine farewell CKS** For Catherine's farewell, *Newcastle Sun*, 8 November 1935; for Mary's farewell, *Brisbane Telegraph*, 8 November 1935. **CKS in hospital** Sydney *Sun*, 25 June 1935; for Sheil's response to CKS's illness, see *Caesar of the Skies*, p. 190. **CKS recommends Sikororsky flying boats and refutes accusations of anti-patriotism** Melbourne *Age*, 13 July 1935. **CKS visits Japanese ship** *Newcastle Sun*, 8 August 1935. **Boulton and Pethybridge in California** Sydney *Sun*, 3 July 1935. **CKS tests *Altair*** Melbourne *Argus*, 26 August 1935. **CKS sails on *Britannic*** Melbourne *Argus*, 23 September 1935. ***Altair* shipped to England** Sheil, *Caesar of the Skies*, pp. 182–85; see also for wrangles over fuel tanks. **Imperial Airways is awarded Trans-Tasman contract** Sydney *Sun*, 29 October 1935. **British Pacific Trust** Adelaide *Chronicle*, 26 September 1935; Sydney *Sun*, 25 September 1935. **Mary in hospital, convalesces in Melbourne** Sydney *Sun*, 22 September and 6 October 1935. Mary is said to have been pregnant – see Mackersey, *Smithy*, p. 342. In the medical practice of that time, it is unlikely that doctors would have removed tonsils during pregnancy, so tonsillectomy may be a euphemism for a miscarriage and curettage. **CKS falls ill, delays take-off** Sheil, *Caesar of the Skies*, pp. 180–98 – see p. 191 for Tommy 'would have flown to hell with the Boss he idolized'; Melbourne *Argus*, 21 October 1935. **Government refuses to pay CKS's fare home** *Smith's Weekly*, 21 December 1935; *FM*, pp. 334–35; interview with Mary, Mackersey, *Smithy*, p. 342. **First attempt, CKS turns back over Greece** Melbourne *Age*, 24 October 1935; Brisbane *Telegraph*, 25 October 1935; Sydney *Sun*, 25 October 1935. **Second take-off** Melbourne *Argus*, 7 November 1935. **Melrose and Broadbent** Sydney *Sun*, 10 November 1935. **Progress of CKS's flight** Brisbane *Telegraph*, 6, 7, 8 and 9 November 1935; 'Further Report on the Loss of the Lockheed *Altair*', A461 D314/1/7 NAA; 'Mr Melrose's Theory', Melbourne *Argus*, 22 November 1935. **Search for CKS** Sydney *Sun*, 8, 9, 10 and 11 November 1935. **Allan, Taylor, Purvis and Stannage join the search** Sydney *Sun*, 12, 13, 14 and 18 November 1935. **Leofric's suggestion** Sydney *Sun*, 20 November 1935. **Flares on Sayer Island** Sydney *Sun*, 24 November 1935. **Wilfrid's theory of the crash** Hobart *Mercury*, 4 December 1935. **Other theories** Sheil, *Caesar of the Skies*, p. 196; for Lawrence Wackett's diagnosis of a faulty supercharger, see Pedr Davis, *Charles Kingsford Smith: Smithy, the World's Greatest Aviator* (Sydney, 1977), pp. 152–54. **CKS's wife and mother cannot accept his death** Mary and Catherine refuse a memorial church service, 9 December 1935, A461 D13/1/7 NAA. **Eulogy in Parliament** Melbourne *Argus*, 7 December 1935. **Wreckage found of** *Altair* Kalgoorlie *Western Argus*, 20 July 1937; *Bowen Independent*, 11 February 1938. **Beautiful islands of the Mergui archipelago** 'Sea Gypsies Deny That Smithy Lives', Sydney *World's News*, 11 February 1939.

On 3 December 1937, at a private ceremony in the Powells' house in Irving Road, Toorak, Mary Kingsford Smith married Alan Tully, chief executive in Australia of Ethyl Corporation. In 1940 Alan Tully was transferred by the Ethyl Corporation to America, where he and Mary spent the rest of their lives. Eight-year-old Charles Arthur Kingsford Smith went with them, and still lives in the United States.

INDEX

PICTURE CREDITS

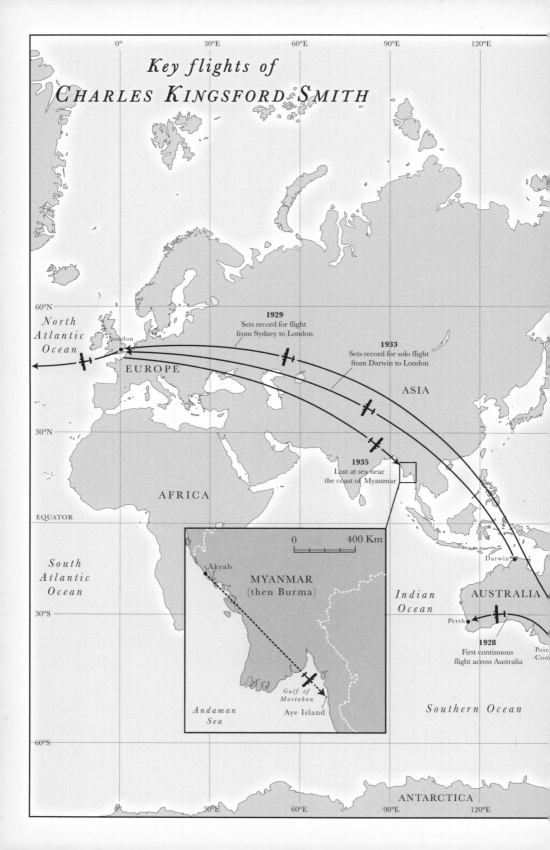

Key flights of
CHARLES KINGSFORD SMITH

0°   30°E   60°E   90°E   120°E

North
Atlantic
Ocean

London

EUROPE

60°N

1929
Sets record for flight
from Sydney to London

1933
Sets record for solo flight
from Darwin to London

ASIA

30°N

AFRICA

1935
Lost at sea near
the coast of Myanmar

EQUATOR

South
Atlantic
Ocean

30°S

0   400 Km

Akyab

MYANMAR
(then Burma)

Indian
Ocean

Darwin

AUSTRALIA

Perth

1928
First continuous
flight across Australia

Poin
Coo

Andaman
Sea

Gulf of
Martaban
Aye Island

Southern Ocean

60°S

ANTARCTICA

0°   30°E   60°E   90°E   120°E